Plant Optimization in the Process Industries

Plant Optimization in the Process Industries

Incorporating Equipment/Assets
in the Decision-Making Process

Marty Moran

WILEY

Published by John Wiley & Sons, Inc., Hoboken, New Jersey.
Published simultaneously in Canada.

Library of Congress Cataloging-in-Publication Data is applied for

Hardback ISBN 9781119707738

Cover Design: Wiley
Cover Image: © Photon photo/Shutterstock

Set in 9.5/12.5pt STIXTwoText by Straive, Pondicherry, India

This book is dedicated to those who struggle with either MS or transverse myelitis. Let's hope science can eventually solve these horrible diseases.

Contents

Foreword by Ron Lambert

It is with great pleasure that I have been asked to write the foreword for this exceptional book on Process Plant Optimization, authored by a distinguished individual who has significantly impacted the industry with his innovative ideas and forward-thinking approach.

I first crossed paths with Marty Moran during our time at Meridium (GE Digital), where we both worked as consultants. Marty's expertise in Reliability, Availability, and Maintainability (RAM) studies made him an invaluable asset to the product team, as he provided groundbreaking ideas and conducted crucial RAM studies using the Meridium product. However, what truly set Marty apart was his ability to see the "Big Picture" and project into the future the benefits gained applying his diverse skill set. It was this exceptional talent that led me to hire him at Sadara for a completely different role – Instrument Reliability Engineer, knowing that his capabilities and contributions would extend far beyond Instrument Reliability and where he made remarkable contributions to our team.

During Marty's tenure at Sadara, he played a pivotal role in several high-stakes projects, using his expertise to solve complex engineering challenges. His unique background in process control and his ability to think beyond single pieces of equipment made him an invaluable resource for our team. I distinctly remember the crucial role he played in determining the root cause of a significant incident within an ethylene furnace, as well as his invaluable contributions applying his advanced multivariable control knowledge to lead the team optimizing the ethylene unit.

Marty's dedication to obtaining data through consultation with those knowledgeable about the unit is a philosophy that resonates deeply with me. The relationships he built displaying shared concern and curiosity with those needing help greatly increased the influence he had in seeing that correct actions would be seen through to their completion and results. His emphasis on using engineering skills and experience to make reasonable assumptions, rather than blindly relying on data, is an approach that is both practical and invaluable in our industry.

Having had the opportunity to review the early stages of this book, I was struck by the innovative concepts presented within its pages, particularly the integration of optimization similar to advanced control algorithms coupled with Monte Carlo reliability modeling. Marty's emphasis on including constraints within all optimization problems, and his unwavering focus on business financial optimization, truly represents the direction in which our industry should be heading. It is my firm belief that this book will make a significant contribution to the field of process plant strategy and optimization, and will undoubtedly shape the way we approach these critical concepts in the future.

In conclusion, Marty's groundbreaking ideas and forward-thinking approach have paved the way for a new era in Reliability and Optimization. It is my sincere hope that this book will inspire and empower the next generation of engineers, and lead our industry to new heights.

Ron Lambert
Engineering Manager
Atlantic Methanol Production Company

About the Author

I grew up in an ethnic Irish neighborhood on the west side of Chicago in the 1960s/1970s and hold both American and Irish citizenship. I attended college in downtown Chicago at the University of Illinois and graduated with a BS in Chemical Engineering in March 1982. I am currently a Chartered Engineer in Engineers Ireland.

At a high level, my career has concentrated on using Advanced Manufacturing Technologies, such as Advanced Process Control/Optimization and Asset Optimization Technologies that helped customers in the process industries increase throughput, reduce manufacturing costs, increase asset availability, and reduce risk with progressive roles in technical, consulting, business, and marketing.

Due to the fact that I worked for Advanced Manufacturing Technology vendors for many years, I have spent considerable time in sales/marketing roles. I have been fortunate in that I have worked in over 60 plants in 25 countries on 6 continents.

Most of my "process" experience has been concentrated in the worldwide oil refining industry. However, I have also worked on projects in midstream, a couple ammonia plants, and in a large-scale petrochemical plant.

While I spent the majority of my career working for vendors, my most recent experience was working directly for a massive petrochemical plant in the Middle East, filling a role in a group of technical experts, leveraged across more than 20 petrochemical plants. During my last year there, I requested and accepted a role to be the technical lead for the Advanced Control/Optimization project. For the three years prior to that assignment, I was a Senior Reliability Engineer providing consulting for the strategic portion of the program, in addition to being used for special projects when a seasoned engineer was required. Ultimately, I got to play significant roles on both the "process" and "asset" sides of the business, in addition to using my marketing and writing skills.

My personal interest has always been in the area of optimization. To that end, I have principally worn two hats in my career. One was to "optimize" the process side of the business using Advanced Process Control/Optimization, while I spent the second half of my career trying to optimize assets using various technologies. The essence of this book is to couple optimization of the processes with optimization of the assets thereby multiplying the benefits to the business.

While I am proud of my technical capabilities, I am business orientated and believe that technology should only be used if it achieves a specific business result. In my mind, business comes first. Technology only exists to further our business needs.

For that reason, the focus of this book has been generally on Optimization and, more specifically, on how Optimization and Monte Carlo simulation can be pragmatically combined together to optimize profitability in the process industries through better asset decisions. This book has walked the balance between business needs and how that can be achieved technically. In addition, it has described what organizational changes must be made for a company to effectively adapt modeling/optimization technologies. It has suggested a Program Master Plan.

To better appreciate the thoughts in this book and how I came to these conclusions, it's important to traverse the path of how I became involved with asset optimization technologies and specific experiences I had earlier in my career.

Process Optimization Experience

While I will concentrate principally on my asset optimization experience in "About the Author," it's important to highlight what I learned during my early years as an Advanced Control and Optimization (ACO) engineer. From late 1987 until the early 2000s, I was an ACO engineer. I worked for SETPOINT until early 1994, where I learned both Advanced Control and Optimization. The latter is especially important as that is the focus of this book.

At that time, we were using closed equation optimization technologies. I was quite familiar with GRG2, which is a reduced gradient, nonlinear optimization, originally developed at the University of Texas. Many of you may have already worked with it, though you probably didn't know it. That's because it's now embedded within Excel as the Solver. That's what we will use for the majority of optimization problems in this book.

When I left SETPOINT, I joined Continental Controls, where I developed their multivariable controller, for which I received a US patent.[ix] I was so impressed with GRG2 during my time working at SETPOINT that I used it as the mathematical engine for this new multivariable controller. What I did was fundamentally write a sophisticated "wrapper" that called the GRG2 mathematical algorithm to perform nonlinear multivariable control.

Those early years working with optimization were important because I became convinced that one essential missing ingredient in asset management was the lack of an emphasis on true optimization. What you'll see in this book is that we will be now combining, and augmenting, system modeling with optimization technologies.

Another thrust during my years optimizing the process side was the emphasis that we placed on driving business benefits. I had the opportunity to lead many process studies where we had to justify our projects based on business benefits. As I moved to the asset side of the business, that justification seem to be lacking, or only haphazardly done.

I should also note that while at SETPOINT, I served as the Hydroprocessing Technology Manager. That will explain why you'll see so many hydroprocessing examples in the book.

Asset Modeling/Optimization Experience

In early 2003, my career changed significantly. In the 15 years proceeding that, I had been focused on Advanced Process Control and Optimization. However, due to a rather dramatic and unanticipated career shift, I found myself as the Director of Sales and Marketing at Clockwork Solutions.

Clockwork was a pure play RAM high-end software modeling company. I was tasked with selling projects to clients to improve their ability to make better design and spare parts decisions using RAM, as well as other standard uses of the technology that we now take for granted. Unfortunately, what I soon learned was that up to this point in time, there had been little adoption of this technology within the process industries. In fact, there had been more projects executed within the military aerospace industry, rather than in refining and petrochemicals. As a new salesperson, this was especially disconcerting, to say the least.

Furthermore, the people attracted to a company like Clockwork were technical types and not nearly as business focused as I had grown accustomed to in my career. Even more troubling was that RAM modeling was still largely a "cool" technology in search of some practical business problems to solve.

In fact, the technology was not even being sold properly. Everything was backward. Normally, one starts with a business problem and evaluates different technologies and their potential unique abilities to solve a particular problem. However, in this case, it was just the opposite. Here, the technology already existed, but it wasn't clear at all as to what business problems this technology could be used to solve. If I were to have any hope of surviving in my new sales role, I needed to articulate that – and quickly.

Of course, from a sale person's perspective, the problems the technology can solve must occur frequently enough that a company would be interested in purchasing software and accompanying services. In addition, the cost/benefit of solving those problems needs to be such that that a company would be interested in spending money on them. Plus, the sales process needs to occur in a timeframe where I could close enough deals so my employer would keep pursuing this business area. I knew I was in trouble.

For all those reasons, I decided to write a paper and give a presentation at the National Petrochemical and Refining Association (NPRA) conference in March 2004 that detailed how RAM modeling driven by an optimization technology could be used by a refiner to increase the quantity of gasoline being produced based on the availability of the different units producing gasoline components, balanced against the cost of various potential improvement projects.[x]

In essence, this presentation showed how to answer an important business question. That is, where should I spend my next dollar to **optimally** increase availability? That entire presentation was eventually republished by Petroleum Technology Quarterly in April 2004.[xi]

As is so often the case with marketing, it takes times for an idea to gain acceptance. Based on the response from that conference, it became clear that the industry wasn't ready for this technology any time soon and that it would likely take years for large-scale interest and/or adoption.

Since Clockwork would not be moving fast enough for my interests, I decided there might be a niche use for this technology that I could exploit as an entrepreneur. To that end, I started writing a VBA-based program on my own using Monte Carlo simulation modeling that would determine the optimal turnaround timing between units based on the reliability performance of different equipment within the units. To garner funding for the idea, I applied for a grant from the National Science Foundation (NSF) for a small company I had formed.[xii] Unfortunately, the NSF review panel split their votes 3-3 and the idea was not approved.

Given that setback, I eventually accepted a position at Meridium where I played numerous roles, For the first 18 months, I taught training classes on how to use their software to solve technical problems and generate business benefits. While I could teach all their classes, I was used primarily for engineering related subjects. For example, I taught the Reliability Analytics class many times, which included a portion dedicated to RAM modeling.

Another role I had was being an internal product consultant for their RAM modeling product. In that job, I worked with the product manager over several product development cycles to make numerous changes to their RAM modeling product. For example, we dramatically improved RAM analytics and added spare parts functionality, as well as tighter integration with other parts of the Meridium reliability suite.

Eventually, I became a functional consultant and I got a chance to implement several RAM modeling projects. One was for an Australian mining company, who were desperately trying to increase capacity to meet iron ore demand from China. Another was for a large refiner evaluating different electrical grid configurations.

While at Meridium, I was selected by the company president to help create the Meridium Value Assurance Program (MVAP).[xiii] which was a comprehensive reliability assessment technology that graded a client in 30 different elements of a reliability program. 10 of the 30 elements were specific Reliability Work Processes, one of which was Reliability in Design. This work process encompassed RAM as one of its principal components. Each element contained approximately 10 different assessment questions ranked against standards for each of 4 different maturity levels.

Another important aspect of the MVAP program was the ability to estimate a client's benefits when making improvements in their maturity levels. For example, I could estimate a client's benefits from improving their Reliability in Design program using just client supplied data and the MVAP program methodology.

Later, when I returned to my personal consulting practice, I created and implemented a large-scale RAM model for a multi-billion-dollar, coal gasification plant in late 2014/early 2015.[xiv] The completed model would show that the plant could not achieve the specified technical and business objectives and would later become evidence supporting legal action brought by the tax-payers against the utility.

At Sadara, I was hired as an Instrument Reliability Engineer, so I never worked directly with their RAM modeling technology or associated data. That program had been used to determine the exact processes, plant configuration, and equipment size and had been completed in Houston years before I started working for the company at their Saudi plant site. (I hadn't started until about 35% of the plant was already running with the remainder nearing 'Mechanical Completion'.)

However, I eventually morphed into being the group 'marketing' person due to my experience writing books[xv] and working in sales/marketing organizations. (Every reliability program should have a marketing person in the group that can help publicize their successes or lest they be forgotten about or be taken for granted.) For that reason, the Reliability Director asked me to write a Sadara success story article that would later appear in Hydrocarbon Processing.[xvi] This was my preferred publication, since I had been an author there frequently in the past. [xvii xviii xix xx] This article would be a typical 'Case Study', of which I had written many during my marketing career.

When I interviewed the Reliability department management for the article, they all felt strongly that Sadara's RAM effort was one important reason why their plant availability was so high upon start up. While I didn't have an opinion since I hadn't participated in the program, nor had access to the specific software or required training needed for the program, their response didn't surprise me since I had known the power of modeling from my previous 15 years of experience with the technology.

What Sadara claimed was what we, as vendors, had been preaching for many years. It was just that Sadara was the first grassroots plant to use this technology on such a grand scale. Sadara's experience and recent changes in the vendor landscape are encouraging because we may have finally gotten to the point where more executives throughout industry might now be open to the idea of using this technology.

As impressive as that effort was, the new intellectual property introduced in this book will move beyond just RAM models by layering optimization on top of it. A useful analogy is that Monte Carlo modeling is like a dial telephone, while the technology introduced in this book more akin to a smartphone. It extends it is new direction.

A Unique Perspective

To summarize, there were four major events that occurred throughout my career that allowed me to provide a unique perspective on the thoughts articulated in this book as noted below:

1) 1996 US Multivariable Control Patent
2) 2004 NPRA Paper
3) 2004 NSF Overhaul Optimization Submission
4) 2014/2015 Plant-Wide Modeling Project

Let's take each of these one at a time to understand how they added value.

The multivariable controller that I wrote in 1994 used GRG2 as it's "heart and soul." Fundamentally, I wrote software around this core optimization technology to implement multivariable control. This application was both a traditional closed equation optimizer and a pure multivariable controller. This is important because those on the reliability/asset management side of the business are not trained to think in "optimization" terms. I had an advantage because those of us who worked on the "process" side of the business had been heavily exposed to this technology.

The 2004 NPRA paper was the first detailed application that combined standard optimization technology with Monte Carlo modeling. Once that groundwork had been laid, it would be easy to extend it to other applications. In fact, that would occur later that same year.

Turnaround Optimization was the second application attempted. While the NSF application in December 2004 for "Overhaul Optimization" was not funded, the intellectual foundation had been created. (In fact, I had already written much of the code for that application, though it was never used in any commercial application.)

The last major experience was the plant-wide model that I had created for a US coal gasification plant in late 2014/early 2015. While I had been using RAM modeling on a number of important projects in prior years, this was my first large-scale attempt at using a "Consulting" approach to generate the required data for the models. The Consulting approach is also the main work process used for this methodology.

These four major experiences laid the groundwork for easily extending optimization to the other applications as cited in the book, as well as the methodologies proposed.

Acknowledgments

I start by thanking all my colleagues in the process industries. Over the past 40+ years, I've had the opportunity to work with so many talented professionals. As my LinkedIn page can attest, the list is long and impressive. At times when I reflect on my career, I realize how blessed I am not only to have had so many unique experiences all around the world but also to have shared that time with so many exceptional individuals. Given how many high-quality people I have interacted during my career, in this section, I will only call out a few key individuals that significantly impacted this book in some manner. Otherwise, this section would become far too long.

While I thoroughly enjoyed learning about chemical engineering while in college, my career got off to a rocky start. I had the unfortunate distinction of graduating at the worst possible time in the history of chemical engineering – March 1982.

I always remember a discussion in a Starbucks a few years ago with a gentleman roughly my age. When he learned that I was a chemical engineer of his era, and was still practicing chemical engineering, he was astonished. He couldn't believe that I could have somehow navigated the layoff minefield of the mid-1980s. It wasn't always fun.

I started working professionally in November 1982 when I accepted a job at a company on the west side of Chicago that manufactured onstream process analyzers for the refining industry. It would be the start of a career anchored in oil refining that would ultimately be exceptionally good to me.

My career blossomed when I joined SETPOINT in Houston in November 1987. While I did well in college scholastically, SETPOINT was a humbling experience because I realized what it was like to a "C" student. That is, I was around so many talented people that I was often in awe of them. It took some years to get to their level, but their mentoring must have worked well because I now knew the industry well. It's the experience gained at SETPOINT that I would leverage during the remainder of my career – and is certainly reflected in this book.

I was taught the business/process side of the refining industry by many people. However, one person stands out: Walter Bare took my aside and taught me the refining industry in a way that few others could. I would not be where I am without his help. SETPOINT was instrumental in many other ways since that is where I really got to practice optimization, which is the central element of this book.

As for this book specifically, I thank Ron Lambert, who has contributed in many ways. Ron and I first met when we were both working at Meridium. While we crossed paths occasionally in those years, we really got to know one another when I joined Sadara, where Ron was the Reliability Department Director.

Philosophically, both Ron and I believed that Reliability had become too enamored with its own technologies/methodologies and often forgot that the goal of a plant was to make product – not care for individual assets. We were both pragmatists, who had a conviction that one didn't need to have perfect data to make good decisions. Rather, we strongly agreed that we should lean heavily on utilizing engineering assumptions as one of the better sources of data for this type of program from the many different people who had intimate knowledge of the unit.

I thank Greg Weber. A former Turnaround and Reliability Manager at one the world's largest integrated oil companies, we first met some years ago through a colleague who had recommended me for a project that Greg was working on. He eventually contracted me to put together a Monte Carlo simulation model for a large plant for one of his better US clients. As we worked together on that project, I educated him on how Monte Carlo technology could pragmatically be used to make better asset decisions.

In fact, Chapter 7 largely describes the process we used during that project to obtain modeling data. I had had this idea floating in my head for years, but this was the first project where we used that idea in a major way. Greg and his colleague conducted the client workshop to generate the data for the model I developed, independent of my involvement. What amazed me was how quickly they achieved that goal and the quality of the information that they gathered during this process. This gave me the confidence to know that the consulting methodology that we highlight later in the book was not only possible, but absolutely the right approach for asset modeling.

Disclaimer

The information and insights presented in this work are based on optimization, combined with Monte-Carlo simulation. As an author, I acknowledge the following: (i) the content within this work has been developed through rigorous analysis, empirical data, and industry expertise. All examples were developed within Excel™ using Excel's Solver as the optimization engine and Excel's data tables to simulate Monte-Carlo principles; (ii) the optimization techniques described herein are not based on any proprietary methodology; and (iii) the ideas developed herein, in general, are based on the author's experience using Advanced Control and Optimization, including a US patent owned by the author.

The publisher and the author make no representations or warranties with respect to the accuracy or completeness of the contents of this work and specifically disclaim all warranties, including without limitation warranties of fitness for a particular purpose. No warranty may be created or extended by sales or promotional materials. The advice and strategies contained herein may not be suitable for every situation. This work is sold with the understanding that the publisher is not engaged in rendering legal, accounting, or other professional services. If professional assistance is required, the services of a competent professional person should be sought. Neither the publisher nor the author shall be liable for damages arising here from. The fact that an organization or Website is referred to in this work as a citation and/or a potential source of further information does not mean that the author or the publisher endorses the information the organization or Website may provide or recommendations it may make. Further, readers should be aware that Internet Websites listed in this work may have changed or disappeared between when this work was written and when it is read.

1

Optimizing a Process Plant

Fundamentally, this book is about optimization. To be more specific, optimizing a process plant.

However, like love, where beauty is in the eye of the beholder, optimization can, and often does, mean different things to different people. It all depends upon your vantage point.

While engineers like to focus on all the technical details in a process plant, and often lose sight of the big picture, it's important to understand that oil refining, petrochemical manufacturing, and specialty chemical, food, and pharmaceutical manufacturing enterprises are profit-seeking businesses.

1.1 High-Level Business Goals

Maximizing profit entails many factors, but for the most part, the high-level business goals of most process plants are well known and straightforward:

- Maximize production, when able to do so, using the existing asset base
- Keep the plant running as long as possible between turnarounds. However, when choosing turnaround timing, duration, and what equipment to be restored balance the increased production of longer run rates against exposing the plant to unacceptable safety or production failure risks
- Carefully manage capital dollars to earn the best return on those investments
- Perform only as much day-to-day maintenance that balances short- and long-term production needs in the most cost-efficient manner
- Maintain enough spare parts to adequately service equipment, especially to prevent any production disruptions, but do not overstock
- When new capital projects are proposed, consider designs within budget constraints, with higher reliability if they can be cost-justified
- Utilize outside resources to supplement internal resources when it is in the plant's business interests to do so

Let's dig a bit deeper into plant profit and its two main components – revenue and costs.

1.2 Profit

When we refer to optimization in this book, we will be focused on helping make decisions using *Advanced Manufacturing Technologies* to maximize the profit of a facility, balanced across all the important time horizons – short, medium, and long term. While this book is principally focused on

Plant Optimization in the Process Industries: Incorporating Equipment/Assets in the Decision-Making Process, First Edition. Marty Moran.

using asset modeling technologies to achieve that goal, we will also compare and contrast that with other advanced manufacturing technologies that have been used in other aspects of the business, principally on the process side, when appropriate.

Profit is defined as revenue minus costs. Obviously, senior management is heavily compensated to maximize profit, and there are many factors at work in being able to achieve that goal.

1.2.1 Revenue

One aspect of maximizing profit is to maximize revenue.

For a process plant, revenue is simply the amount of product produced multiplied by its sales price. A plant will always have production targets that it must meet chosen based on market demand, current product margins, plant configuration, and potential feedstock availability. If there is sufficient market demand, there may also be an incentive or opportunity to maximize production, beyond so-called nameplate capacity, or what has been achieved in the plant in the past.

To generate revenue, there are two major components that are necessary to achieve that goal.

The first is process technology, whereby raw materials can be converted into final products. In an oil refinery, this might be a fluid catalytic cracking (FCC), alkylation, or a hydrocracker process unit. Sometimes, the process technology might have been developed internally, but more often in today's world, it's licensed directly from a process technology company. Process technology includes the process flow, catalysts, high-level equipment requirements, and process "know-how" of how to generate products in the real world.

The second part necessary is the actual real-life equipment itself – pumps, compressors, distillation towers, heat exchangers, instrumentation, etc.

Thus, to drive revenue, process technology, and "know-how" are required in the right combination with the proper collection of assets. Neither exists independent of one another. Both are necessary. In many ways, that is the thrust of this book and probably more important that you can appreciate at this point. However, as the chapters proceed, it will become more obvious why it matters so much. Also, that statement has major implications for the technology solutions that we will ultimately employ.

1.2.2 Costs

Costs that influence plant profitability originate in many different departments, such as operations, maintenance, engineering, reliability, process control, and HR. These include costs for personnel, raw materials, utilities, capital projects, and spare parts, among others. Since these span such a wide spectrum, to remain focused, we will only consider those that advanced manufacturing technologies could influence or provide guidance upon. There are definitely places where these technologies can help, but many others where they cannot.

For example, these technologies won't be able to tell you what the right salary range should be for maintenance personnel. However, once we know what that figure is, we can use it to determine what the proper maintenance interval might be to optimize a given task. In addition, once costs are known, the optimization might determine that we can do more frequent maintenance since labor costs are lower. Or it could be just the opposite.

These technologies won't be able to tell you what the proper spare parts costs should be. However, once we know those figures and normal delivery times, they will be able to determine the optimal quantity of spare parts to hold.

It would also be able to determine whether we should consider purchasing more reliable equipment because the incremental production is significantly greater than the increased asset cost. Using these technologies, we could also determine when it makes economic sense to spend money on an improvement project, and which is the best choice from a pool of potential projects.

These examples all demonstrate how optimizing assets can improve plant profitability. While these focus on how better asset decisions could improve profitability, there are also examples from the process side where advanced manufacturing technologies add value and increase a plant's bottom line.

For example, in a similar thought process to what has already been presented, advanced manufacturing technologies won't be able to determine what utility costs should be, but they will likely be able to determine what the optimal utility utilization should be for a given situation that balances increased production against increased utility usage.

1.3 Each Plant Is Unique

You might have noticed a pattern from the previous discussion. That is, these optimization methodologies can't determine the actual prices of raw materials, spare parts, or capital equipment, but rather how we should react to a certain price environment. That's important because each plant is unique and operates in its own specific environment.

For example, local labor markets differ between plants. Thus, wages will vary from plant to plant as the availability of different skill sets varies, and the costs for those skill sets within a given geographical region. In addition, while many plants will be using the same large vendors that most of us are familiar with, there will always be other significant local supplier variations that influence a plant's economics.

The bottom line is that the optimal way to manage plant costs on the US Gulf Coast is almost always going to be totally different from how we manage plants in the Middle East. Moreover, even within a given region, such as the US Gulf Coast, there will be differences among plants literally just down the street from one another.

In other words, everybody's situation is different. Understanding that nuance is important. Those who can adapt to that reality and exploit its advantages will thrive and benefit. On the other hand, those who can't take advantage of their unique environment will be less profitable. That is why optimization can play a significant role.

1.4 Plant Optimization Nirvana

True plant optimization should, by definition, consider all variables that influence either plant revenue or costs and allow those variables to be manipulated within the optimization such that we eventually find that set of values whereby profit is maximized. However, given the large number of variables that do influence plant profitability, it hasn't historically been possible to frame the problem in such a manner since the problem becomes too unwieldy and overwhelming. So, to make the problem easier to solve, we have normally bucketed it into its major categories and solved it independently of one another.

This simplification approach has pluses and minuses. While it makes the problem easier to solve, it also compartmentalizes the problem more. This implies that one group, who is trying to do the right thing, may not know about what another group is doing. This makes it possible, maybe even likely, that they will be working at counter purposes to one another without ever knowing it.

For example, during my career, my observation is that there seems to be an imaginary brick wall between those responsible for the "process" side of the business and those who are focused on the equipment side. They have different worldviews and are not necessarily comfortable in each other's world. It's been my impression that it's like one group speaking French and the other Swahili. They often talk right past one another.

That's a major problem. Ultimately, it doesn't benefit either side – neither the equipment people nor the process people, and certainly not the business. We all know that you can't make a product without the corresponding equipment. On the other hand, we also know that equipment doesn't exist on its own. It's there solely to make products.

True plant optimization nirvana would be to meld those two worldviews into one overarching comprehensive view, where you could perform the optimization of all relevant variables from both sides seamlessly in one technology package. Unfortunately, we are not there yet – and won't be anytime soon. It will take time and there will be intermediate stepping stones on the way to that goal.

To borrow an analogy, a comprehensive optimization vision like that outlined above would have been similar to people talking about a Smart Phone back in 1984. Those on the bleeding edge might have been able to envision it, but that futuristic product wasn't going to be in the hands of practical people anytime soon.

So, since we won't be able to get there in one fell swoop, a reasonable compromise in today's era might be to perform an optimization that isn't as integrated as we might like, but at least one based on profit, which includes some intelligence from everybody's world.

For example, when I manipulate process unit availabilities for the different process units in my plant to find the best choice for each of them, I don't make choices that alter the optimal choices that have already been made on the process side.

There's an important principle at stake. That is, optimal asset choices don't upset previous optimal process side choices. In other words, when I perform an optimization on one side of the business, I don't do anything that runs counter to the other side. As in medicine, the first philosophical principle is "Do No Harm."

Even better, when I perform an optimization on one aspect of the business, I use that information to further improve the optimal value on the other side of the business. For example, in the previous discussion about choosing optimal process unit availability targets, I recalculate yields and total production to drive the plant toward optimal profit.

Figure 1.1 expresses both those ideas under the overarching theme of optimization. The process and equipment sides contribute in different ways, but they always maintain a balance between the two.

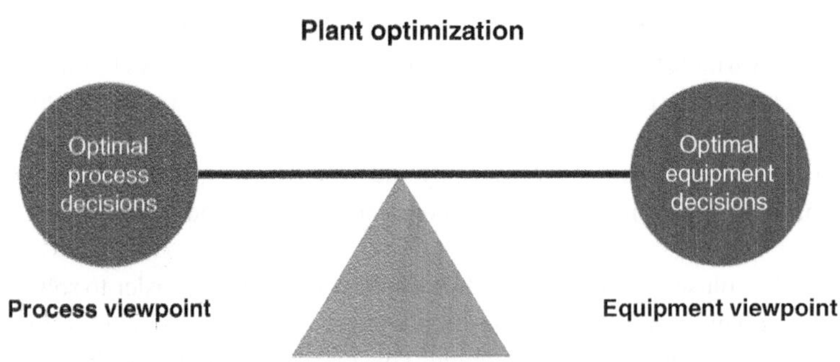

Figure 1.1 Plant Optimization – Process/Equipment Relationship

What this discussion has exposed is that the process and equipment views of the business have not historically been aligned with one another in the manner that best suits the business. Let's explore that topic a little bit further.

1.5 Process/Asset Views of the Business Need Alignment

I always remember a conversation in 2005 when I was talking with somebody who was working at one of the largest oil refining companies. At the time, I was working at a prominent asset management software provider.

He said, "We understand the process side of the business quite well and don't feel we need help there. However, if you have suggestions on the asset/equipment side, we are all ears since we know that we have significant room for improvement."

The longer I worked on the asset side of the business, the more I realized how true that statement was. On the process side, as will be explained in the next section, planning/scheduling/process control/optimization technologies are quite mature. They are "must have" and most companies have already adapted them. The process side of the business has largely been "optimized." Is there room for improvement? Sure, but there's always ways to make improvements. However, it's more in the realm of diminishing returns rather than major gains.

In my opinion, the reason that there has been significant progress in optimizing the process side of business is that large oil refineries and chemical companies are run principally by chemical engineers. Production and product slate decisions are their main goals. Those numbers are highly visible within the organization and generate a lot of attention. There's always significant interest and benefit in optimizing the production/product slate. Plus, that's chemical engineers "sweet spot." They enjoy solving those types of problems.

The equipment side of the business is totally different. It's historically been dominated by a different group of people – mechanical engineers, who have a different set of skills, focus, and worldview than chemical engineers.

It's probably not surprising that chemical engineers are not nearly as passionate about the equipment side – and even less so about maintenance. I am not trying to discount the importance of maintenance, just stating what I have observed during my career that many chemical engineers consider it more to be a necessary evil.

Let's be realistic. When you bring your car into the shop, it's not because you were looking forward to a day with the car mechanic. I think that most would agree that this is not nearly as much fun as hitting new production records. After working on both sides of the business, I've concluded that the two different groups of people spoke in different languages. Worse yet, they often talk past one another.

While the process side may be interesting and glamorous, the reality is that the process industries are a highly capital-intensive business. The equipment in our plant consumes an enormous quantity of resources, dollars, and attention. Given that, it's important that the asset side of the business be just as optimized as the production and product slate currently is.

So, for the time being, we won't be able to fully align those views. However, it's still in our best interest to do so as much as possible. As you will see, as this book proceeds, we will make significant strides toward that goal with the suggested technology and methodologies.

Given that introduction, let's determine how well advanced manufacturing technologies has done in optimizing the two major sides of our business – the process and equipment sides. We'll start with the process side since it has had a significant head start. As we explore that area, what we will realize is that there are learnings from that area that we can leverage to improve the asset side of the business.

1.6 Optimization Technologies on the Process Side of the Business

Over the last couple of decades, many advanced manufacturing technologies have appeared that are now used to optimize product slate and total production. In fact, many of these advanced manufacturing technologies on the process side have become truly "must have," and most plants would not think of running their facilities without them. For example, all plants undergo a planning and scheduling process, almost always using industry-standard tools.

Linear programs (LP) used for planning are of particular significance. These are especially important in refining applications, which allows planners to optimally choose what feedstocks they will run and the final product slate. LPs consist of literally thousands of equations that describe the unit yield structure in a reduced form, as well as simple unit constraint expressions. LPs are typically run for time periods of 30, 60, or 90 days.

The LP is the most important tool used to optimize a refinery's economics. While other industries might not be nearly as complex, most businesses still go through a similar process – even if the planning process is done in a spreadsheet. Planning is important in all process industry businesses.

It's important to note that the planning function does not determine what will be run on a specific day. Rather, it concentrates on determining the different feedstocks that would be needed over certain time frames, as well as the amount of product produced.

To create more granularized schedules, most plants also use specific vendor-supplied scheduling programs that translate the LP plan into daily or weekly schedules. In other words, scheduling takes the plan and puts it into an actionable form for operations.

While some companies use vendor-supplied technologies, others rely on a sophisticated spreadsheet to perform their scheduling. Whether in a spreadsheet or using vendor-supplied tools, these programs are complex and have usually been honed for that particular plant over many years.

Up to this point, we have described two advanced manufacturing technologies – planning and scheduling – that have been critical to maximizing profit through their ability to optimize product slate and production. However, there is another technology that also plays a significant role – advanced control and optimization (ACO). Planning and scheduling determine what products we will make. ACO optimizes that product selection, dynamically, and in real-time.

As an ACO engineer, one of our more important goals was to safely run the process unit up against its true operational constraints. That's not to say that operations personnel haven't always been trying to do the same thing to the best of their ability. However, they operate under significant constraints that ACO does not.

That is, it's been estimated that the average operator has ultimate responsibility for as many as 400 control loops. With so many loops to manage, operators have learned to build in "safety margins" to prevent products from going off spec. They know that with a built-in margin, if a variable starts drifting toward a limit, it will give them enough time to react and keep the plant out of trouble. For example, if the debutanizer overhead C5+ composition is 1%, then operations will keep the actual product spec below 0.5%, just to be safe. If I were in their job, I'd do the same. Unfortunately, this is not an optimal plant operation.

ACO optimizes a plant by dynamically allowing it to run much closer to its actual constraints. Using the power of the computer, ACO can quickly recall the history of all past moves, perform many calculations based on that history, project where critical variables will be in the future, and then calculate unit setpoints that optimize production – and easily do this every minute of every day.

Using ACO, the large "safety margin" operators typically build in can be eliminated enabling the plant to run much close to actual plant constraints, which is almost always the most profitable operation point. With so many variables to consider, there's simply no way any human could effectively compete against the computer. ACO does this reliably each minute of each day, 7 days a week, 52 weeks a year, without vacations or calling in sick. In addition to running the plant up against its true constraints, ACO is also able to maximize production by safely increasing the feed rate, when there is room to do so.

So, by using just three advanced manufacturing technologies – LPs, scheduling, and ACO – we are able to produce the optimal product slate, maximizing production when advantageous to do so, minimizing product giveaways, and minimizing energy use – all at the same time. The combination of these inherently maximizes revenue – and to a certain degree profit. So, from the process and operational standpoint, our manufacturing technologies have been doing a great job in helping us maximize profit.

Now, let's turn our attention to the asset side to see how well we have been using advanced manufacturing technologies to optimize our asset base.

1.7 Optimization Technologies on the Asset Side of the Business

Optimization of the asset side of the business can be based on either a "single asset" or "collection of assets" viewpoint. As opposed to process optimization, which was largely concentrated on pure production, where we didn't need to consider costs nearly as much, optimization of the asset side will, by necessity, entail significant emphasis on the cost side. It does and can influence production. We shouldn't forget that. However, costs are going to play a more major role within asset management, unlike the case with the process side.

Since this can be such a broad discussion, let's narrow the focus to a handful of important areas to make it more relevant. Even though it might appear limited, our chosen areas will cover a lot of important problems that affect most plants grapple with. These include the following:

- Turnaround timing/duration, as well as what equipment should be restored in each turnaround
- Proper selection of improvement projects (and ultimately process unit availability)
- Task optimization
- Spares optimization
- Maintenance plan options optimization
- Design optimization

In addition, all of these optimization problems should incorporate knowledge of how feed quality and process conditions affect equipment performance.

1.7.1 Turnaround Considerations

Turnarounds are expensive. They not only either dramatically reduce production or curtail it completely for a period of time, but they also incur significant costs in terms of personnel and equipment repairs.

Turnaround timing and duration are chosen for many reasons. Common reasons include necessary catalyst replacement, critical inspections that need to occur within a certain time frame, a capital improvement project that needs to be implemented, or supply issues, among others. Since turnarounds are so complex, there will always be "art" in terms of how they are managed. However, the more science that can be injected into the process the better.

Whereas the turnaround critical path may be any of the aforementioned items, one area where we could be doing a better job is determining when equipment availability might be a bottleneck in an upcoming planned turnaround or some future turnaround. This is a classic optimization problem that helps us determine when a turnaround should be taken if equipment availability is the critical path.

On the other hand, even if equipment availability is not the critical path, we can still determine the optimal choice of what equipment should be serviced in this turnaround or what could wait until a future turnaround so that we can optimize total production and costs over the long haul.

Today, this is art. However, optimization could play a more decisive role in how we manage turnarounds. There is a lot of room for improvement.

1.7.2 Improvement Project Selection – Determining Optimal Process Unit Availability Target

Most companies have a bad actor evaluation process. One of the outputs of this process typically highlights potential improvement projects for consideration. Sometimes, it's obvious when an improvement project is needed. However, it's not always trivial to figure out which improvement project provides the most incremental value for a limited amount of improvement dollars in an integrated facility with several interconnected process units.

Even if we haven't thought about it in this manner, what improvement projects we choose will ultimately determine each process unit's availability. That's because each unit contributes to production in different ways, and how to incrementally increase production can vary significantly from unit to unit depending on improvement project options and cost.

What this lesson ultimately teaches us is that it's not necessary for every process unit to run at the same availability target. Even if it seems counterintuitive, there's no need to keep "improving" the target availability, unless there's a business reason to do so. Ironically, maintaining the lowest availability target consistent with meeting the business needs often results in maximum profit.

How to prioritize improvement projects is a classic optimization problem. By extension, what each process unit's availability target should be is also an optimization problem.

Today, neither of these problems is being optimized in any significant fashion. There's also considerable room for improvement here.

1.7.3 Task Optimization

There are many different maintenance routine tasks executed daily in every process plant. The timing for each of those tasks has historically been chosen once and only rarely reevaluated thereafter. This is clearly suboptimal.

There are many reasons why task timing should be periodically reevaluated. For example, technology may have changed, which could now allow an interval to be extended without any production or equipment impact. However, this rarely happens. Not only does this potentially waste labor resources, but it also consumes spare parts for some equipment because every time that equipment is broken into certain spares must be replaced. Rings are the classic example. Thus, task frequencies ultimately affect what spares must be held.

Task optimization that can balance production, labor, and spare costs is an area that is ripe for improvement.

1.7.4 Spares Optimization

How many spares to retain is often determined based on a vendor's initial recommendation and never altered. This leads to excessive spare parts spending.

There are many reasons why spare parts holdings should be periodically reevaluated. First, operational reasons may be driving increased failures that we may not have considered previously. For example, there may be more impurities in a process line that are damaging seals. That might necessitate a reason to hold more of that type of seals. It could also be the opposite, where operations had been able to correct a past problem, and there is no longer the need to hold as many spares.

Also, there may be many supplier reasons why one would want to increase spare holdings. For example, if a reliable vendor that the plant relied upon in the past was suddenly unable to supply parts and we now needed to rely on other vendors with longer lead times, we'd want to reevaluate our holdings.

For many reasons, there's significant room for improvement in optimizing a facility's spare parts.

1.7.5 Maintenance Plan Options Optimization

During turnarounds or for more complicated repair jobs, there are often different maintenance plan options that could have significant cost, schedule, and production consequences. Sometimes, what option is best is obvious. However, in other situations, it's not as straightforward and could represent a classic optimization problem.

While this is done to some degree today, there's plenty of room for science to play a better role in optimizing maintenance plans.

1.7.6 Asset(s) Design Optimization

During an improvement, debottlenecking, or grassroots design project, many design decisions need to be made for each asset or collection of assets within the project.

Typical questions designers wrestle with might be, "What level of reliability do I need to this particular piece of equipment? What if I can find the same type of asset at a higher price, is the incremental production worth the extra investment? When do I need parallel equipment trains or redundancy?"

Many of these design decisions are excellent candidates for optimization.

1.7.7 Storage Tankage Determination/Optimization

Most plants are faced with the question of what is the proper amount of storage tankage? Operations generally argues for more, but tankage costs capital to build and maintain, plus it can have tax implications in many locales. However, there's no doubt that it certainly helps buffer "unreliability" – especially in those cases where an upstream unit has a poorer inherent availability than the downstream unit that it feeds in series. What's the best balance between these differing objectives?

Optimization can determine the proper amount of tankage or how to allocate that storage capacity throughout a plant. For practical purposes, this is not being done today in the process industries.

1.7.8 Feed Quality and Process Condition Considerations

Decisions made on the process side often have implications for how we manage the asset side of the business. For example, refineries often run different crude feed slates, which can have dramatic and significant effects on equipment performance. To make better business decisions, we should include those asset effects in our optimization problem definitions.

To the best of my knowledge, this is not being done at all in the process industries, or if so, not nearly as well done as should be the case.

All of the items discussed in this section show that optimization of the asset side has significant potential and promise. However, most of these items are not being optimized to anywhere near the same degree as the analogous problems on the process side. The remainder of this book will explore those problems, and more, to show how we can make significant progress.

1.8 Conclusion

This book is primarily about plant optimization, which, by definition, is about maximizing plant profitability – the difference between revenue and costs. Since there are many different aspects of plant operation that can affect profitability, this book will only focus on advanced manufacturing technologies and methodologies that directly influence plant performance.

There are two main "sides" of the business – the process and equipment aspects. In theory, plant optimization should consider both simultaneously because a process needs equipment to make a product, and equipment doesn't exist in isolation. Rather, equipment is an essential element among an interconnection of many other different pieces of equipment, there to help manufacture saleable products.

Unfortunately, the process and asset sides of the business have not historically been optimized simultaneously for a variety of reasons. One reason is that different people with different worldviews and skill sets manage the process and asset sides of the business. In addition, the technologies and methodologies on each side of the business have developed separately and at different times from one another. Plant optimization nirvana would be to combine those aspects of the business together, so that each person is cognizant and supportive of the other.

The industry has put into place tools, largely adapted worldwide, to perform optimization on the process side. Those are considered "must have" and most plants use them in one form or another. The situation on the asset side of the business is much less mature, with only spot success stories.

We view plant optimization proceeding in two distant stages. The first would be analogous to moving from the push-button phones to the Blackberry era of the mid-2000s. It would be to improve our ability to perform optimization by making better asset decisions. This optimization would be based on maximizing profitability for a handful of important asset-related problems:

- Turnaround timing/duration, as well as what equipment should be serviced in each turnaround
- Task Optimization
- Spares Optimization
- Proper Selection of Improvement Projects
- Process Unit Availability Determination
- Asset or "Collection of Assets" Design Optimization

This will move the industry forward in a significant way. Whereas this will create value where it never existed before, this is an intermediate step, but a necessary step, in moving to final optimization nirvana, which seamlessly integrates the process and asset sides of the business together into one comprehensive whole. That would be equivalent to the analogy of moving from the Blackberry to the smartphone era. However, we are nowhere close to that today.

1.9 Future Chapters

This book consists of 20 chapters and 8 logical sections, which are listed below:

- Reliability Programs – Past and Future
- First Optimization Problem/Solution
- Program Foundational Information
- Remaining Optimization Problem/Solutions
- Organizational Considerations
- Conclusion
- About the Author
- Appendices

Generally, we start with a history of how reliability has evolved over time and where we hope it will head. Then, we introduce the first optimization problem and its corresponding solution. In many ways, this is like your introductory chemistry or physics class. It has all the elements that we use throughout this program. We then step back to detail some important foundational elements, before proceeding to detail the remaining optimization problems/solutions. Once the reader has command of all the different problems the methodology can solve, we give a detailed view of how to choose the first problem for each company along with all the major organizational changes that would accompany a program like this. Last, we summarize what we have learned.

1.9.1 Reliability Programs – Past and Future

In Chapters 2 and 3, we will discuss how reliability programs have evolved over the years. Chapter 2 will discuss the first generation of reliability technology, whereas Chapter 3 will discuss Gen 2, which is the subject of this book – using asset methodologies to optimize process plants.

1.9.2 The First Optimization Problem/Solution

In Chapter 4, we present the details of our first optimization challenge – How to Select the Best Improvement project (which also dictates the optimal process availability targets for each process unit). This example contains all the important elements of our program. It's complicated enough to understand the nuances of the approach, but not so much that one becomes overwhelmed.

Besides the problem and solution itself, the reader is introduced to the two main technologies discussed in the book – Optimization and Monte Carlo simulation.

1.9.3 Program Foundational Information

After the introduction of the first optimization problem and solution in the previous chapter, we step back to introduce more detail on important aspects of the problem. These include:

- Chapter 5 – Monte Carlo Simulation Overview
- Chapter 6 – Optimizer Overview
- Chapter 7 – The Consultation Process - The Main Work Process

Chapter 5 will give the reader an introduction to key aspects of the Monte Carlo simulation, with more details supplied in an appendix.

Chapter 6 will provide an overview of how to structure optimization problems. After a review of some material from Chapter 5, it will provide optimizer definitions for all remaining optimization problems discussed in the book.

Chapter 7 will introduce an important aspect of the asset optimization methodology – the details associated with the consulting process that will drive the development of the data and a sanity check on the models that are developed. This is the major work process for this methodology.

1.9.4 Remaining Optimization Problems/Solutions

This section will be the bulk of the book. It consists of a detailed explanation of all the remaining optimization problems/solutions. Topics covered include:

- Chapter 8 – Turnaround Considerations
- Chapter 9 – What About Process Conditions?
- Chapter 10 – Opportunistic Maintenance Optimization
- Chapter 11 – Spare Parts Optimization
- Chapter 12 – Task/Resource Optimization
- Chapter 13 – Tankage Determination/Optimization
- Chapter 14 – Improving Availability
- Chapter 15 – Equipment Reliability Optimization
- Chapter 16 – Plant Optimization within the Design Process
- Chapter 17 – Combined Optimization

Chapter 8 introduces one of the more complicated chapters in the book, but likely one of the most valuable – Turnaround Considerations. There, we look at how we could use this technology to help determine when we might take a turnaround, its length, and what equipment we could possibly restore within that turnaround.

In Chapter 9, we introduce a mechanism for how to incorporate feed quality, process condition changes, or contaminant levels and how they might influence asset performance decisions.

Chapter 10 shows how we could optimally evaluate different maintenance options for a process unit or across a group of process units.

Chapter 11 is about how we can optimize spare parts inventory to balance the need for lower cost, but also ensure enough parts are available so production is not impacted.

Chapter 12 shows us how we can optimize maintenance tasks for a process unit or across a group of units.

Chapter 13 examines how we can optimize intermediate storage tankage. This can often be used when it is not practical to improve an upstream unit's reliability in a cost-effective mechanism.

Chapter 14 isn't about optimization per se, but rather a general discussion of different options to improve the availability of a process unit or group of units.

Chapter 15 introduces the idea of how the user can evaluate different equipment design options where there are reliability and cost differences between different equipment choices. This is a necessary concept for the chapter on design optimization.

In Chapter 16, we show how to optimize the design of a simple improvement or debottlenecking project. We limit the scope of this discussion to no larger than a single process unit.

Chapter 17 may well be the most important chapter in the book since it ties together different individual optimization problems into a combined optimization problem.

1.9.5 Organizational Considerations

This section shows how we can implement the program within our specific organization.

The first chapter in this section, Chapter 18, Mapping Models to Optimization Problems, presents a generalized approach to how each specific plant can determine the best opportunity to begin its program. On the other hand, Chapter 19, Creating a Program Master Plan, provides many ideas on the various organizational aspects that the plant will need to consider for successfully sustaining a program.

1.9.6 About the Author

This chapter summarizes the author's journey and how he gained the experience to write this book based on many years of Advanced and Control Optimization, along with extensive time spent working for reliability software vendors with an emphasis on Monte-Carlo simulation modeling.

1.9.7 Appendices

There are two appendices. The first, Appendix A, provides further details on Nuts and Bolts of Monte Carlo Simulation, while the second, Appendix B, is a detailed explanation of the refinery example used throughout the book.

Let's now take a look at the evolution of asset management in the process industries over the last couple of decades. The next chapter will provide significant insights into the state of the art today and eventually, where we'll need to head. Once we have that background under our belt, we can lay out the vision of where we need to go to manage our assets truly to maximize plant profitability.

Let the journey begin!

2

Gen 1 – Transitioning from Reliability to Asset Management

If you observe how any technology changes over time, what you will notice is that it tends to move forward in fits and starts. That's normal. It was true of computers, cell phones, and has been true of Asset Management as well. Progression is what's important – even if it isn't always in a straight line.

Thus, this chapter will discuss reliability's first generation of technology – both its successes and areas where it could still improve. That will set the stage for Gen 2, the main subject of this book, which will be discussed in the next chapter.

After a brief historical introduction of its early beginnings, we will discuss how Gen 1 Reliability has morphed into Asset Management. Next, we will detail how reliability was organized in its early years, as well as how leading organizations are now changing that structure as reliability departments have become more common and sophisticated.

Then, we will discuss where Gen 1 has fallen short and what shortcomings we need to address before we can successfully move into the next phase. As part of this improvement discussion, we will also focus on one particular technology – Monte Carlo simulation technology, and what has held it back. That's critical because it is one of the key technologies needed for Gen 2.

Last, we will try to get everybody excited about the future as we paint the vision of what's possible with the next generation of technology. A key part of that discussion will be to demonstrate how similar technologies on the process side evolved from their nascent beginnings to becoming "must have" technologies nowadays. We can achieve the same with new asset optimization methodologies as those other technologies did over the past 30 years.

2.1 Reliability's Early Days

When I first started in this business over 30 years ago, reliability departments didn't exist. Now, they are common. That's significant progress.

When I was in college, it wouldn't have been possible to obtain a degree in Reliability Engineering. While still not offered at all mainstream colleges, it's now possible to find universities that offer those programs – if one is motivated enough. That's even more progress!

As Reliability Engineering became a true discipline, there was a convergence on what topics were essential. These included criticality analysis, strategies, reliability-centered maintenance (RCM), failure modes, effects, and actions analysis (FMEA), and reliability modeling. The modeling aspect can include both individual equipment modeling, such as Weibull analysis, and system modeling. If one obtains a degree in Reliability Engineering, these are the types of classes that one would be studying.

Plant Optimization in the Process Industries: Incorporating Equipment/Assets in the Decision-Making Process, First Edition. Marty Moran.

Most reliability departments were initially organized as adjuncts to existing maintenance departments. That made sense originally since reliability initially tended to be focused on equipment, and most of the people who staffed the original reliability departments transitioned from existing maintenance departments.

However, as time progressed, it became widely recognized that reliability encompasses more than just maintenance. In fact, reliability and maintenance are fundamentally different in many key respects. For example, reliability is inherently strategic, while maintenance is tactical. At its core, reliability is about "managing risk." That's a unique concept that few thought of while working in maintenance.

This led to the understanding that the original management structure of reliability reporting to maintenance has become confining. In fact, it has often resulted in reliability and maintenance working at cross purposes to one another. Like a child that first transitions into a teenager and ultimately an adult, it's time for reliability to grow up and become Asset Management.

2.2 Rebranding Reliability to be Asset Management

We also come to realize that the word "reliability" is tainted because most people already have a preconceived notion of what reliability means. Usually, that implies something akin to maintenance, though its exact meaning will likely vary from individual to individual. Unfortunately, a common impression is that reliability personnel are just maintenance people with a different colored shirt.

There are many issues with people associating reliability with maintenance. First, it pigeonholes the concept. Speaking as a chemical engineer, when something is associated with maintenance, I quickly conclude that it's not my job. So, I tune it out. Yet, Asset Management should be part of my job – and everybody else's job for that matter.

Another problem is that most people have a negative view of maintenance. When I have to take in my car for "maintenance" is that a good day or bad day? Reliability, by simple association with maintenance, has a negative connotation.

Plus, associating reliability with maintenance limits it to reducing losses. The perceived value of improvements becomes the percent reduction in the small piece of the manufacturing time – downtime.

As anybody who has worked in marketing can tell you, when your audience already has a preconceived notion of a concept, that's a problem because they will rarely change their mind. It doesn't matter how much effort you put into it; once their mind is made up, it's made up. It's like trying to remold cement after it has set. Forget it. Why bother even trying?

But it's an altogether different story if you have a blank slate to work with. When you do, you can plant the seed and let it grow into the concept that we want it to be.

That's why, to be effective, we are going to have to broaden this area to make it more positive, inclusive, and applicable to all aspects of the organization. For that reason, we need to rebrand this area as Asset Management, since so few people have preconceived notions of what that entails. That enables us to define it how we want.

Who can be against Asset Management, and by extension optimization of that asset base?

For those reasons, we will rarely use the word "Reliability" from this point forward. Rather, we will now be talking about Asset Management.

2.3 Changing the Reliability Management Structure

As noted, our reliability departments are still too closely tied to maintenance – organizationally, philosophically, and technically. That's an issue because, by definition, optimization of an asset base, normally on the order of several billion dollars, must actively and organizationally involve more than just the maintenance department. We must change the common mindset that managing assets is solely a maintenance function.

In fact, managing assets of this order of magnitude needs to be an essential part of senior manufacturing, as well as production and operations management. Fundamentally, it's everybody's responsibility and should be managed at the highest levels of the organization.

In addition, Asset Management must evolve into the asset equivalent of our modern-day process safety departments – independent and accountable only to senior management. Otherwise, the fox ends up guarding the chicken coop – and we all know that usually ends up.

Fortunately, some leading organizations have already started to reorganize in this fashion.

2.4 Where Did Gen 1 Fall Short?

While reliability made many important strides in Gen 1, it didn't get everything right. That's not surprising. So, we'll need to make some mid-course corrections to increase the odds of success in Gen 2.

First, most reliability approaches have lacked a business focus and are almost always equipment-focused, usually single equipment-focused. We must never lose sight of the fact that equipment exists only to make products. Equipment is solely a means to an end. Thus, reliability professionals must broaden their footprint.

Second, reliability engineers don't seem comfortable making assumptions, and all too often don't want to attempt problems if they don't have excellent data. We can't constrain ourselves in that manner.

Third, reliability departments have become too enamored with technology by themselves and all too often are missing the bigger business picture. At the end of the day, reliability tools and methodologies exist to improve business outcomes. They are only a means to an end.

In particular, there's one technology that has struggled, which is important to any future asset optimization program – Monte Carlo simulation. We need to understand why, so we don't repeat those mistakes.

We will attempt to correct all those shortcomings in the next generation.

2.5 Adoption of Monte Carlo Simulation Technology Has Struggled

Monte Carlo simulation is a key core component of Reliability Engineering's system modeling technologies. We will eventually discuss its particulars at length throughout different chapters in this book. But for now, let's skip the technical details, and just note that most organizations have struggled to use it effectively, even though it has been around for over 25 years.

That can be partly attributed to the fact that organizations were justifiably spending a significant amount of time on issues that were more foundational and important for the long-term development of their programs. In other words, Monte Carlo simulation was a more advanced technology and could wait until other aspects of an organization's program had matured. Plus, even when they had time, most organizations didn't really know what problems to apply it to.

In addition, vendors need to be assigned a portion of the blame. It's been my impression that all too often, Monte Carlo simulation technology has been sold as an ad-hoc software tool, rather than a solution to an operating company's business problems. Unfortunately, it's often been a software tool in search of a business problem to solve.

The downside of being a tool that can be used to solve many different problems is that, well, you can be used to solve many different problems. For the Monte Carlo simulation, trying to be everything to everybody usually means that you end up being nothing to most organizations.

In the business world, you need to narrow the focus to the most important problems. Only then can you begin to optimize those business functions. In other words, focus brings clarity. That's why, in the next chapter, we will show how we can apply Monte Carlo simulation technology to solve the most important business and technical problems. We won't try to make it applicable for all problems. We will focus on the most important problems that most operating companies are dealing with.

Last, even where Monte Carlo's value has been recognized, it hasn't been integrated into a company's main work processes like it should have been. Unfortunately, many projects that have been implemented have been "one of a kind" – done once, and the organization quickly forgets about them after the project has been completed. That makes it difficult to sustain a program.

All these issues will be addressed in future chapters, since Monte Carlo simulation will be one of the two principal technologies discussed in this book, along with true optimization technology.

2.6 Asset Optimization Nirvana – The Future

Up to this point in time, we have examined a brief history of Gen 1 Reliability, discussed how reliability continues to evolve into Asset Management, as well as important organizational changes underway within the industry, highlighted some general Gen 1 shortcomings, as well as specific technology limitations within the adaption of Monte Carlo simulation.

So, what can we expect once we have implemented Gen 2 technology and achieved excellence in Asset Management – and the optimization of those assets?

The most important thing is that we will fundamentally be focused on maximizing production while minimizing cost – and achieving both objectives at the same time. We will realize that we can't separate the production side of the business from the asset side. They are one and the same and must work together to achieve a common goal.

Here's what nirvana might feel like once it has been implemented:

- Senior executives could challenge their organization as to where their next dollar should be spent, and how much, to increase availability so they could meet production targets
- Managers could ask their staff to delineate exactly what projects, and in what order, need to be executed to achieve incremental production goals
- Instead of issuing blanket edicts as to what a process unit's availability should be, each individual unit's availability could be chosen based on science and data instead of through an arbitrary process
- When making decisions regarding when to shut down for turnarounds, and for how long, one could optimally determine what equipment to include in this turnaround, and what could wait until the next cycle, which allows one to maximize profit over several turnaround cycles, while always properly managing risk
- We could determine the optimal maintenance task intervals, balancing the increased cost of the maintenance tasks themselves, against the possibility of lost production at the unit level, or more broadly across a group of units

- We could be assured that we are not overspending on spare parts, yet ensure that we will always have enough when we really need them
- When undertaking a major debottlenecking project, we could have confidence that we are choosing equipment with just enough reliability (but not too much) that minimizes total lifecycle cost (capital now, as well as operational and maintenance later on), yet still allowing the plant to meet capital budget and production goals
- We would be able to determine the adequate amount of storage tankage, as well as its optimal allocation

If these objectives could truly be achieved, the business benefits would be significant.

Not possible?

I would argue that there were a lot of things that weren't possible when I first started in this industry 40 years ago – many of which have now become commonplace. In fact, I have seen, and been a part of several of those trends.

Just to show you how much has changed, I got to witness history when I was a young engineer. That's when I visited one of the last true "board-operated" plants in existence. It's always interesting to watch how young people react when I tell them about that experience.

To even begin contemplating that idea, one must do a Google search to find a graphic of a board operator graphic. You should see a young person's eyes light up when they realize that an operator had to literally "walk the board."

While that concept is now safely in the museum, there have been plenty of other developments that have been more enduring. For example, during my early years, DCS systems were just being installed. Process historians followed suit a few years later. Advanced process control was in its infancy, and multivariable control had yet to be invented. Real-time optimization was still just a mere dream for process and operations engineers. Linear programming planning technologies have not yet been invented. However, all of these are now a reality – and have been for many years.

For example, it's now estimated that there are over 10,000 multivariable controllers operating on six continents today. They have become so dominant that plants don't dare to operate without them. While that may be true nowadays, it wasn't always that way.

So, if the technologies now optimizing the product slate and overall production are standard fare, we need to ask ourselves some tough questions about the asset side of the business:

- Why can't we now easily determine the optimal availability for each process unit that maximizes profit?
- Why can't we easily determine the optimal maintenance tasks across unit(s)?
- While turnaround planning will always involve some art, why haven't we been able to inject more science into the turnaround planning process?
- Why are so many spare parts decisions based on what the vendor recommends, rather than on a true cost/benefit evaluation?
- When will the day come when we can consider these technologies to be as "must have" in the same manner that the industry treats LPs, Scheduling, and ACO?

The answers to those questions should make us uncomfortable. Today, none of these are being routinely done – and certainly not through any automated optimization process.

While our industry has been making some progress in certain select areas over the last 20 years, it hasn't been fast enough. Even today, too many important decisions are based on gut feel and lack the rigor that we could, and should, be applying.

I will always remember a conversation back in 2004 with a very senior engineer at one of the local Houston refineries. He had asked me if I could help him use science to decide whether they should repair or replace a crude tank. He had remarked that if we couldn't help him, a bunch of senior guys would just sit around a table and pull the answer out of thin air. (He actually used a more graphic phrase, but we can't reprint his exact words.)

We were able to help him use science to solve that problem by using a few different technologies in combination. Which technologies we used were less important than the fact that the ultimate decision was based on science – and data. As Edwards Demings famously said, "Without data all you have is an opinion." Unfortunately, too many of our Asset Management decisions are still more opinion, than fact.

Let's change that, and more, by introducing Gen 2 – True Asset Optimization. Fundamentally, it builds upon what reliability departments have done well over the last 20 years but supplements that progress with ideas that have been successful in other aspects of our business.

2.7 Conclusion

The first generation of Asset Management technology originated when reliability departments were first formed. They were generally organized as subsets of the maintenance department, reporting ultimately to their management, and staffed with former maintenance personnel who had shown significant competence in their area of expertise.

Over time, reliability evolved into a separate discipline and has generally included criticality analysis, strategies, reliability centered maintenance (RCM), failure modes, effects, and actions analysis (FMEA), and reliability modeling. In fact, it's now possible to receive a degree in Reliability Engineering, with special coursework in these areas layered on top of traditional engineering areas.

As time progressed, reliability groups are now sometimes referring to their discipline as Asset Management to distinguish how they are separate and distinct from maintenance. In addition, there are now instances where groups have been organized to either report directly to Operations or to senior management. Let's hope that these trends continue.

While a significant step forward, the first generation of Asset Management has fallen short in certain key areas. It has been too focused on single equipment with not enough regard for the process side. The reality is that equipment only exists to help manufacture products.

Too often, reliability departments have been using technology for technologies sake. We must realize that technology only exists to solve real-life business and technical problems, ultimately resulting in increasing plant profitability. A better use of technology would be to only create enough complexity so that we can solve the business problem at hand.

In addition, a real, or perceived lack of data has limited the ability to generate adequate solutions to some real-life problems. We sometimes have to make imperfect assumptions, so we can answer a problem, rather than just punting on the issue.

Last, Monte Carlo simulation technology, a key component of true plant optimization based on improved asset decisions, has struggled and hasn't been implemented in many organizations. This is due to other priorities, vendors' inability to articulate their true business value in a succinct manner, and most organizations' failure to understand their business value, and then institutionalize those gains.

Given the pace of technology development in general, and more specifically in this area, it is expected that Asset Management will evolve further to address these issues. This will be discussed at length in future chapters.

3

Gen 2 – Plant Optimization Using Asset Modeling Methodologies

The worldwide refining and petrochemical industries are capital-intensive businesses with Replacement Asset Values (RAV) estimated in the trillions of dollars. If any technology can save even a small percentage of that value, it could be worth tens of billions of dollars. This chapter will provide an overview of a methodology that has this potential.

Optimization of the major operating variables within a traditional process unit is proven technology. Established in the 1980s, it is now a mature technology that has been generating billions of dollars over many years. It has become a "must-have" technology that is simply necessary to run a modern plant.

Unfortunately, optimization of the large asset base in these industries has lagged considerably behind that counterpart on the process side. While there have been some inroads with modern reliability theory, asset management platforms, advances in condition monitoring techniques, and the introduction of Monte Carlo simulation technologies, the success stories are too few and far between. We can do better. There has simply not been the breakout success on the asset side compared to what occurred on the process side when LP planning, Advanced Control, and Optimization technologies were first introduced.

Thus, this chapter will propose a combination of ideas that are already working on the asset side and then marry them with what we learned as we introduced advanced manufacturing techniques on the process side.

We'll refer to this as Gen 2, the era when we started to apply asset modeling methodologies as an integral part of Plant Optimization, as compared to Gen 1, which was grounded in Reliability Engineering principles.

Gen 2 consists of three major aspects – a philosophical approach, two technologies, and foundational methodologies.

3.1 Gen 2 Philosophy

The major philosophical points within Gen 2 are the following:

- Technology does not exist in a vacuum. Rather, it exists only as a means to solve business problems. Therefore, we will be using Gen 2, first and foremost, to drive significant business benefits. We will not be using technology solely for technology's sake.
- The emphasis will be on creating solutions only as detailed as necessary to produce quantifiable business benefits. Simpler is better.

Plant Optimization in the Process Industries: Incorporating Equipment/Assets in the Decision-Making Process, First Edition. Marty Moran.

- We will never have perfect data, so we must rely on asset experts for a significant portion of our input. Asset data will be used as a backup and/or as an adjunct to expert input.
- The focus will not be on assets in isolation but rather on their combination since the ultimate aim of a plant is to produce a saleable product, which comes from the collection of assets.

Those four points are significant and have consequences, as will be shown as we proceed in this chapter and throughout the book.

3.1.1 Gen 2 Technology

Since we are interested in the "combination of assets," we will normally be using a technology that easily accommodates that requirement. Monte Carlo simulation fits that need, has now become more accepted within the reliability world, and works well. Not all problems will require Monte Carlo simulation, but the majority likely will. While it's one of the significant pieces of the technology puzzle, it's not the only one.

A second technology, equally powerful and necessary, is true optimization technology. This will iterate to the optimal financial solution, usually profit, by repeatedly adjusting variables to the Monte Carlo engine (if we are using Monte Carlo) until it finds an optimum. How these two technologies work together to achieve this objective are shown below in Figure 3.1.

While Monte Carlo was definitely part of the Gen 1 technology tool kit, what truly separates Gen 2 from Gen 1 is the reliance on the optimizer. Having the optimizer as the top block is important because it moves us from a focus on technical results to financial metrics, since we will usually be trying to maximize profit. Previously, most of the emphasis in the reliability world has been on technical requirements, and all too often on single equipment. So, putting on our new asset management hat and adding the optimizer fundamentally shifts our thought process. It moves us from the maintenance management mentality to how executives in the business suite think.

Another key aspect of optimization, which is not shown in the graphic, is the ability to incorporate constraints into our problem definition. In the real world, plants encounter many different constraints, and any realistic technology needs to seamlessly incorporate those into successful problem solutions.

This structure is remarkably similar to that which has proved extraordinarily successful on the process side. Advanced process control and optimization, now "must have" technologies in the

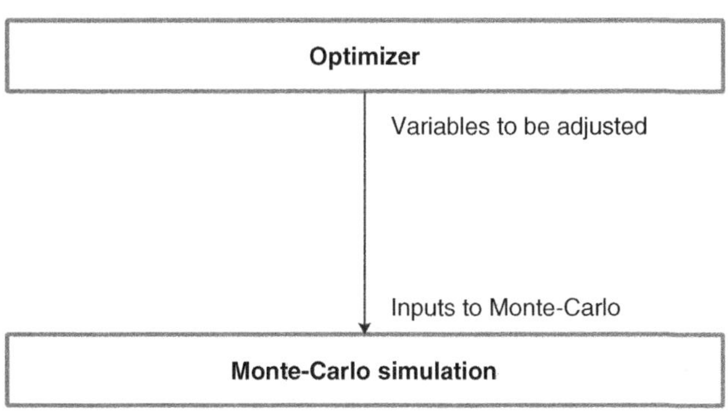

Figure 3.1 Optimizer/Monte Carlo Hierarchy

process industries, have a similar structure. Replace the words “Monte Carlo Simulation” with Multivariable Control, and you have exactly the same, remarkably successful structure.

The ideas behind this approach were first articulated by the author in a presentation at the 2004 NPRA (National Petrochemical and Refiners Association) Annual conference in San Antonio, Texas.[1] It was also subsequently published in a PTQ article the next month.[2] Chapter 4 will detail that example using a slightly modified problem.

The structure shown is a conceptual framework. In real life, getting this to work seamlessly will require some work from industry product vendors so each layer can talk to one another properly. This should be doable and would make it much easier for clients to use.

To remain product agnostic, for purposes of this book, we used the Solver within Excel as the optimization engine for the book’s specific examples. Monte Carlo was never used directly, but the probabilistic ideas behind it were simulated using Excel’s Data Tables, and the resulting data were regressed into equations before being used within the Solver.

3.1.2 Gen 2 Methodology

The methodology is also important. Gen 2.0 will rely on experts to make predictions and will not “punt” if there is a lack of data or if that data is not perfect. If engineers had always waited for perfect data, we’d still be living in the hunter-gatherer culture.

Our Asset Experts will make experience-based predictions in two key areas.

The first is to explicitly generate failure and repair data, when that does not adequately exist, or verify that existing data is accurate and can be used. A Consultation process will be introduced in a later chapter that will drive the initial data-capturing exercise for these key inputs.

This Consultation process will have other positive side benefits, besides just generating the input data. That is, it will enable a healthy, ongoing “conversation” between all major groups within the plant that either doesn’t take place today, or not to the degree that it should.

However, there’s a second area where our equipment experts will play a crucial role. To illustrate that, let’s add a third box to our Monte Carlo/Optimizer diagram, as shown in Figure 3.2.

You’ll now see that there is a box in between the choice of optimization variables and input to the Monte Carlo program. The reason for that is there will be times when we set up our optimizer to vary an input variable, but that variable doesn’t have a direct counterpart in the Monte Carlo program. So, what we will need to do is perform an intermediate calculation that will transform the

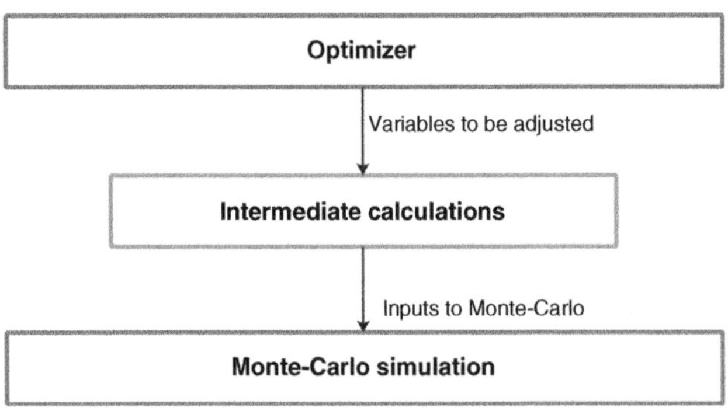

Figure 3.2 Intermediate Calculations

initial variable chosen by the optimizer into variables that can be used directly by the Monte Carlo program. Determining these intermediate calculations will be the province of equipment experts.

In fact, Intermediate Calculations, or dependent variables as they will be referred in later sections give us additional flexibility and can be used to include other aspects of the problem outside the scope of Monte Carlo. For example, they could end up being budget, labor, or production constraints – all important in real-world problems.

3.2 Gen 2 Asset Optimization Applications

So far, we have introduced the basics behind the Gen 2 approach. Now, let's paint a vision of important business/technical problems that it can be used to help solve. The following are some of the more important problems that most plants face where this technology is applicable and will drive major benefits. However, this list is not exhaustive. Nonetheless, it will give you a feel for the power of the approach.

- We all know that turnaround intervals and what equipment is serviced within a turnaround are determined based on many factors. In addition, we also know that the longer our units run, process unit availability is bound to decline. How can this technology help us make an informed judgment as to whether equipment availability will be a turnaround bottleneck, and how can we better choose the equipment to be serviced in the turnaround, the interval, and the duration based on those factors?
- Given a limited budget, what are the best improvement projects we could choose to optimally increase overall production, and hence profitability?
  - (As we make that determination, we'll see that this naturally sets each process unit's availability target).
- How do we include the effect of feed quality changes and different process conditions on equipment degradation as we make these important asset decisions?
- Once optimal process unit availabilities have been determined, how can we optimize tasks at a process unit level, or across a group of plants?
- What quantity of spare parts optimization is needed to minimize spare parts spending without needlessly sacrificing production?

3.2.1 Optimizing Turnaround Interval/Equipment Restoration Decisions

Turnaround intervals and durations are chosen for many different reasons. For example, the main turnaround driver may be to regenerate catalyst. Or we may need to shut down to implement a given capital project. On the other hand, in some countries, shutdowns may be dictated by the requirement for mandatory inspection of key equipment.

Obviously, a given process unit does not exist in isolation, so turnaround timing on a unit can be driven by factors external to that unit. For example, a turnaround may be required due to product supply limitations.

The general trend within the industry is to increase the length of time between turnarounds. However, the longer that we run a unit before we turn it around, the more likely it is that we will lower our availability below a minimally acceptable economic threshold. In other words, our unit becomes uneconomical to operate beyond a certain time. Since turnarounds take extensive time to plan, nobody wants to be surprised by an early shutdown, which could be catastrophic economically.

So, it makes sense that we evaluate whether equipment failure will be a key driver of turnaround timing at some point in the future. Since equipment can often operate for a very long period of time before failing, we need to be cognizant of the fact that the equipment under consideration might not fail before the next turnaround but fail in the middle of the next production run cycle – at exactly the moment when we can least afford it. Therefore, we need to potentially look out multiple cycles, for certain equipment, before making any final decisions. The approach described here will allow us to simulate this idea on any number of equipment and any number of turnaround cycles in the future.

The optimizer can be structured to choose the length and duration of several future turnarounds and what equipment is restored to proper functioning during each turnaround. Appropriate constraints can be included in the calculations that bracket the start and end of the turnaround, as well as what equipment can, or cannot, be restored during a particular turnaround.

The objective function will be structured to maximize profitability over several cycles. This will require calculating process stream availability, hence yields, from the turnaround(s) in the problem definition, as well as including the cost to make equipment restoration repairs during each run.

To perform these calculations, the Monte Carlo simulation will need to be run several times with different sets of data within a particular optimizer iteration. Figure 3.3 shows the structure of the problem when considering two turnaround cycles. The optimizer will choose when each of the turnarounds occur, the length, and what equipment should be repaired in each turnaround.

Availabilities are calculated for each run to generate production volumes, as well as the costs to service and restore the equipment. The optimizer can then determine incremental profitability because it has already calculated the revenue and costs. This process will repeat itself many times until the optimizer finds the optimal values of occurrence, length, and what equipment to restore in each turnaround.

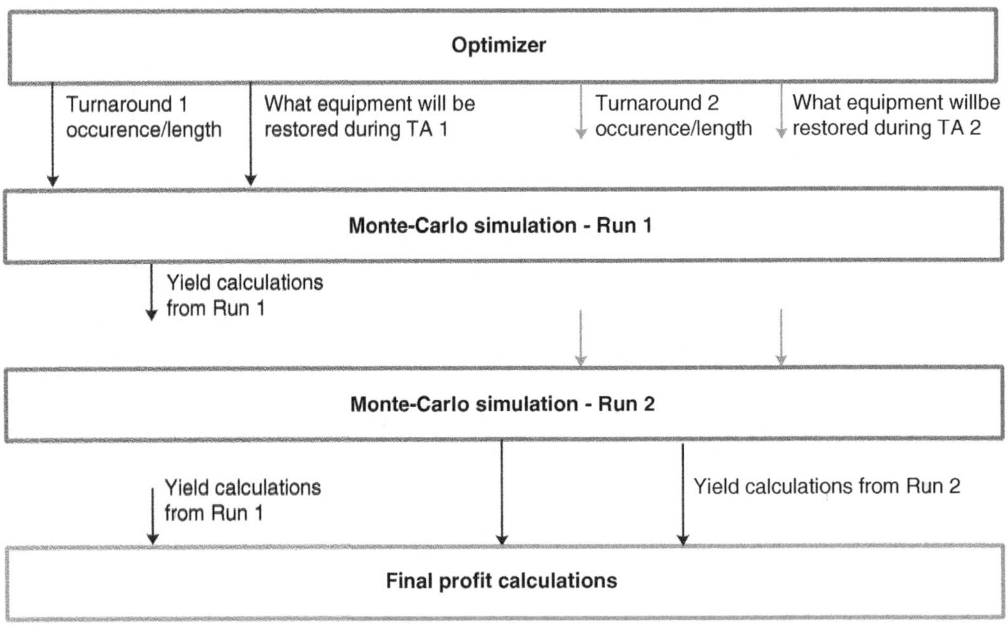

Figure 3.3 Turnaround Optimization Overview

The ideas for this optimization problem were initially developed by the author in 2004. Late that year, as part of his company, Onstream Optimization, he applied for a grant to the National Science Foundation to fund this idea under the title, "Financial Quantification/Optimization of Maintenance Decisions."[3] Unfortunately, the committee split 3-3, and the idea was not funded. However, the good news is that with modern technologies, it's now possible to solve this same problem in a slightly modified fashion.

3.2.2 Choosing the Best Improvement Project – and Ultimately Optimum Process Unit Availability Targets

With limited improvement dollars to spend, an important business question facing many senior managers is, "Where do I spend my improvement dollars to get the biggest bang for my buck?"

That's not a trivial question because not all improvement projects are created equal. For example, spending $1,000,000 for one improvement project on a certain unit may yield a certain production gain, while spending the same amount of money on a different process unit generates significantly less. So, there's clearly an optimum.

As we ask that question, we end up inadvertently challenging the conventional wisdom about setting process unit availability targets. Today, there's a belief, rarely challenged, that process unit availability over time should always be getting better for each unit. In fact, if you have felt like the target values for each process unit's availability seem to be set arbitrarily with little business consideration, you are not alone. That's because they generally are.

So, let's challenge that wisdom.

Does it really make sense to set the target availability for each unit equal to one another? In addition, does it make sense that each unit must keep improving? Or let me phrase it differently: do we need to keep spending money on that unit to improve availability? Or is there a point of diminishing returns for every process unit?

As one delves into the business and technical details behind those questions, the cracks start to appear. It turns out, as we will see shortly, that the reason that we don't want to set all availability targets equal to one another is that different process units contribute to profitability in different ways, given where they are on the availability curve.

In other words, each unit will have a "sweet spot" where we will want to operate from an availability standpoint, and that availability will not be the same for each process unit. In fact, it's intimately related to which improvement projects we choose to implement.

The same is also true about availability always improving over time for every unit. Sometimes, that is the right answer. Other times, it is not. It also depends on which improvement projects are a possibility and those that we decide to implement. More on that in the next chapter.

The following graph will illustrate the concept for solving this problem through optimization, using an example from the refining industry. In this particular case, a refiner was trying to maximize profit through increased gasoline production. Figure 3.4 shows the optimizer choosing the availability target for each of the major units that influences gasoline production and then determining its effects on all the major gasoline component streams. (The example refinery used here and in the remainder of the book will be introduced in more detail in the next chapter and is detailed fully in Appendix B. Since the example here is an overview, one can either accept it as is or turn to Appendix B to become more familiar with the refinery process units used in the example.)

As we mentioned earlier, a key component of the Gen 2 approach is having equipment experts make critical assumptions in order to fill knowledge or data gaps. For this particular problem, those familiar with the unit would need to provide high-level estimates of what discrete projects

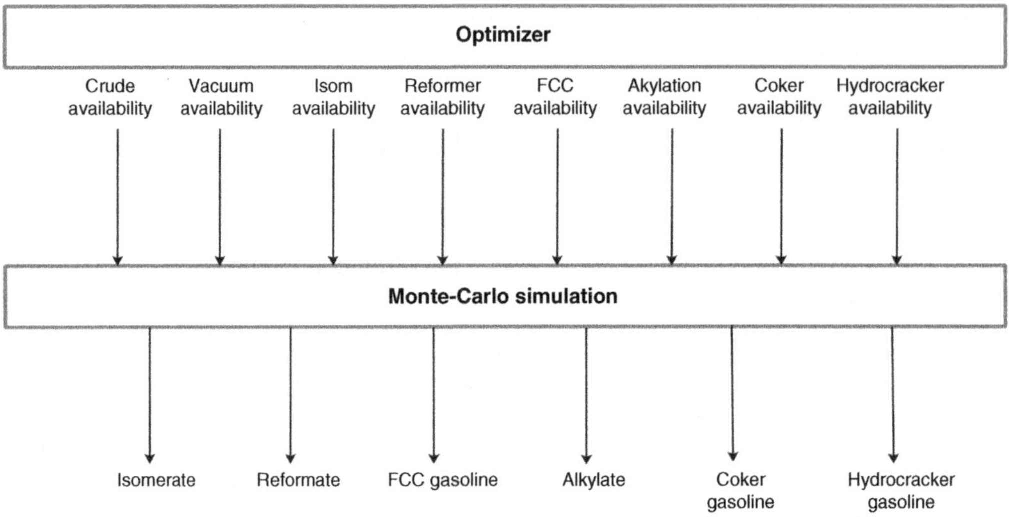

Figure 3.4 Process Unit Target Availability Optimization

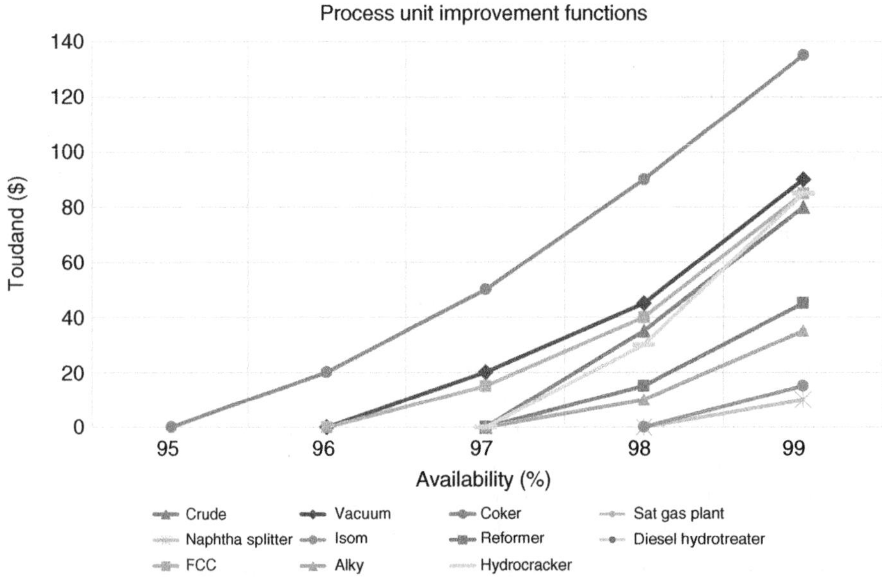

Figure 3.5 Process Unit Total Project Cost vs. Availability

would be implemented as part of a prioritized list that would be necessary to improve availability. Over time, these assumptions will be replaced with data, as predictions evolve into reality.

Of course, there is a cost associated with each of those projects. The discrete project values for these individual projects would be converted into smooth functions for the optimizer to work with, as shown in Figure 3.5. These graphs should be interpreted as the cumulative spending necessary to improve a particular process unit's availability from its current value to some future value.

In this example, the optimizer would start by choosing the availability for each major process unit. Using the chart, the optimizer would understand how much it needs to spend to generate that

Table 3.1 Original vs. Optimal Availability

Process Unit	Initial Availability (%)	Optimal Availability (%)
Crude	97	99.75
Vacuum	96.5	98.32
Coker	95	97.3
Sat Gas Plant	98	99.75
Isom	96	99.75
Naphtha Splitter	98	98
Isomerization	98	99.75
Reformer	97.5	99.75
Diesel Hydrotreater	97	97
FCC	96.5	98.86
Alkylation	97	97
Hydrocracker	97	97.96

incremental availability. But it also knows how much incremental gasoline can be produced, so it can now calculate profit as the difference between the revenue generated from incremental gasoline minus the value of the improvement projects. The optimizer will try many different sets of individual process unit availabilities and eventually iterate to a solution.

In our refinery example, Table 3.1 shows what the initial process unit availabilities were and what the optimizer determined to be the optimal values for each process unit.

If you didn't quite appreciate all the details related to this example, don't worry. By the time you finish the next chapter, you will understand those concepts well, as the next chapter is devoted solely to that problem.

3.2.3 Incorporating Feed Quality and Process Conditions into Our Asset Decisions

In some industries, such as oil refining, there can be cases where changes in feed rate, process conditions, or even contaminant levels can significantly influence equipment performance. If it is possible to predict how equipment failure rates could change given those disturbances, those effects should be included in the optimization structure to enable us to make more realistic asset decisions.

This is another example where our equipment experts can play a significant role. In the particular case shown in Figure 3.6, an oil refinery is trying to decide the best choice of three different crudes over a given time frame that would optimize yields, while keeping equipment degradation within certain limits.

The optimizer chooses the amount of each crude to run, within operator-entered limits, and then calculates the effect on equipment failure rates. Since crude feed types are changing, it is also necessary to calculate how the yield structure changes, in addition to availabilities. The optimizer will keep iterating varying the crude slate until it finds the proper combination of crudes that maximizes profitability.

While this structure is shown for an oil refining example where different crude feeds are run in varying quantities, the idea applies equally well to all other industries where either different

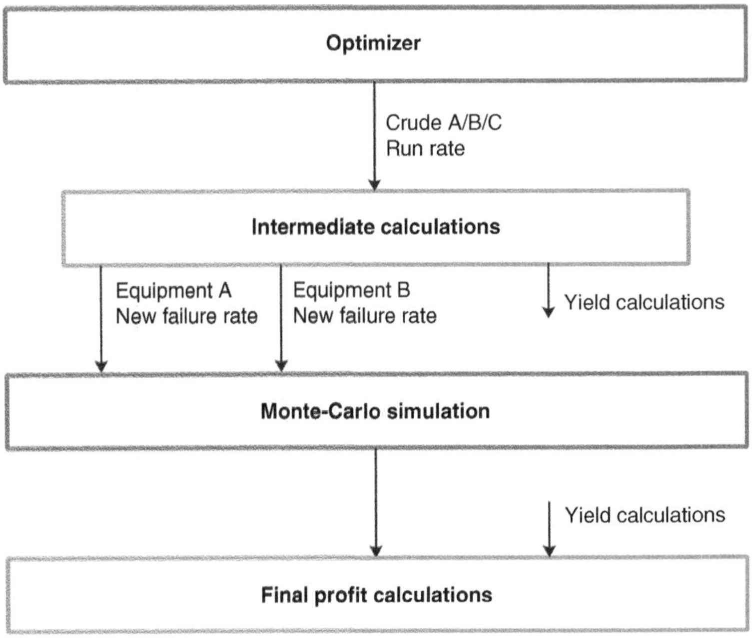

Figure 3.6 Incorporating Process Conditions into Optimization

process conditions, feeds, or some contaminants could change, and equipment experts would be able to reasonably estimate those effects.

In other words, this structure provides a mechanism for varying equipment failure rates (or, for that matter, any other input variable into the Monte Carlo program). It could be used separately but would often exist as a part of any of the optimization problems described in this book. Examples such as these will be presented in the chapter on combined optimization.

3.2.4 Optimizing Spare Parts

When equipment fails, every plant hopes that they have enough spare parts to allow the equipment to be brought online as quickly as possible. However, buying, storing, and maintaining spare parts is a considerable expense.

Spare parts are a lot like Goldilocks. Not too hot, not too cold. Or for spare parts – not too much, not too little. Easier said than done. At either the process unit level or plant, an optimal quantity of spare parts needed to be purchased and held to help mitigate equipment maintenance and reliability issues.

Gen 2 technology can easily be used to determine optimal spare parts at many different levels. Figure 3.7 shows a simple example of a refinery fluid catalytical cracking unit (FCC), where we are interested in determining the optimal quantity of spare parts for certain rotating equipment in the unit. While it is shown for rotating equipment, it can easily be extrapolated to any quantity of different equipment.

While one can choose optimal spare parts values for individual equipment, it often makes sense to perform that analysis on numerous equipment together, since spare parts budgets are aggregated together. The total budget value can easily be incorporated as a constraint in the optimizer.

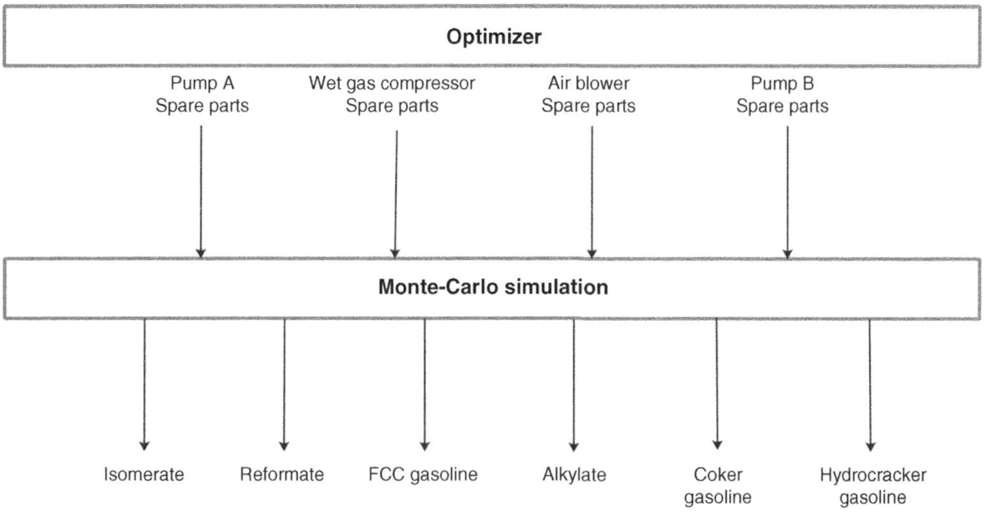

Figure 3.7 Spare Parts Optimization

3.2.5 Optimal Tasks within a Process Unit

Once we have determined what the availability should be for each process unit, we can focus on further optimization at the unit level. Assuming that we will be working with the equipment that already exists and not making any asset replacements – the case for most existing plants – generally, we will want to minimize costs while maintaining the optimum availability targets just determined. Any task within a process unit is potentially fair game for optimization. It should be noted that this optimization will be an ongoing exercise, executed periodically.

Generally, there's a tendency for task intervals to be set once and never be reevaluated. That doesn't make sense for several reasons. First, equipment ages over time, and the optimal set of tasks for equipment, especially given resource constraints that work on that type of equipment, will often change over time. Second, the technology associated with that particular functional area may have become more efficient with the benefit that task frequency might be able to be decreased with no additional risk.

Let's use an example to illustrate the task optimization concept. In this particular case, we would like to optimize oil change intervals for all pumps on a given process unit.

Let's say that there was an incident on the Fluid Catalytic Cracker (FCC) a few years back, where a piece of rotating equipment was implicated in an incident. One of the Root Cause actions from the incident was that oil would now be changed at a six-month frequency for all pumps in the unit to ensure that such an incident never occurred again.

For this problem, the optimizer is structured to choose oil change intervals separately for all X pumps on the unit. Of course, spreading out the oil change interval decreases maintenance costs, but increases the risk of an additional failure.

To solve this problem, we will also need to estimate how failure risk changes over time. That's where our equipment experts come in. They would need to make estimates for each piece of rotating equipment in the unit, similar to what is shown in Figure 3.8.

In essence, the optimizer would choose individual oil change intervals for each pump, which would allow it to calculate a new failure rate using the intermediate calculation provided by the equipment experts. The Monte Carlo engine would then be run with these failure rates to determine a new availability.

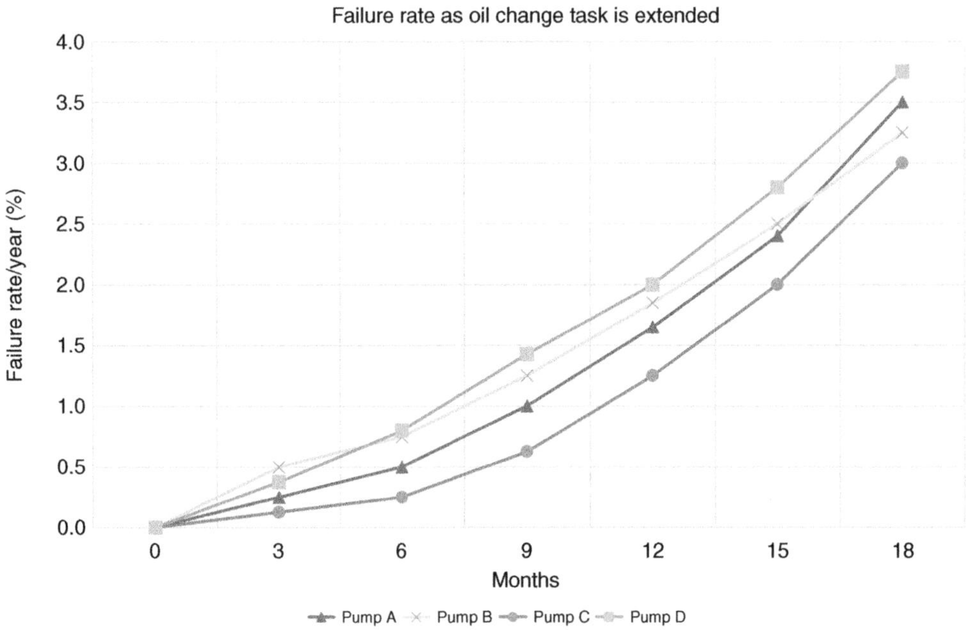

Figure 3.8 Pump Failure Rate vs. Oil Change Interval

We would now know the product distribution based on the availabilities of each pump and the maintenance costs for the oil changes. Those two combined together are the profit. The optimizer would then keep adjusting the task intervals for each pump until it found the combination of oil change intervals that maximized profit. It's important to note that the oil change interval would likely be different for each of the pumps because each pump likely contributes to profit in a unique way.

The structure of the problem is shown in Figure 3.9.

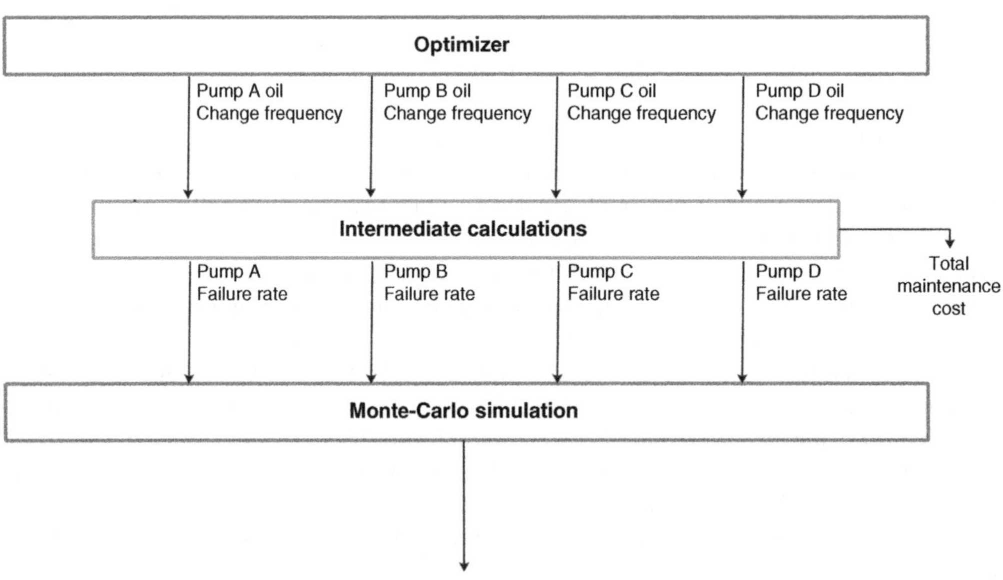

Figure 3.9 Pump Oil Change Interval Optimization

While this is admittingly a simplistic problem, it was chosen to illustrate the concept. However, we should not let its simplicity detract us from the usefulness of the approach. Each unique process unit will have its own more, complicated tasks, where task optimization would be desirable.

In addition, while the example only showed one task type, there is no reason that it needs to be limited to only one task. In fact, most of the time, it probably would not. We could use any combination of tasks where we wanted to find an optimum. Since most tasks are mutually exclusive, leaving tasks out of the problem is not an issue.

3.2.6 Other Potential Problems

The five problems presented so far are not the only problems that can solved with this technology. Not hardly. For example, maintenance optimization, storage tank optimization, equipment reliability optimization, and optimization within the design process will all be detailed in subsequent chapters.

But this still doesn't do justice to what can be accomplished with this approach. Everything highlighted so far is relatively "pure" in the sense that it only focuses on one aspect. For that reason, the last example in the book will be "Combined Optimizations," where we mix some of the purer problems together. In fact, the nirvana of this approach is to combine as many of these together as make business sense. This would better mimic how the real world works, especially when incorporating all the different constraints that one encounters in real life.

3.3 Conclusion

This chapter has introduced the main ideas behind Gen 2 Asset Optimization, which are centered in three main areas – philosophy, technology, and methodology.

At a high level, the philosophy is to stay centered on business benefits, only abstract problems to the level of detail necessary to answer the business question at hand and realize that we will never have perfect data.

Two important technologies that Gen 2 will focus on are Optimization and Monte Carlo simulation. The methodology will utilize equipment experts as often as necessary to fill knowledge and data gaps and create simple relationships where necessary.

This chapter has also introduced five main uses of the technology that are certain to drive business benefits. These are what equipment should be restored during a turnaround, as well as the optimal turnaround interval and duration, based on equipment considerations, how to choose the best improvement project, and hence optimal availability for each process unit individually, task optimization, spare parts optimization, and how to include feed quality or process condition changes into the problem structure.

It should also be noted that there will also be other problems described in later chapters that were not included in this chapter. For example, maintenance optimization, storage tank optimization, equipment reliability optimization, and process unit design.

While this chapter presented the main examples and vision of what's possible, subsequent portions of this book will detail the roadmap and "the how to" so your program can progress. Organizations may not get there in one fell swoop, but there are plenty of nuggets that will be presented to put a program in place.

Let's now delve deeper into how to determine the best improvement project, and hence the optimal process unit availability for each process unit. This is a great example that illustrates all the important concepts.

4

Selecting the Best Improvement Projects – Optimal Process Unit Availability

Most process plants have more than one process unit. In fact, usually, there are many. So, the first business question from an asset standpoint that we must ask ourselves is, "What should be the target unit availability for each unit in our plant?"

To answer that question, let's introduce the example plant that we will use throughout the book. It will be based on a refinery with the typical units found in an American refinery configured to produce maximum gasoline quantities.

A refining example was chosen for several reasons. First, it's the process area that I understand best. So, I can easily delineate its details. Second, it has significant complexity. Certainly, not all plants have this level of complexity. However, if you start with examples that are too simple, you risk not being able to deal with complexity as it arises. However, it is much easier to dial back that complexity, to a simpler example, rather than go the other direction.

Figure 4.1 shows a refinery with 11 different process units:

1) Crude
2) Vacuum
3) Saturated Gas Plant
4) Naphtha Splitter
5) Isomerization Unit
6) Catalytic Reformer
7) Diesel Hydrotreater
8) FCC
9) Alkylation
10) Coker
11) Hydrocracker

These units, in combination, would produce maximum gasoline.

If you are not familiar with the refining industry, you can refer to Appendix B for a complete process description of how this refinery would operate, the products it would produce, and a complete material balance for each unit. For those even moderately familiar with refinery operation, you can probably just skip the Appendix, and keep reading on.

So, back to the most important asset management problem question facing us. As we look at our 11 different process units, what should be their target process unit availability averaged over a given time period, say 20 years, for each of the 10 individual process units?

Plant Optimization in the Process Industries: Incorporating Equipment/Assets in the Decision-Making Process, First Edition. Marty Moran.

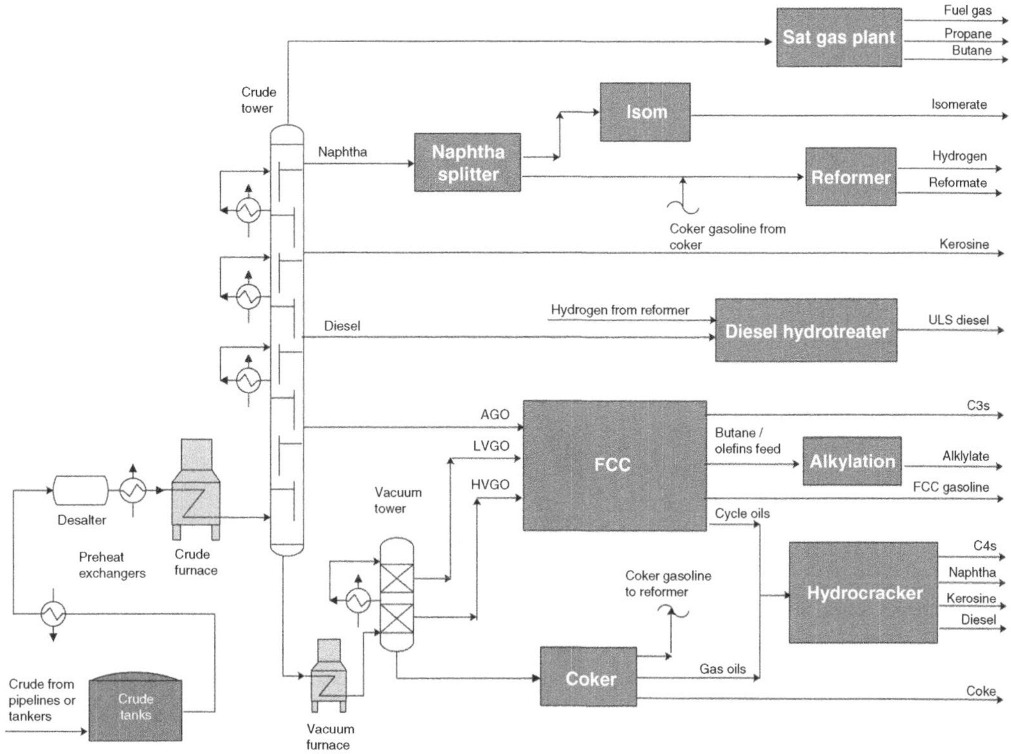

Figure 4.1 Example Refinery

Should we just state a company policy that all units should have 98% availability? Or better yet, 99% availability? Or if we are truly shooting for "world class reliability," how about 99.5% availability for each unit? If we've gone that far, what prevents us from shooting for 99.99% availability?

On the other hand, horror of all horrors, could we settle for a smorgasbord of availabilities like what is shown in Table 4.1, in which some units have availabilities significantly less than 98%, and that's actually the right solution for our plant?

From a business standpoint, we must realize and make peace with the fact that aiming for higher availabilities requires more resources, capital, maintenance, etc. In fact, it usually becomes significantly nonlinear, and much more costly, as we approach 100% availability. So, given these realities, are higher availabilities really the right answer if we can obtain our business goals with a lower availability?

So, back to our original question. What availability is optimal for each of the 11 process units?

The correct answer is the availability chosen for each unit should be the lowest value consistent with meeting our business goals over the short, medium, and long term. Unfortunately, there's no easy way to generalize what the process unit availabilities should be for a given plant. You have to do the analysis, and the math.

So, it depends. It depends on the plant, its unique business situation, the material balance of the feeds and products, the process unit and plant configuration, local economics, the amount of tankage, etc. What is the "optimal" availability for the Alkylation unit for a large refiner in Mississippi may not be the same as for a smaller refiner in Utah, let alone a world-class operator in The Netherlands or Singapore.

Table 4.1 Possible Optimal Process Unit Availabilities

Unit	Availability (%)
Crude	99.5
Vacuum	99.5
Sat Gas Plant	99
Claus Unit	99.5
Naphtha Splitter	99
Isom Unit	97
Reformer	98
FCC	98
Alkylation	96
Coker	96
Hydrocracker	97

So, how do the vast majority or plants decide what should be the individual unit availabilities today? If you work at the typical plant, an availability target is just chosen, often plucked out of thin air, and then the plant's entire maintenance and improvement programs are driven with a metric, which normally does not have a meaningful business basis. Other times, managers consider "World Class Performance" based on global benchmarking as their target without considering their own business situation. At least, this is better intended, but both of these approaches are misguided and can result in a significant amount of money being spent that is wasted. Even worse, money that could be beneficially spent, never happens.

Fundamentally, this occurs because business goals were never the primary driver of those decisions. In fact, it's the reverse. The tail is wagging the dog, with the poor dog getting whiplashed because it is being wagged so hard.

Sadly, the most important decision in our entire asset management program starts off on the wrong foot.

What's a better way?

4.1 Industry Challenge

In 2004, I challenged the process industries to rethink the answer to that question during a presentation at the National Petrochemical and Refining Association Executive Conference.[4] The presentation attempted to answer the following question – "What is the best methodology to determine where I should spend my next asset improvement dollar to optimally improve gasoline production in a refinery?" While that question was targeted toward a refining audience, it could have easily been generalized to any industry.

In fact, that question could be simplified to, "Where do I spend my next dollar to improve availability to optimally increase production?" At an even higher level, you might rephrase it as, "Where do I spent my next asset dollar to get the biggest bang for my buck?"

Unfortunately, the answer to that question is complex and involves many different facets of plant operation – process, logistical, or equipment issues, etc. It is almost certainly specific to the particular plant.

To address the question, I created a refining example in the original paper. While we will use the same general refinery layout for our example in this chapter, the specific data are different and totally fictitious.

Ultimately, we will discuss three different cases – gasoline production based on the current availabilities (they were exaggerated in the original paper, but done to make a point), a case where every unit is mandated to have X% availability (we'll use 98%), and the last case, which is the determination of each unit's actual optimal availability.

Before we discuss those ideas, we'll need to introduce a couple of important concepts. Let's talk about how to estimate the incremental availability improvement and associated cost for a series of projects that we might contemplate executing for each unit. Then, we will have a brief overview of the two key technologies that form the basis behind the approach outlined in this book – Monte Carlo simulation and optimization technology.

4.2 Improvement Projects

It should be possible for the Operations team within most process units to identify individual projects that could increase availability, if implemented. In fact, every plant should challenge its Reliability Group, as well as others that have responsibility for the unit, such as the Operations, Maintenance, and Engineering, to identify the exact list of projects they could implement if management asked them to increase unit availability.

Ironically, it doesn't need to be very detailed. In fact, all that is required is a simple listing of what projects they would implement, using "back of the envelope" cost estimates, along with an educated guess of how that project would increase availability. What's important is the projects must be prioritized in exactly the order that they would be executed. (The optimization will use this information later.)

So, let's say that a process unit currently has an availability of 97%; such is the case for our Hydrocracker. The list of improvement projects to increase availability might be something similar to that shown in Table 4.2. (Notice that there are no project specifics. Those were intentionally left off since we don't want to get lost in the details that are not significant enough to understand the gist of the example.)

A couple of comments about the list.

First, since it is a prioritized list based on what the teams feel should be executed, the level of difficulty and practicality plays a strong role in the order of the projects. It shouldn't be based on what project might have the greatest increase in availability or the least cost. That might be the case, but it doesn't have to be that way.

Table 4.2 Hydrocracker Project Costs and Incremental Availability Increase

Project	Starting Availability (%)	Final Availability (%)	Estimated Cost ($)
Project 1	97	97.4	10,000,000
Project 2	97.4	98.0	20,000,000
Project 3	98.0	98.6	30,000,000
Project 4	98.6	99.0	25,000,000

For example, in our list, Project A increases availability by 0.4%. However, it may be a relatively easy project to execute (for example, it doesn't require a turnaround). Each of the next two projects increases availability by 0.6%, but those projects may be considerably more difficult to execute outside a turnaround.

Now, we need to extend this idea to each project unit in the plant. The same rules apply.

What we need to understand is that these lists don't need to contain significant detail in order to be useful. We all know that engineers like to do a deep dive into the details before they commit to numbers, but the point here is that we want to know what our options are across a wide array of units. It's OK to make assumptions.

The combined lists allow senior management the ability to understand the big picture and what it can and cannot do. Believe it or not, this combined list will play an outsized role in our asset optimization program.

Each of these lists can then be conceptualized, as shown in the following graph. The individual, discrete steps are the set of projects that will increase availability by a given percentage, if executed. Figure 4.2 depicts our Hydrocracker projects.

One should be careful when reading the graph, since availability is on the vertical axis. The actual individual projects are vertical lines in the graph. For example, after a $10,000,000 investment, availability will increase from 97.0% to 97.4%. After another investment of total investment of $20,000,000, there will be an additional availability increase of 0.6% from 97.4% to 98.0%.

To use these more effectively in the optimizer, it is often a lot easier to translate the discrete projects into smooth functions. Mixed integer optimization can be used for discontinuous functions, but many optimizers strongly prefer continuous functions. While converting discrete projects into a continuous function is not strictly correct, it is generally good enough for optimization purposes. If, after examining the optimization results, we feel that we need to refine the problem, it's easy to rerun the problem. Figure 4.3 shows the Hydrocracker improvement projects now shown as a continuous function.

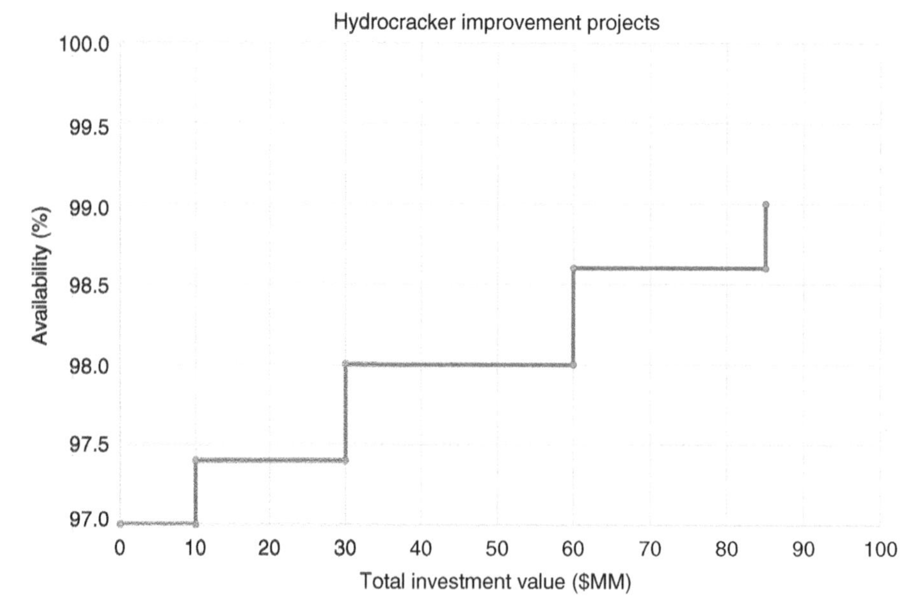

Figure 4.2 Discrete Hydrocracker Improvement Projects

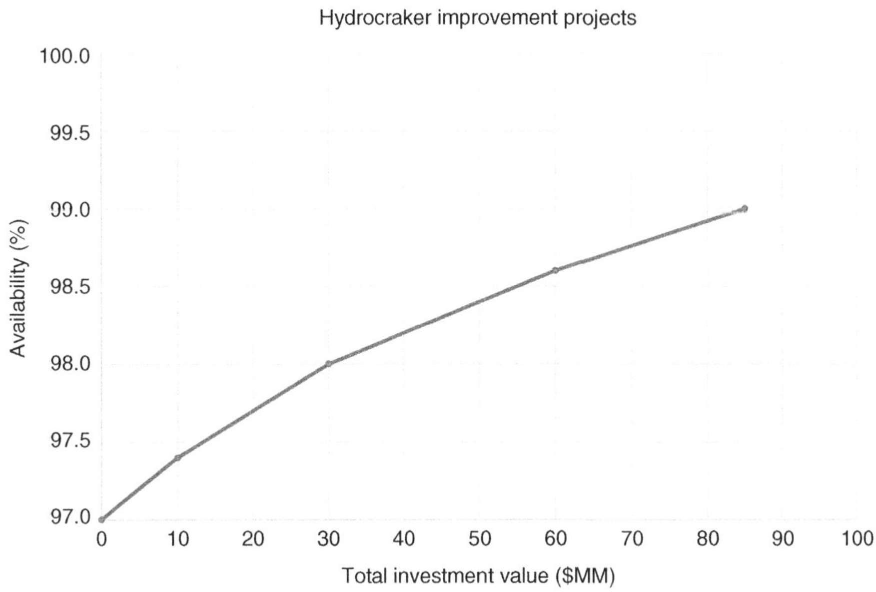

Figure 4.3 Regressed Hydrocracker Improvement Projects

Table 4.3 Incremental Project Costs per Availability

Base	Unit	95	96	97	98	99
97	Crude				$35	$45
96.5	Vacuum			$20	$25	$45
98	Sat Gas Plant					$15
98	Naphtha Splitter					$10
98	Isom					$15
97.5	Reformer				$15	$30
97	Diesel Hydrotreater				$10	$25
96.5	FCC			$15	$25	$45
97	Alky				$10	$25
95	Coker		$20	$30	$40	$45
97	Hydrocracker				$30	$55

After performing this analysis for one unit, we would need to perform the same analysis for each process unit. Table 4.3 below shows the incremental project costs that would need to be incurred to improve availability for our remaining units and their order. Let's assume the Base values as shown in the table, which are quite low by industry standards. While exaggerated, they are chosen to illustrate the concepts. Let's assume the Base values as shown in the table, which are quite low by industry standards. While exaggerated, they are chosen to illustrate the concepts.

We can now take those discrete values and transform them into continuous, cumulative functions, as shown in Figure 4.4. These functions will be later used in our example optimization problems.

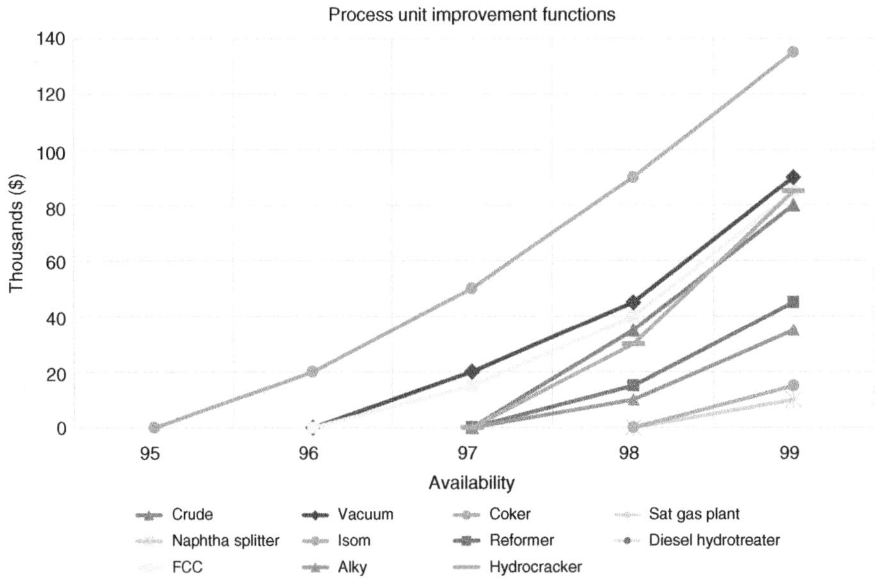

Figure 4.4 Process Unit Total Project Cost vs. Availability

4.3 Asset Optimization Technologies

Now that we understand how to incorporate improvement projects into our problem definition, let's step back and introduce our two Asset Optimization technologies.

4.3.1 Monte Carlo Simulation Technology

The first of these technologies is shown as the bottom box in Figure 4.5. It's Monte Carlo simulation technology. It was largely developed by the American government after World War II when the Americans were working on nuclear bomb technology.

One might remember from school that Mathematics is usually thought of as "deterministic." Thus, if we know the variables on the right-hand side of an equation, we should be able to group them in a way such that we can directly calculate those on the left-hand side. For example, A = B + C. If we know B and C, we can add them together to arrive at the value of A. That's deterministic math.

Unfortunately, some problems are so complicated that they cannot easily be arranged into this format. That's why Monte Carlo simulation technology was developed. Using Monte Carlo

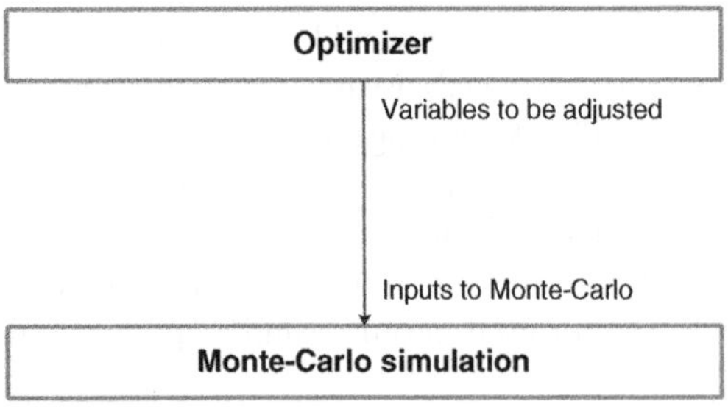

Figure 4.5 Optimizer/Monte Carlo Hierarchy

methodology, we don't need to know the exact deterministic equation. Rather, we can just guess at important variables, perform some intermediate calculations, and then determine if we can chart a complete path from a starting point on a Reliability Block Diagram through to a final point. If we can traverse that path from our initial point to the final point, then our unit is available. If not, it is unavailable. This process is then repeated hundreds, if not thousands of times, until a statistically significant set of results is generated. The plot is a distribution of likely results over a given time period. (For more details on how Monte Carlo works at a detail level, please refer to Chapter 5 and Appendix A.)

The Monte Carlo simulation technology used in our industry is often referred to as RAM modeling, an acronym that stands for Reliability, Availability, and Maintainability. Since there's so much technical detail behind all those "-ability" words, the reader is directed to Chapter 5 and Appendix A for a more complete discussion. What's important for now, is that Monte Carlo simulation technology and RAM are terms that are often used interchangeably.

In this book, we will largely refer to this as Monte Carlo simulation technology rather than RAM, primarily because we want to elevate the discussion beyond Reliability. As noted in Chapter 2, the word Reliability is tainted and has often been a barrier to the wide-scale adoption of asset management. We want to change that, so let's focus more on the business side.

What's more important to realize is that Monte Carlo simulation technology is nothing more than a sophisticated mathematical engine. Nothing more, nothing less. The good news is that if we structure our problem correctly, it can represent how assets perform over time and allow us to perform sophisticated "what if" analyses of what options we have for improvement. As with so many technologies, what's important is the data used in the analysis.

Since this book is primarily concentrated on optimization, we don't want to get the reader too hung up on the details behind RAM modeling. If one desires more technical information, first start with Chapter 5 and Appendix A. If that isn't sufficient, then you may want to consider vendor courses as your next course of action.

While we are discussing Monte Carlo analysis, I want to caution the reader it's easy to fall into the typical Monte Carlo black hole. That is, one loses sight of your primary objective – solving an important business problem. It's a hole that many a good engineer has become trapped in – never to see the light of day (worse yet, never to solve problems of importance). I point out two areas that have been problematic in the past.

The first is, that far too often, engineers have conceptualized problems to an excruciatingly high level of detail. Not only is this often unnecessary, it requires much more work and underlying data to solve the problem. Worse yet, it doesn't ultimately improve the business decision. Let's make a promise to ourselves that we will keep our problems as simple as possible.

The second is that many think they need perfect data to do the analysis. Nothing could be farther from the truth. We are engineers, and we must make decent assumptions. Chapter 7 details the Consulting approach that will be central to how we obtain data for our analyses.

Now that we have had a brief introduction to Monte Carlo technology, let's introduce our other major technology focus – true optimization.

4.3.2 Optimizer

The second technology that will play an outsized role in our asset optimization journey is true optimization technology. It will be structured to optimize an objective function, usually based on maximizing profit or revenue. It sits on top of the Monte Carlo simulation technology and sets key variables in the simulation.

All optimizer technologies involve the following aspects:

- Independent variables
- Dependent variables
- Constraints
- Objective Function

Each of these is explained in more detail below.

Independent variables are variables that the optimizer chooses.

For example, in our example refinery problem, the optimizer will adjust each of the unit availabilities.

Dependent variables are those variables, often calculated through a complex set of calculations, that can only be determined after the independent variables have been chosen by the optimizer.

In our example, the Monte Carlo simulation will run after the optimizer chooses the new set of process unit availabilities. These availabilities will then be used to calculate the final values of all component streams, which will be used for revenue and profit calculations.

Constraints can be applied to either independent or dependent variables.

For example, since crude unit availability is one of our independent variables, we can limit the optimizer to only choosing values between 96% and 99%, if we so choose. That's easy to do since they are simply variables that the optimizer chooses at points during a particular run.

Independent variable constraints, such as the process unit availabilities in our example problem, will always be honored.

However, we can also constrain dependent variables. For example, we could limit the total amount of gasoline produced, which we might need to do since we have contractual limitations, where we can't sell any more beyond a certain amount. So, each time that the optimizer runs, it chooses the independent variables, calculates the dependent variables, and then the optimizer tests to see if it has violated those limits.

Dependent variable constraints will be honored – if it is possible to do so. However, sometimes, if the independent variables' limits are chosen too narrowly, it may be possible that the optimizer can't honor the dependent variable constraints.

The **Objective Function** is that mathematical equation that the optimizer is trying to either maximize or minimize. In our case, we are usually trying to maximize profit, which is defined as incremental gasoline revenue minus the increased capital investment needed for the individual units.

4.3.2.1 General Optimizer Solution Approach

Given that high-level introduction, it makes sense to discuss how the optimizer proceeds through its calculation sequence. Figure 4.6 shows the optimization structure and the sequence of events leading to convergence.

The process begins when the optimizer chooses a set of independent variables, always honoring the independent variable constraints. Then, any dependent variable equations are calculated. The optimizer now checks to see if any constraints have been violated. If it has, the process starts again with the optimizer choosing another set of independent variables. If not, it will calculate the objective function. It then checks to see if it has arrived at a final optimum. If so, the last set of independent variables are the optimum choices. If it hasn't reached an optimum, the process repeats by choosing another set of independent variables. This process will repeat itself, potentially thousands of times, until an optimum is reached.

In our example, each time that the optimizer iterates, it recalculates the streams produced from each unit based on the new unit's availabilities. Based on current pricing, we can easily calculate the incremental revenue from increasing the availabilities.

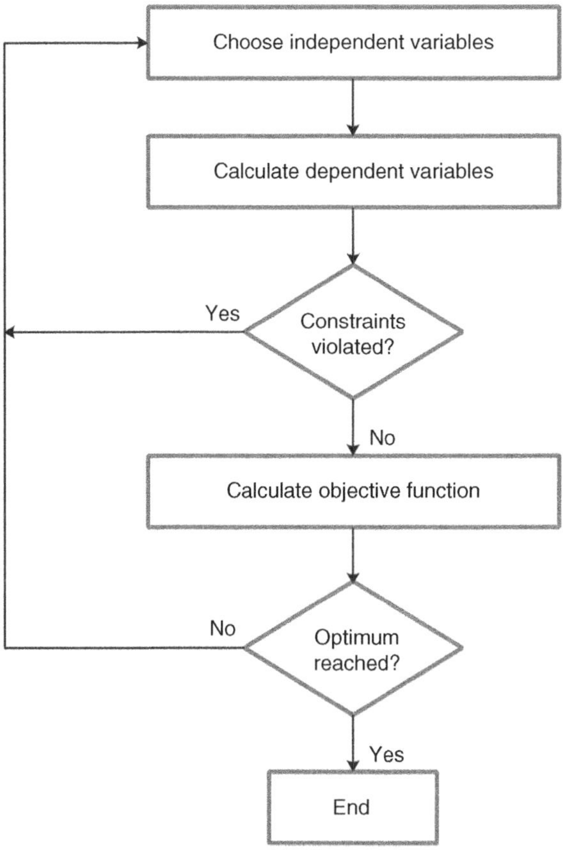

Figure 4.6 Optimization Sequence Steps

However, the only way to obtain that incremental production is through an investment, which can be determined using the improvement functions detailed earlier. That is the cost of our objective function. The optimizer would calculate the incremental profit as the difference between the incremental revenue and the investment costs necessary for that incremental production.

This process is repeated many times within a given optimization run by choosing new sets of availabilities and calculating the new profit for that set of variables. Eventually, the optimizer will iterate to a set of availabilities that maximize profit.

4.3.2.2 Choosing an Optimization Search Direction

The discussion above showed the general solution approach for any optimization algorithm. As one can probably imagine, internally, it's considerably more complicated than what was just described. Let's add some important details.

When the algorithm first begins, it must determine a search direction. To do so, the normal approach is to adjust each independent variable a tiny increment one variable at a time, while holding the other independent variables constant. Then, it calculates all the dependent variables and the objective function. Using this information, it can then determine the partial derivatives for all the dependent variables and objective function relative to the independent variables it just moved. It then repeats this process for all the independent variables.

Now, the algorithm has enough information to more accurately determine how to execute an efficient search. It will then start moving all independent variables simultaneously and will perform many iterations using that approach. However, it may need to occasionally stop and redetermine all the partial derivatives. This is done so that it can do a better job of finding a true global optimum and not just settle for a local optimum.

4.3.2.3 Examples of Optimizers – Solver within Excel

We are probably more familiar with optimization than we may realize. For example, if you have ever used the Solver within Excel, it has three different options. One of those options is a nonlinear optimization technology, GRG2, which was developed at the University of Texas. It uses a user-friendly frontend that allows the user to enter what cell is the objective function, the independent variables, and the constraints.

While the NPRA paper used another commercially available, popular, optimization technology,[5] I personally used GRG2 as the mathematical engine for the multivariable controller that I developed at Continental Controls, for which I received a US patent.[6] This book will remain agnostic as to which optimizer technology is used within your program, as long as it meets the criteria for successfully arriving at the true optimum. However, the problems in this book will be solved using the GRG2 option within Excel's Solver.

4.3.3 Combining the Technologies

Now, we have introduced the two main technologies, Monte Carlo and Optimization, that we will be using for the majority of problems throughout this book.

What's interesting is that this combination is layered similarly to what is found in Advanced Control and Optimization (AC&O). In AC&O, a dynamic multivariable controller technology is on the bottom, similar to Monte Carlo simulation engine in our combination, with a true Optimizer sitting above it that is trying to maximize profit by manipulating a set of independent variables. So, for those already familiar with AC&O, asset optimization will have a similar look and feel to it. In fact, the Optimization aspect in both areas is almost exactly identical in function.

As is the case with AC&O, there are times when the top optimization function will not be needed or used, but for our more important problems, we will want it. Unfortunately, many of the industry's RAM modeling vendors have not yet recognized the need for the optimization function and will need to adapt. However, it's an easy step to marry the two technologies together, as we did in the 2004 presentation.

From a philosophical point of view, it's important to include the optimization layer because it keeps us focused on financial-related metrics, which is ultimately what our process industry businesses are intended to do in the first place – maximize profits. Too many have ventured into the Monte Carlo modeling world only to have gotten lost in lower-level details, missing the big picture. In addition, it allows us to introduce constraints into the problem definition which are a part of all real world problems.

4.4 Optimizer Definition

Now that we have been introduced to optimization principles, let's define the independent and dependent variables, applicable constraints, and the objective function for our example problem where we are trying to determine the most desirable improvement project (which, by definition, sets each process unit's availability target). Figure 4.7 shows the high-level problem structure.

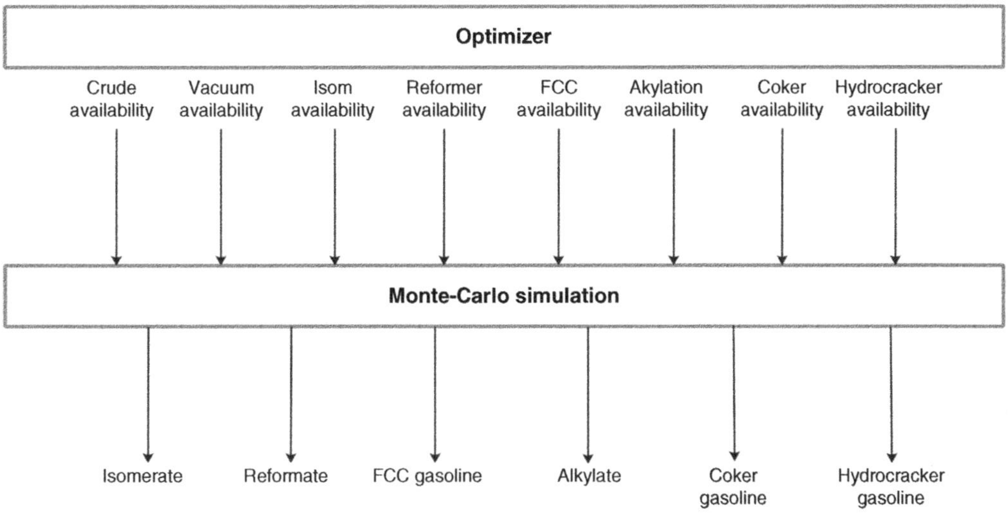

Figure 4.7 Process Unit Target Availability Optimization

4.4.1 Independent Variables

In our example, the independent variables are each process unit's availability target. While we choose to include all the availability targets in our example, there's no explicit requirement to do so. We could have easily included only the crude and vacuum unit's availability and then let the optimizer calculate their optimal values. At the end of the day, it's the user requirements that drive this choice. However, for this problem, let's stick to using all process unit's availability targets.

4.4.2 Dependent Variables

There are four main dependent variables for this problem:

1) Each unit's improvement function, as highlighted earlier
2) Total product volume increase
3) Individual gasoline component totals
4) Total improvement project expenditure

As noted earlier, each time that the optimizer chooses a process unit availability target, it will use the improvement project graphs to determine how much it needs to spend to achieve that new availability target. There is one equation for each different process unit.

The refinery material balance, shown in Appendix B, allows us to calculate gasoline component production for each new availability target. While this is done for each process unit individually, it's actually more complicated because of intermediate storage tankage.

The final gasoline component streams are then added together and multiplied by the wholesale gasoline price to determine the total revenue increase from the new availability targets.

The total values for each of the individual gasoline component streams can also be included as constraints, since the planning department can estimate the maximum/minimum values that it needs to blend gasoline successfully later on.

For those not familiar with the refining industry, gasoline bought at the pump is actually a blend of many different individual components. Refinery product blending is a complex blending batch operation and is beyond the scope of this book. For that reason, we'll model that by just using a simple constraint for each component that the planning guys will supply.

The last major calculation is the total improvement expenditure. This will be used in two different places. First, it will be one of the major dependent variable constraints. In addition, it will be one aspect of the objective function calculation.

4.4.3 Independent Variable Constraints

Independent variable constraints are the upper and lower bounds that the optimizer can move a particular independent variable within. For example, the maximum/minimum crude unit availability in this example will be set to 99.75% and 97%, respectively. This will give the optimizer the ability to move that independent variable anywhere within that 2.75% range, but not outside of it. These are important constraints, and the user needs to use judgment on how they are set.

When evaluating problem results, one should always check to see if a particular independent variable is hitting either its maximum or minimum limits. If so, the user should ask themselves if those limits are actually appropriate, or if they could be widened to give the optimizer more room to move. If the limits are real, that's OK, but always check that when evaluating any optimization results.

From this point forward, we will always assume that every independent variable automatically has a maximum and minimum associated with it. (In fact, all the major optimization packages automatically provide a high/low limit for each of those variables.)

4.4.4 Dependent Variable Constraints

Dependent variable constraints are an important part of virtually every optimization project. They are any constraints the user would like included in the problem definition, which are not already independent variable constraints. There is no limit to how many can be included.

As noted in the previous section, we will include constraints for total improvement project expenditure, as well as individual gasoline component production values.

Remember that just because a dependent variable is defined doesn't mean that when the optimizer runs, it will bump up against that constraint. Which constraint it hits, if any, depends on many factors, one of which is the independent variable ranges. However, if you do run up against a limit, ask yourself if it is because independent variables associated with it don't have enough room to move.

Also, remember that sometimes the optimizer can't honor all the dependent variable constraints. When this happens, it often has to do with the fact that the independent variable limits are too tight or the problem is not defined properly.

4.4.5 Objective Function

The Objective function is the equation that the user is trying to either maximize or minimize. In our case, we would like to maximize profit, which is simply the difference between incremental production revenue minus incremental improvement project spending, subject to both independent and dependent variable constraints.

A couple of important notes. Since dependent variables are a result of independent variables, at the end of the day, the objective function is fundamentally only a function of independent variables.

It's important to note that each independent variable should have some discernable effect on the value of the objective function over at least part of its defined range. If not, the optimizer won't adjust it. That's not a big deal. However, what it means is that this variable has no effect on the objective function and shouldn't have been included as part of the problem definition anyway.

4.5 Optimization Example

We will proceed through three different scenarios. In the first case, we will determine production quantities with availabilities set to the historical average for each of the process units. In the second case, we will set all availabilities equal to 98%, simulating a case where there is a plant-wide process unit availability mandate. The third and final case will be a series of true availability optimization runs.

4.5.1 Optimization Approach

All these examples were simulated in Excel using the GRG2 option of the Solver. There were many supporting calculations that determined production totals based on the yield structure for Crude A, in addition to assumed process unit availability targets. In addition, there were also regressed equations where each unit's project improvement costs were a function of its availability target. Of course, there were numerous equations that totaled production streams, as well as calculated improvement project totals.

A shortcut was taken in this example whereby we decided not to include tankage as a part of the problem definition. In essence, we assumed that tankage was large enough not to be a limiting factor.

Once that decision was made, there was actually no reason to use Monte Carlo as part of the problem definition/solution. Thus, it was actually possible to solve the optimization problem directly.

Since project costs are a one-time expenditure and production occurs on a daily basis, one needs a reasonable method to combine those two together to compare/contrast the results. Hence, we totaled five years of production (as well as crude feedstock cost) as the revenue source and then subtracted the total cost of improvement projects.

4.5.2 Base Case

In the base case, we ran the simulation engine with the historical availabilities for each of the process units as shown in Table 4.4.

Assuming the following wholesale market prices shown in Table 4.5, the simulation results are shown in Table 4.6.

The results show that given the unit availabilities, our example refinery was able to generate a profit of almost $168 MM. In this case, no improvement projects were implemented, so there were no incremental dollars spent. This is our base case.

4.5.3 All Process Units with Same Availability Target

Now, let's proceed to the next case, where every unit will be mandated to implement improvement projects to bring each unit up to 98% availability. Achieving that availability will require a significant expenditure as dictated by our process unit improvement functions shown earlier in this chapter. In fact, we know exactly what we must spend – $275 million. This is broken down in Table 4.7.

Table 4.4 Base Case Availabilities

Process Unit	Availability (%)
Crude	97
Vacuum	96.5
Coker	95
Sat Gas Plant	98
Naphtha Splitter	98
Isomerization	98
Reformer	97.5
Diesel Hydrotreater	97
FCC	96.5
Alkylation	97
Hydrocracker	97

Table 4.5 Assumed Market Prices

Stream	Price
Crude	$45/bbl
Propane	$42/bbl
Butane	$42/bbl
Gasoline	$55/bbl
Kerosine	$55/bbl
Diesel	$55/bbl

Table 4.6 Base Case Results

Stream	Price
Crude Cost	$45,000,000
Propane Production	1,929 bbls
Butane Production	4.798 bbls
Gasoline Production	54,162 bbls
Kerosine Production	12,583 bbls
Diesel Production	11,608 bbls
Propane Revenue	$81,001
Butane Revenue	$201,498
Gasoline Revenue	$2,978,898
Kerosine Revenue	$692,069
Diesel Revenue	$638,420
Profit	$167,691,845

Table 4.7 Improvement Projects Costs at 98% Availability

Process Unit	Improvement Project Spend
Crude	$35 MM
Vacuum	$45 MM
Coker	$90 MM
Sat Gas Plant	$0 MM
Naphtha Splitter	$0
Isomerization	$0
Reformer	$15 MM
Diesel Hydrotreater	$10 MM
FCC	$40 MM
Alkylation	$10 MM
Hydrocracker	$30 MM

Table 4.8 Results – 98% Availability

Stream	Price
Crude Cost	$45,000,000
Propane Production	2,004 bbls
Butane Production	5,111 bbls
Gasoline Production	57,734 bbls
Kerosine Production	12,713 bbls
Diesel Production	11,727 bbls
Propane Revenue	$85,006
Butane Revenue	$214,641
Gasoline Revenue	$3,175,382
Kerosine Revenue	$699,204
Diesel Revenue	$645,002
Profit	$307.9 MM

The simulation results are shown in Table 4.8.

The results show that given the unit availability of 98% for each process unit, our example refinery was able to generate a profit of almost $308 MM. This is approximately $140 MM more than our base case.

The question then becomes if this is the best we can achieve.

4.5.4 True Optimization

In this section, we will determine the optimal set of availabilities for each process unit. The optimization will be performed in Excel using GRG2, which is a nonlinear, constrained optimization technology.

Before we detail the exact results, let's introduce a format that we will use for all optimization problems in the book. Per the discussion earlier in this chapter, we will define the specific independent and dependent variables, as well as any constraints and the objective function.

4.5.5 Independent Variables

The independent variables for this problem are:

- Availability target for each process unit in our example refinery

4.5.6 Dependent Variables

There will be one dependent variable equation for each process unit that regresses the discrete improvement projects into one equation that will allow the optimizer to calculate the improvement cost required to achieve a given availability. In essence, these are the curves shown earlier in this chapter.

There are also a significant number of dependent variable equations that define the refinery material balance based on the yield structure for Crude A, as described in Appendix B. In addition to the yield structure, the production totals are also a function of the process unit availabilities.

Profit is calculated as the five-year yield production totals for each stream multiplied by the appropriate sales price minus the total of all improvement projects required, as well as subtracting the feedstock costs, which are also calculated over a five-year time horizon.

4.5.7 Constraints

There will be model runs where cost and gasoline production capability are constrained.

4.5.8 Objective Function

The goal of this optimization is to determine the maximum profit.

4.5.9 Results

In total, 37 optimization runs were conducted. The first optimization run determined the optimal availabilities given the prices and project improvement costs, as detailed in the previous section. However, additional model runs were conducted to test the optimization's sensitivity to other variables. Those runs consisted of the following:

- Effect of gasoline pool price
- Sensitivity to crude price
- How changing improvement project costs for these units individually affected the results:
  - Coker
  - FCC
  - Hydrocracker
  - Crude

- Then, one case was conducted that combined all the previous project improvement costs together
- A series of cases were run that changed the total improvement budget limit
- A few cases were run where gasoline production was constrained

Let's now examine those results.

4.5.9.1 Initial Optimization Run

Using the original improvement project costs and product price/crude costs from a previous section, the optimizer determined the following availabilities for the process units, as shown in Table 4.9.

We see that the availabilities determined from the optimizer range from a minimum of 97% all the way close to 100%. They are definitely not values that could easily be determined manually.

All process unit availability manipulated variables were bracketed by an upper and lower independent variable limit. In all cases, the base value was the current availability. On the other hand, the upper value was chosen to be 99.75%. That's somewhat deceiving and inaccurate because the improvement projects were only estimated based on their current availability to 99%. So, we were extrapolating beyond that to 99.75%, which strictly speaking is incorrect. For the purposes of this book, we'll assume that this assumption is OK. However, in real life, we'd want to estimate what projects would actually be required to achieve this kind of availability level.

Table 4.10 shows the projects that the optimizer decided to spend money on. Two columns are shown. The first is the value that the optimizer determined, which was based on a continuous function that models the individual discrete projects. The second column is the actual discrete projects closest to those values.

The discrete column is significantly lower than the continuous column, but that can be attributed almost exclusively to the assumption that the process unit availability can exceed 99%. If the upper bound is clamped at 99% availability, then the answers are virtually identical. Let's leave the upper bound to be 99.75% because, in a real-world refinery, we'd want some of our

Table 4.9 Optimal Process Unit Availabilities

Process Unit	Availability (%)
Crude	99.75
Vacuum	98.023
Coker	97.3
Sat Gas Plant	99.75
Naphtha Splitter	99.75
Isomerization	98
Reformer	99.75
Diesel Hydrotreater	97
FCC	98.86
Alkylation	97
Hydrocracker	97.96

Table 4.10 Optimal Process Unit improvement Project Costs

Process Unit	Improvement Project Spend (Continuous Values)	Improvement Project Spend (Discrete Values)
Crude	$125.7 MM	$80 MM
Vacuum	$54.85 MM	$45 MM
Coker	$61.5 MM	$50 MM
Sat Gas Plant	$26.25 MM	$15 MM
Naphtha Splitter	$17.5 MM	$10 MM
Isomerization	$0 MM	$0 MM
Reformer	$67.5 MM	$45 MM
Diesel Hydrotreater	$0 MM	$10 MM
FCC	$76.2 MM	$95 MM
Alkylation	$0 MM	$0 MM
Hydrocracker	$28.41 MM	$30 MM

units to achieve that level of availability to maintain economically viable plant profitability. All that means is that we would need to refine our equations by adding project estimates to achieve that level of availability.

It's not always trivial to figure out why the optimizer chose the values that it did because that has a lot to do with process configuration, rates, and how expensive various improvement projects are relative to other projects. However, we can make some educated guesses as to likely reasons why.

- It maximized crude unit production capabilities. Normally, if the crude unit improvement projects aren't outlandishly expensive that makes sense. It's the first unit in the train, so lower availabilities will eventually negatively affect all downstream users. However, we'll see later in this chapter that maximizing crude unit availability has its limits, if improvement projects become too expensive.
- The vacuum unit is fed to the FCC and Coker, so it makes sense that it would spend some money here. How much depends on the cost of the improvement projects.
- The Coker started with a low availability, so it chose to spend money on some relatively cheap projects. That's beneficial for the Hydrocracker since the Coker feeds that unit.
- The optimizer spent money on both the Naphtha Splitter and Sat Gas plant. That's likely because they were cheap projects. In the case of the Naphtha Splitter, it feeds two conversion units – the Isom and Reformer, so you'd like this unit to have a higher availability.
- The FCC is the heart of the refinery, driving substantial gasoline production, so it's not surprising the optimizer spent a considerable amount of money there.
- It spent somewhat less on the Hydrocracker, but feedrates there are significantly less than for the FCC.

All in all, the results make sense upon reflection.

The economic values from the simulation results are shown in Table 4.11.

What we can see is that the optimizer achieved a significantly higher level of profitability than with the mandated 98% process unit availability case. In fact, it's $188 MM more than the 98% availability case.

Table 4.11 Optimum Case Results

Stream	Price
Crude Cost	$4,500,000
Propane Production	2,130 bbls
Butane Production	5,347 bbls
Gasoline Production	60,736 bbls
Kerosine Production	12,940 bbls
Diesel Production	11,937 bbls
Propane Revenue	$89,447
Butane Revenue	$224,557
Gasoline Revenue	$3,340,456
Kerosine Revenue	$711,690
Diesel Revenue	$656,520
Total Project Improvements	$457.9 MM
Profit	$495.9 MM

4.5.9.2 Effect of Gasoline Pool Price

Now that we have completed our first optimization run, let's see how sensitive that optimum is to other variables. For example, since gasoline is the principal product consisting of 60% of the total products, it makes sense to ask how the optimum values would change if gasoline price was varied. Thus, four additional model runs were made with gasoline product prices varying from $45/bbl to $65/bbl with the results shown in Figure 4.8.

What one sees is the optimum for four process units changes values throughout the price range – Vacuum unit, Coker, FCC, and Hydrocracker. The most pronounced change is for the Coker, where the optimum moves from 96.5% availability at $45/bbl gasoline price all the way to 98.5% at $65/bbl. That's likely due to its poor initial availability and the relatively cheap project improvement costs over this availability range.

There are also other units that change as well showing gradual increases in optimum availability values as price increases. These are shown as other curves on the graph. What's noteworthy is that their changes are not nearly as significant as for the Coker. Other process units were not shown on the graph since they had minimal movement in their optimal values over the price range.

Model runs were also conducted for all the other product prices, but none showed any significant differences in the ultimate optimal values. That's likely since they are such a small fraction of the overall product mix.

4.5.9.3 Effect of Crude Price

Model runs were conducted, changing crude feedstock prices over a $10 range. The results were somewhat surprising, since your intuition would likely have predicted some change. However, optimal values didn't budge from the original optimal values. However, that's why you conduct the model runs. The results aren't always what you think.

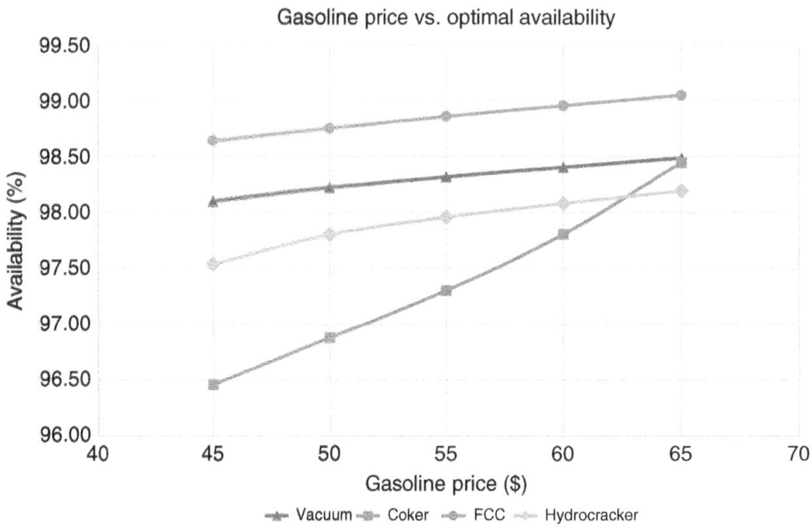

Figure 4.8 Gasoline Price vs. Optimal Availability

4.5.9.4 Coker Improvement Project Change

At this point, it makes sense to determine how sensitive the optimization results are to improvement project cost changes. Therefore, we conducted model runs for four different units to see that effect – Coker, FCC, Hydrocracker, and Crude unit.

We start first with the Coker. For this model run, we reduced the Coker improvement project costs by half, as shown in Table 4.12.

Originally, the optimizer had determined the optimal Coker process unit availability to be 97.3%. However, with these much cheaper improvement project costs, the optimizer drove the optimal availability up to the maximum limit of 99.75% over the entire gasoline price range. In other words, the improvement projects were so cheap that it made sense to implement all of them. All other process unit optimal availability targets remained at the values determined during the original optimization run.

4.5.9.5 FCC Improvement Project Change

For this model run, we changed the Coker improvement values back to the original values and instead reduced the FCC improvement project cost values, as shown in Table 4.13. Similar to the Coker example, we cut those values in half.

Originally, the optimizer had determined the optimal FCC process unit availability to be 98.9%. However, with these cheaper improvement project costs, similar to the Coker case just shown, the optimizer drove the optimal FCC availability to the maximum limit of 99.75% for all gasoline price values. Once again, the improvement projects were so cheap that it made sense to implement all of

Table 4.12 Coker Project Cost Changes

Coker	96	97	98	99
Original	\$20	\$30	\$40	\$45
New	\$10	\$15	\$20	\$23

Table 4.13 FCC Project Cost Changes

FCC	97	98	99
Original	$15	$25	$45
New	$8	$13	$23

Table 4.14 Hydrocracker Project Cost Changes

Hydrocracker	98	99
Original	$30	$55
New	$45	$68

them. The higher availability allowed the plant to maximize gasoline production. All other process units remained at the values determined from the original optimization run.

4.5.9.6 Hydrocracker Improvement Project Change

In this model run, the FCC improvement project cost values were returned back to the original values. Instead, we increased the Hydrocracker improvement project costs by 50% over their original values, as shown in Table 4.14 to see that effect.

As opposed to the two previous cases, this change did not drive the Hydrocracker optimum value to a limit. Rather, the optimum value for four different process units – Vacuum, FCC, Hydrocracker, and Coker were a function of the gasoline product price, as shown in Figure 4.9. The effect was quite pronounced for the Coker, yet relatively mild for the other three process units. All other process units not shown on the graph did not show significant changes to their optimal availability values.

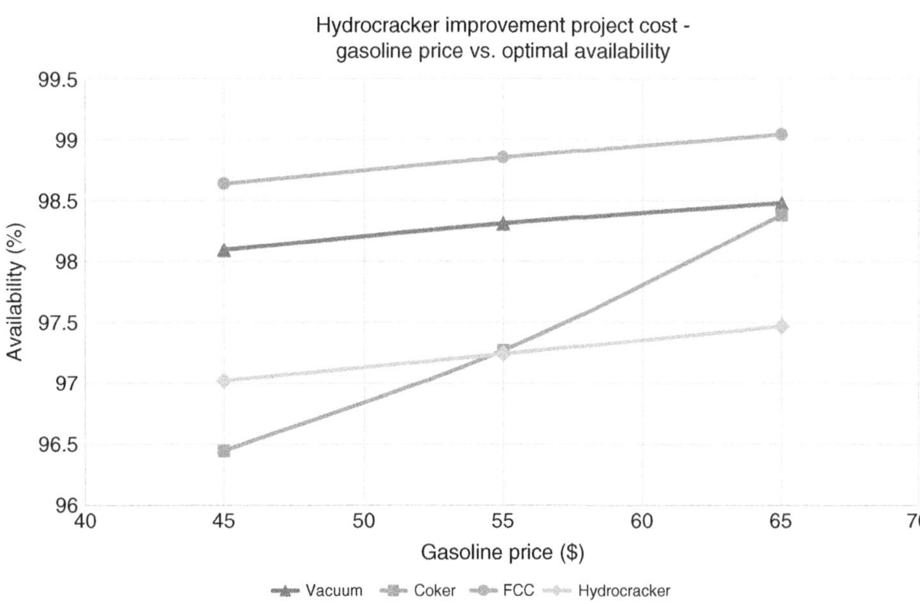

Figure 4.9 Optimal Availability – Increased Hydrocracker Project Costs

Table 4.15 Crude Unit Project Cost Changes

Crude	98	99
Original	$35	$45
New	$105	$135

4.5.9.7 Crude Unit Improvement Project Change

In this case, we wanted to test if there came a point where crude unit improvement project values could become expensive enough such that the optimizer did not drive the crude unit availability level to its maximum limit. The New improvement project values were triple the original values as shown in Table 4.15.

Figure 4.10 shows that the optimal crude unit availability is a strong function of gasoline product price. Only as the product price approaches $65 does the optimal crude unit availability hit the upper limit of 99.75%. At lower prices, the crude unit optimal availability is significantly less.

Since the crude unit is the first unit in the processing chain, its optimal value has a significant effect on what the optimal values are for other related units – especially those process units that are conversion units.

4.5.9.8 Four Improvement Project Changes Combined

In the four previous cases, we examined individual improvement project cost changes for one unit at a time. In this case, we combine all of those changes together in one case to determine the net effect. Figure 4.11 shows those changes where the gasoline product price is $55/bbl.

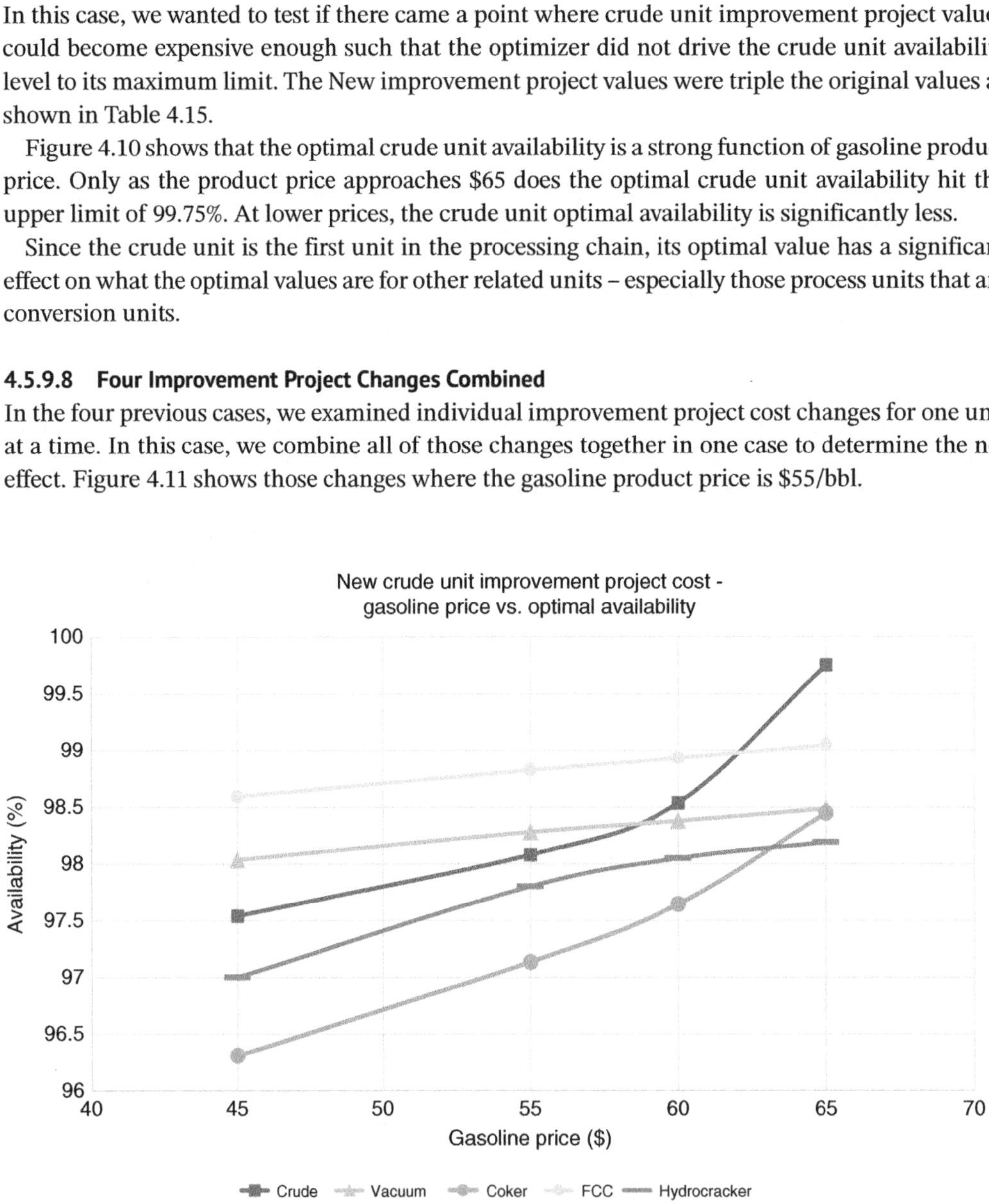

Figure 4.10 Optimal Availability – Increased Crude Unit Project Costs

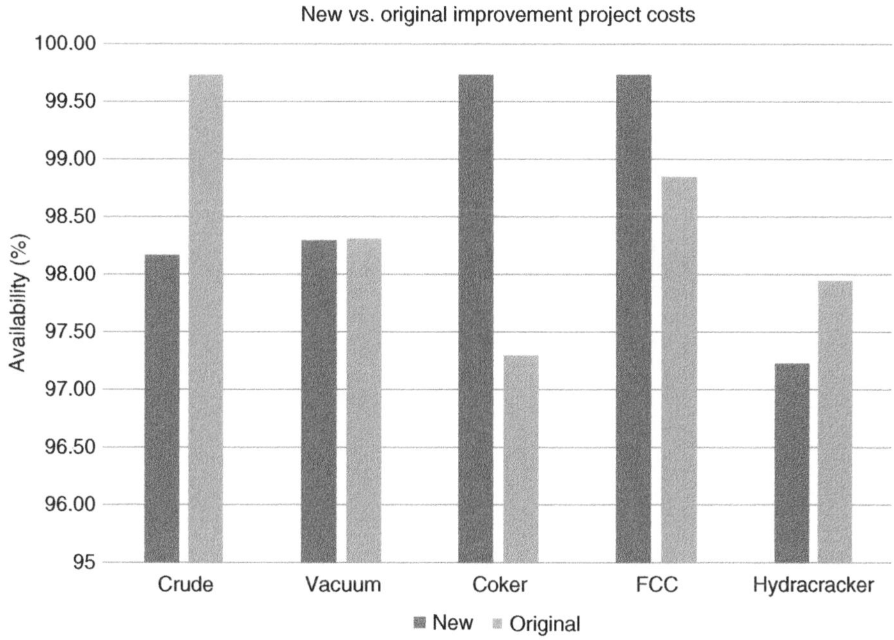

Figure 4.11 Optimal Availability – Combined Increased Project Costs

Individual effects from the four independent changes highlighted in the previous examples seem to predominate. Coker and FCC optimal availabilities are driven higher, while the Crude unit and Hydrocracker optimal availabilities are lower. On the other hand, there is no net change for the Vacuum unit.

4.5.9.9 Constrained Total Improvement Project Costs

In this example, we show how changes in the total improvement project cost budget (constraint limit) affect optimization results. These results presented in Figure 4.12 are remarkably interesting and allow a particular plant to determine the exact order in which projects should be implemented based on the plant's budget.

The best way to read the graph is to move from left to right (lowest budget to highest) and watch how various units jump from one value to another. With the exception of the Coker, which shows a more gradual movement across the budget range, all the other units show a fairly abrupt change at a certain point.

We analyzed the results and produced Table 4.16, that summarizes the best order in which projects should be implemented. The lighter shaded cells are projects that should be implemented, with the numbers in the cell reflecting the order in which we should implement the projects. Darker shaded cells are projects that should never be implemented because plant economics just don't support them.

What this table shows is that we should implement all the crude unit projects first, followed by the first couple of FCC projects. Then, the first Coker project, and so on. In many ways, these results are not all that surprising. The crude unit is the first unit and affects all downstream units. The FCC is the most important conversion unit, so spending money there is quite typical. However, these results could have been dramatically different based on how expensive the improvement

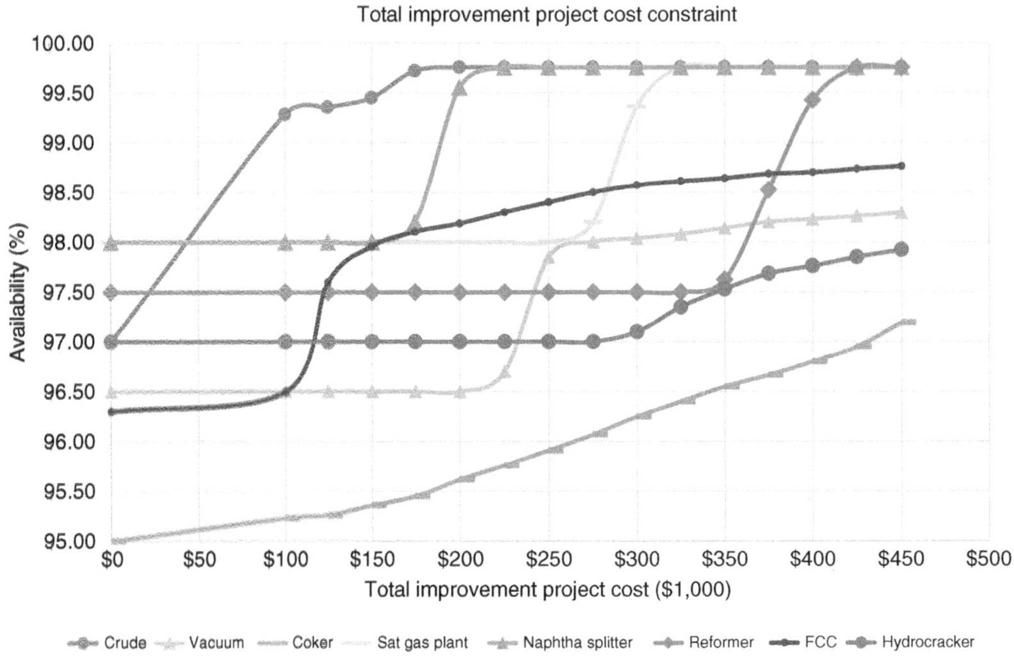

Figure 4.12 Optimization – Total Improvement Project Cost Constraint

Table 4.16 Project Execution Order

	96	97	98	99
Crude			1	2
Vacuum		7	8	N
Sat Gas Plant				10
Naphtha Splitter				6
Isom				N
Reformer			11	14
Diesel Hydrotreater			N	N
FCC		3	4	9
Alky			N	N
Coker	5	13	N	N
Hydrocracker			12	N

projects actually were. If we had run this example using the figures from a previous section where the crude unit improvement project costs were triple what they are now, the results would have been significantly different. So, you do have to perform the analysis to get the proper results.

How many projects we ultimately implement is based totally on our budget. If we have unlimited funds (in this case, over $460 MM), then we can execute all 14 projects. If our budget is less, then we know which ones should be removed first (highest numbers). We would then keep removing projects until we were back within our budget.

What is also interesting from this discussion is that there are eight projects using this example that should never be implemented because they cannot be justified economically.

4.5.9.10 Constrained Gasoline Production Quantities

In the example shown in Figure 4.13, we show the effect of how constraining gasoline production can change optimization results.

In many ways, the results are quite consistent with the discussion in the previous section. In this case, we read the graph from the right side to the left (higher gasoline production yields back toward lower values). Thus, Reformer projects should be eliminated first, followed by the Hydrocracker, Naphtha Splitter, FCC, and the crude unit last. Similar to the last example, the Coker shows a more gradual change over the gasoline production range.

4.5.10 Optimization Example Conclusions

There are important conclusions that we can draw from the series of optimization discussions just completed. They are detailed below:

- The optimum availability for each process unit isn't always obvious because it's a complex problem that involves many different facets – what various improvement projects cost, how units interact with one another, product prices, etc.
- Changes in improvement project costs can dramatically change the final results.
- It's possible to determine the exact mix of projects based on what your budget is.
- However, even with unlimited funds, some improvement projects, while possible, should never be implemented because the underlying economics don't support them.

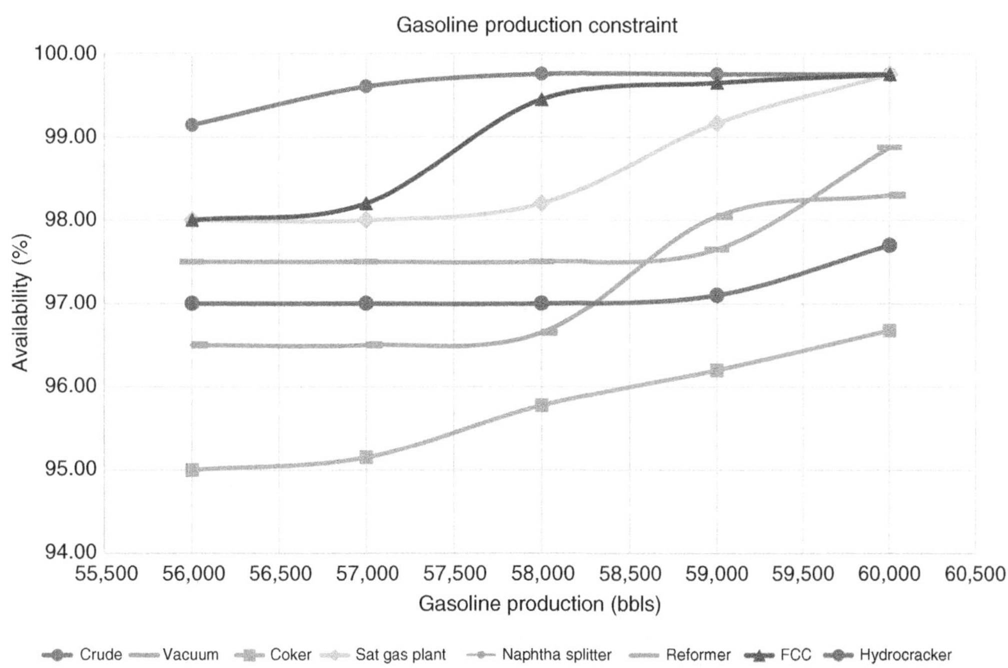

Figure 4.13 Optimization – Gasoline Production Constraint

4.6 More General Optimization

A few comments concerning the optimization results that were just presented.

There's nothing that says that we must use all availability targets as our optimization variables. We could mix them up in any combination that suits us. For example, we might know that we aren't going to be implementing improvement projects on certain units – e.g., the Vacuum, Coker, and Alky. Therefore, we could drop them out of the optimization and only use the other units.

Even right now, we have only focused on availability, and didn't include other types of variables, but there's no reason why the thought can't be generalized even further. You can include any variable as something that the optimization will adjust that you desire. For example, later in this book, we will attempt this very same problem, but will include crude mix as another optimization variable.

The only practical requirement for every variable chosen by the optimizer is that there must be dependent variable equations that show how the change in the optimization variable affects unit availability, the product material balance, or costs in some fashion. Normally, that would require some equation similar to the Process Unit Improvement functions shown earlier.

4.7 Does Reducing Availability Make Sense for Any of Our Process Units?

The methodology shown in this chapter so far is for determining the optimal allocation of capital among a group of process units to improve production. By definition, since we are talking about improvement projects, the optimizer may decide that it makes financial sense to improve the availability of some units, while not choosing to alter the availability of other units. So, at the very least, we are assured that the availability will never DECREASE for any of the process units.

However, are there any possibilities where that might make financial sense? I know that I'm venturing into territory when people are going to question my sanity, but there truly is a method to my madness. Before I even attempt to answer that question, I must relate an experience from the past when one considers situations where a mathematical optimizer might consider moves that challenge the conventional wisdom.

Early in my Advanced Control career, we were doing a project at a refinery in Oklahoma. This particular project included Advanced Control on the crude and FCC units, along with an optimizer for the FCC unit. The optimizer could adjust FCC unit feedrate, riser temperature, and other variables to maximize yields.

I will never forget one day, when our client looked us straight in the eye and said, "I sure hope that your optimization will never tell us to reduce FCC feedrate to improve yield. You'd be laughed out of the refinery and driven straight to the airport."

In other words, we should never violate the golden rule of refining, which states that feed is to be maximized at all times. The good news is we were saved from that possibility. I'm fairly sure that the optimizer never did reduce feedrate.

So, I am prepared for how most will react to my suggestion. I only raise the question to keep us thinking about the big picture and not allow our thinking to become too pigeonholed.

Now, let's return to the question I brought up. For the most part, I admit that it's not a very practical scenario, and might not occur very often, because whatever process equipment exists right now in a process unit is going to remain there. So, it's not the reverse of an improvement project, where we are physically adding equipment. (We don't intentionally remove equipment (at least, not that I can ever think of).)

In fact, it only makes sense to consider it if we can reduce some activity that we are currently doing that would make it financially worthwhile. From a mathematical standpoint, what it means that the improvement functions shown earlier don't have to start at zero. Rather, it could extend into negative territory (which would allow availability to actually decrease). However, to be practical, one would have to understand the relationship between how cost reductions would affect availability. How easy that might be is questionable, which might render the idea moot.

Once again, I don't think this is a very practical scenario, but I do want to challenge the conventional wisdom that says that we must always be improving availability on each and every unit. There might actually be some cases when reducing availability makes more financial sense. However, getting somebody to consider that is likely the equivalent of getting my former client to consider reducing FCC feedrate.

4.8 Conclusion

This chapter introduced the first details behind Monte Carlo simulation and true optimization, which will be the principal technologies used throughout this book.

Monte Carlo simulation is a probabilistic technology that determines system availability based on the failure/repair characteristics of a unit or equipment within a unique process configuration. On the other hand, the optimizer determines the incremental profit of a system based on manipulating independent variables and eventually iterating to a solution. The optimizer has a provision for honoring both independent and dependent variable constraints.

Many process plants set the same target availability for each process unit within their facility. However, in this chapter, we showed, using a refining industry example, that this idea is often suboptimal. Three cases were presented – one using current unit availabilities, another with the same availability target for each unit, and a series of cases where the optimizer determines the proper individual targets.

The optimizer ultimately chose the optimal availabilities after evaluating possible improvement projects for each different process unit. The improvements were first detailed as a series of discrete projects, then smoothed into a continuous function for the optimizer to use. The last case, true optimization, resulted in the largest profit among the three scenarios considered.

The optimization runs taught us that the optimum availability for each process unit isn't always obvious because it's a complex problem that involves many different factors. In addition, changes in improvement project costs can significantly alter the final results. It also shows that some improvement projects should never be implemented because the underlying economics don't support them. Last, this process can help a plant to determine the exact mix of projects that should be implemented based on an allocated budget.

There are many other types of optimization problems that can be implemented using this general approach. However, before we consider those topics, we will use the next couple of chapters to further discuss our fundamental technologies – Monte Carlo simulation and true optimization – in more detail.

Then, we will take a detour to discuss the Consultation Work Process, which is a fundamental part of this methodology since data for the analyses is so important.

5

Monte Carlo Simulation Overview

Monte Carlo simulation technology is an important part of our asset optimization methodology. (While this chapter is meant as an introduction to Monte-Carlo simulation, a more complete description can be found in "Appendix A - Nuts and Bolts of Monte-Carlo Simulation". This includes a simple work process and its explanation, which is, more or less, common sense. In addition, the real bottleneck to implementing Monte-Carlo simulation is obtaining high quality data. Therefore, Chapter 7 is devoted to that topic - "The Consultation Process - The Main Work Process".)

Since Monte-Carlo simulation is foundational, it makes sense for us to step back and spend enough time so that one grasps how it works conceptually. Instead of starting with detailed mathematical explanations that would likely make your eyes glaze over, we will start with a process example and slowly add detail.

Figure 5.1 depicts one of the process units in our refinery example – the semi-regen Reformer.

There are 11 important pieces of equipment in the unit – three furnaces prior to each of the three reactors, the recycle gas compressor, reformate feed/effluent and cold water exchangers, a main separator, and the debutanizer (which contains numerous pieces of equipment to effect its separation).

In our particular refinery, there is a Naphtha Splitter ahead of the Reformer, which splits naphtha feed from the crude unit into a C6 aromatics and a heavier portion, which becomes Reformer feed. Lighter C5/C6 overhead material from the Naphtha Splitter ends up as Isom feed.

Before entering the charge furnace, a feed/effluent exchanger preheats naphtha feed, where the material is further heated to about 980 °F. In the reactor, naphthenic material in the feed is quickly converted to higher octane aromatics. Since the reaction is endothermic, the reactor outlet temperature drops. Thus, the first Interheater furnace is needed to reheat the material back to approximately 980 °F before entering reactor two to restore reaction rates back to reasonable levels.

Reactors 2 and 3 are similar in that they principally try to convert isoparaffins to naphthenes, which are then quickly converted to aromatics. Each reactor size becomes progressively bigger, since the isoparaffin reaction rates become slower with each subsequent reactor. A typical catalyst loading is 20% of the total catalyst in the first reactor, 30% in the second, and the remaining 50% in reactor three. Depending on feed composition, some refinery reformers contain four reactors, instead of the three shown here.

Reaction products are separated into a hydrogen-rich gas, which is compressed and used throughout the remainder of the refinery, as well as a high-octane reformate product. A portion of the hydrogen is recirculated to the reactor to reduce catalyst coke formation. Liquid product from the separator is fractionated in the debutanizer into a small quantity C3/C4 fraction, in addition to the main stabilized reformate product.

Plant Optimization in the Process Industries: Incorporating Equipment/Assets in the Decision-Making Process, First Edition. Marty Moran.

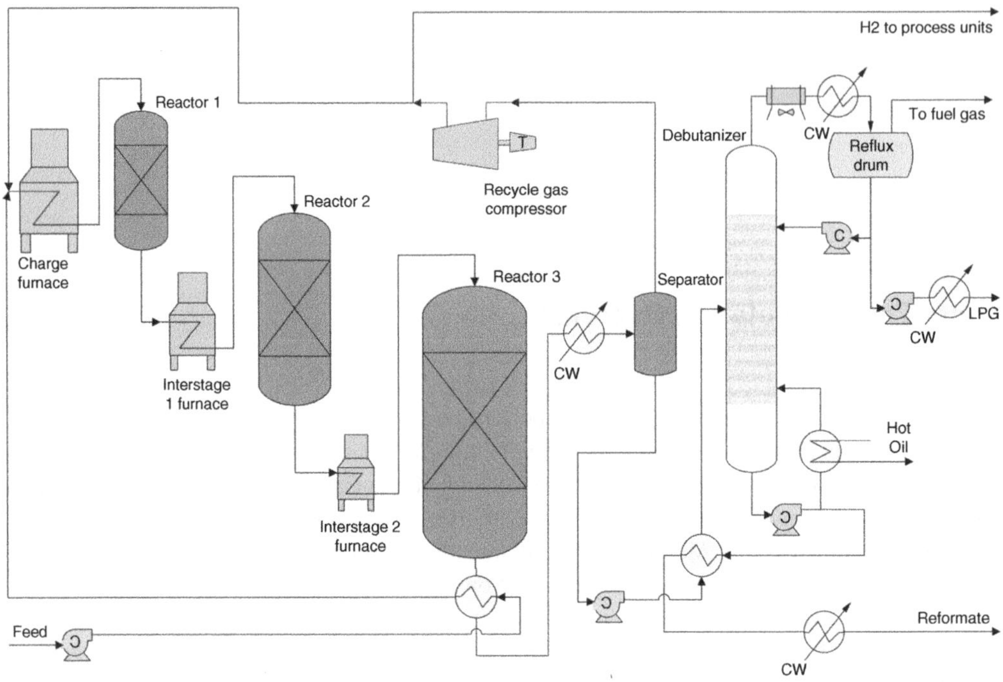

Figure 5.1 Semi-Regen Reformer

5.1 Reliability Block Diagram (RBD)

As noted previously, when one examines this unit from a process perspective, it's obvious that all 11 major pieces of the semi-regen Reformer unit must be operational for the unit to be considered "Available."

To show how availability is calculated, we'll need to introduce the first major concept within the Monte Carlo simulation. That is the idea of Reliability Block Diagrams (RBD). For the Reformer, the most basic RBD is shown in Figure 5.2.

This diagram shows the 11 major pieces of equipment in the unit, along with the starting and ending points. At any given moment in time, each of our 11 pieces of equipment can be in only one of two states – Available or Not Available.

If all 11 pieces of equipment are "Available," then one can begin at the starting point and continue all the way through to the ending point. If that's possible, then the system, or process unit, is considered to be Available. In other words, all the equipment is working properly, and we are producing product at design rates. Later on, we will see that we can also have partial production, so it doesn't have to be 100% design rate or nothing, but that's a complication that can wait until later.

The RBD shown is considered to be a "series" configuration since all the blocks are connected in series. There's also the concept of "parallel" configurations, which is used for spare pumps, or true parallel trains of processing equipment. More about that later.

However, what's really important to note about series and parallel configurations is that they can be quite different from the way that process engineers think of series and parallel – especially when viewing it in a typical process simulator. In other words, the flow path within an RBD and the typical process flow in the unit itself don't necessarily need to align with one another, since they are actually different concepts.

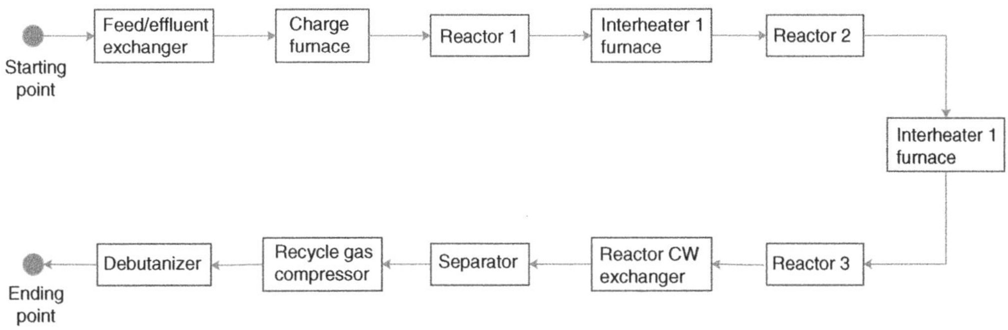

Figure 5.2 Reformer Reliability Block Diagram

The important differentiator in the Monte Carlo world is what equipment needs to be Available in order for the "system" or product unit to be "Available." For example, normally only one pump of a typical pump with a spare arrangement needs to be operational for the pumping function to be considered to work properly. Thus, this would be represented by a parallel arrangement.

Right now, we are representing the debutanizer as one block, when in reality, it consists of numerous pieces of equipment – feed pump, tower itself, overhead condenser, overhead vessel, reflux, product pumps, reboiler, etc. We'll see later on that if we want to expand our debutanizer and detail it at the equipment level (and we will want to do that), then we will use the concept of submodels. Once again, we'll explain that later. However, for now, let's stick to this simple representation so that we can talk through our Monte Carlo simulation, which actually determines whether the semi-regen unit is actually up and running.

5.2 Rolling the Dice

As we talked about earlier, each of our 11 major areas can be Available or Not Available. How can we determine which state it is?

First, we must provide data to the mathematical engine in the form of failure and repair data. Don't worry about the terms for now; rather, just the concepts.

Since Monte Carlo is a probabilistic algorithm, each time it encounters 1 of 11 blocks, it rolls the dice and asks whether that item is Available. Mathematically, that's done by "sampling" a reliability distribution, which is a mathematical representation of the likelihood of that item being Available.

Prior to building the RBD within the Monte Carlo simulation, we would have reviewed data for the debutanizer as a complete unit and determined the percentage of time that it had been down over a period of many years. We then could build a "reliability distribution" from that data.

That could have been in the form of a normal distribution, which would be the mean with some uncertainty associated with it. The idea of uncertainty is important because we never know from day to day which equipment will actually fail, so there will be some uncertainty with all equipment. Most of the time, RAM modelers will actually use what's referred to as Weibull distributions, but let's leave that aside for now. What's important to understand now is that each block in our model will have a mathematical representation of how often it is operational. Each time that Monte Carlo encounters that block, it rolls the dice and determines whether the block is Available or not based on its distribution.

If that block is Available, then it moves on to the next block in the series and asks the same question using the next block's reliability distribution. The process then continues in the same manner until either a block fails, or we hit the ending point. Let's assume in this particular example that all 11 pieces of equipment are "Available," which means that the system is "Available" at our design production rate.

The usual time frame for a Monte Carlo simulation is over a 20-year time span or something similar. Normally, Monte Carlo conceptually breaks down the process into much smaller chunks (the math does it differently, but what is described is conceptually correct).

Let's say that it samples our 11 blocks at the start of each day, and it can connect all equipment from the starting point to the ending point; then, the process unit is considered to be "Available." It then begins the same process again tomorrow. It will continue this process each day until it rolls the dice one day and comes up with a block that has failed. What happens then?

In addition to a reliability distribution, which determines if a block is Available, each block also has a repair distribution. The repair distribution concept is similar to the failure distribution, but in this case, it's a mathematical representation that informs us how long it would likely take to "repair" the block back to its normal working state.

If any 1 of our 11 blocks enters this state, then it would not be possible to form a complete connection from the starting point to the ending point, so the system would become Not Available. Furthermore, it would stay this way until that block was repaired. At that point, the original process would start all over again by sampling the blocks to determine if they are Available. This process would then continue across all 365 days × 20 years to determine the system availability.

5.3 Histories within a Model Run

The process just described could be graphically depicted by what is shown in Figure 5.3.

A state of 1 represents the process unit being Available, while 0 means that it is Not Available. The time axis is in days, which will extend us to 8766 days over a 20-year time frame. The way to interpret that is that the unit was continuously Available from day 0 through roughly day 270. Then, it was down for a few days. Then, it was continuously Available from day 300 through almost day 475, and down again for awhile. If we sum all the days that it was in state 1 and divide by the total number of days, 8766, we can arrive at the system availability for the unit. However, this run only represents one particular iteration within the Monte Carlo engine. We will actually repeat this process, potentially thousands of times.

For example, run 2 might look similar to that in Figure 5.4.

It's important to note that it's different from the first run. That's normal. It's a statistical process.

On the other hand, run 3 might look like that in Figure 5.5.

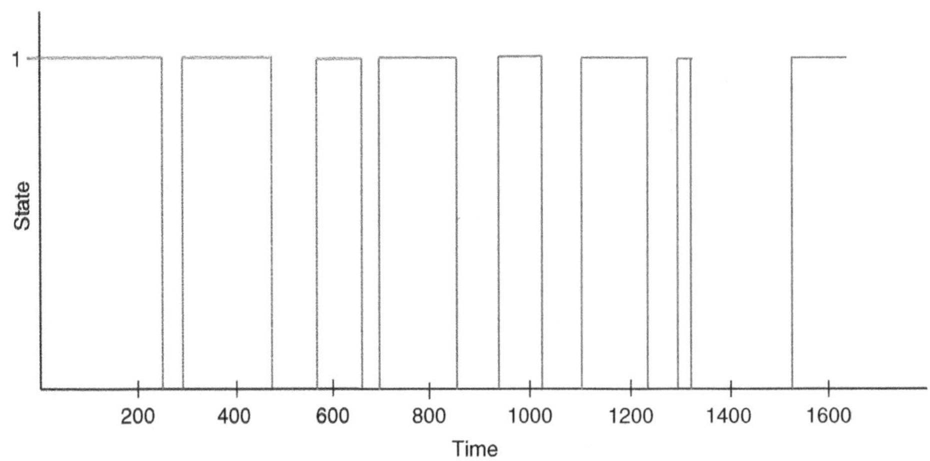

Figure 5.3 History Run 1

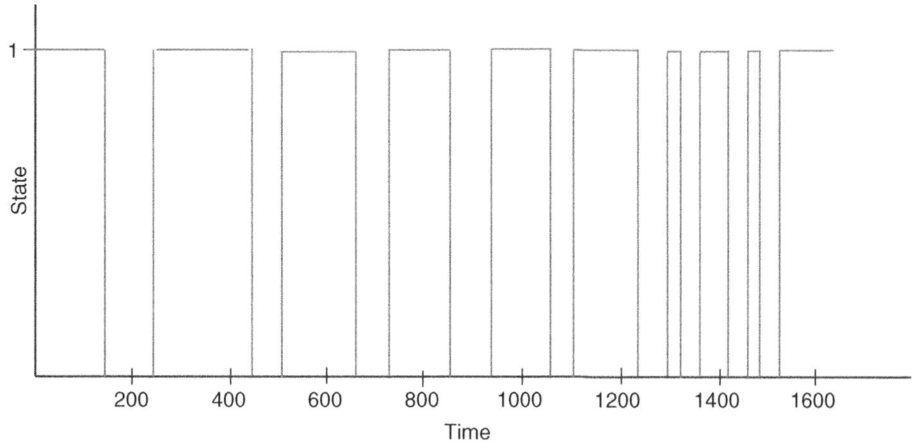

Figure 5.4 History Run 2

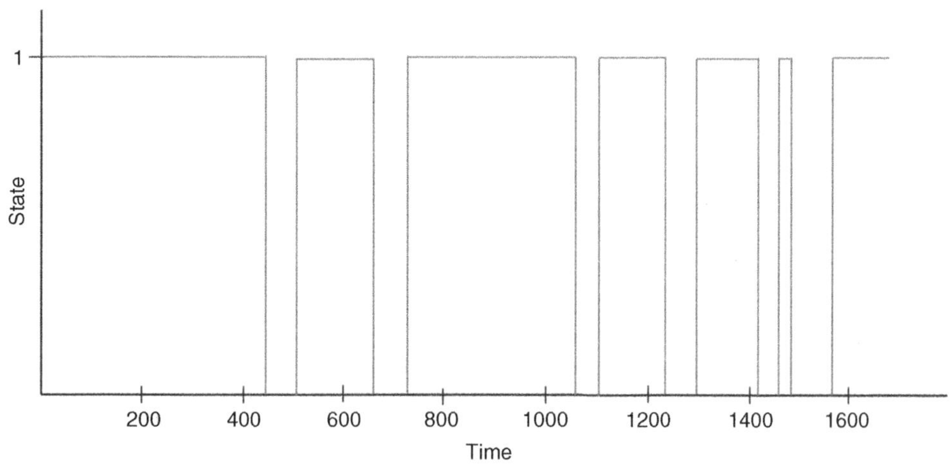

Figure 5.5 History Run 3

5.4 Results

A Monte Carlos simulation will repeat the runs just described many times – perhaps as many as a thousand times. With so much data, the math engine will be able to generate excellent statistical data. Result data comes in many different forms, but we'll only concentrate on the two that are most important – system availability probability distributions and Pareto results for the blocks that are the biggest contributors to unavailability.

5.4.1 Probability Distributions

The most important Monte Carlo result is the system availability probability distribution. Commercially Available packages report not only the mean system availability, but also the shape of the curve, which is usually NOT a normal distribution. Both the mean and curve shape are important in being able to assess if this system will meet the needs of the organization.

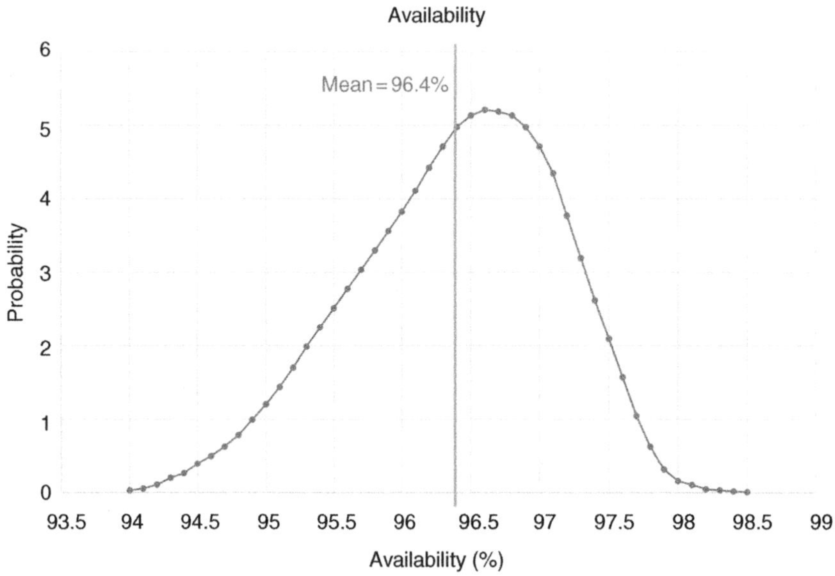

Figure 5.6 Availability Probability Distribution

Figure 5.6 shows what the availability distribution for the Reformer might look like. The mean availability is 96.4%, which should be interpreted as the most likely availability. However, the probability curve is not a normal distribution and is clearly skewed toward lower availabilities. Thus, it is statistically possible that the availability could range anywhere from 94% all the way up to 98.5%.

Thus, the availability will never drop below 94%, but it also means that if you had wanted 99% availability, you wouldn't be able to achieve that with the current equipment, process/equipment configuration, and operating/maintenance practices within the unit.

A skewed probability distribution is quite normal. As your system approaches higher availability numbers, it becomes increasingly more difficult to improve availability. So, the math confirms what we already know as engineers.

To summarize, the three important numbers from this analysis are the mean availability of 96.4%, as well as the absolute minimum of 94%, and the absolute maximum of 98.5%.

The probability distribution tells us if the system can achieve the desired availability target. On the other hand, the Pareto chart shown in Figure 5.7 provides users with knowledge of equipment that are the greatest contributors to "unavailability." In other words, if one needs to increase availability, what equipment should we concentrate on? The largest contributors to unavailability are always shown on the far left-hand side of the chart. As one moves toward the right, a given equipment's contribution to unavailability decreases.

Upon examining the chart, one can see that equipment A contributes 0.7% to unavailability. If it's practical to improve this equipment's reliability, then total system availability could be increased by that amount. In a similar fashion, if equipment B's issues could be adequately addressed, then system availability could increase by 0.45%. It's important to note that there is no requirement that one start with equipment A before proceeding to equipment B.

In fact, sometimes, it's impractical to correct issues with a piece of equipment furthest to the left, yet it is possible to correct equipment somewhat farther to the right. For example, it might

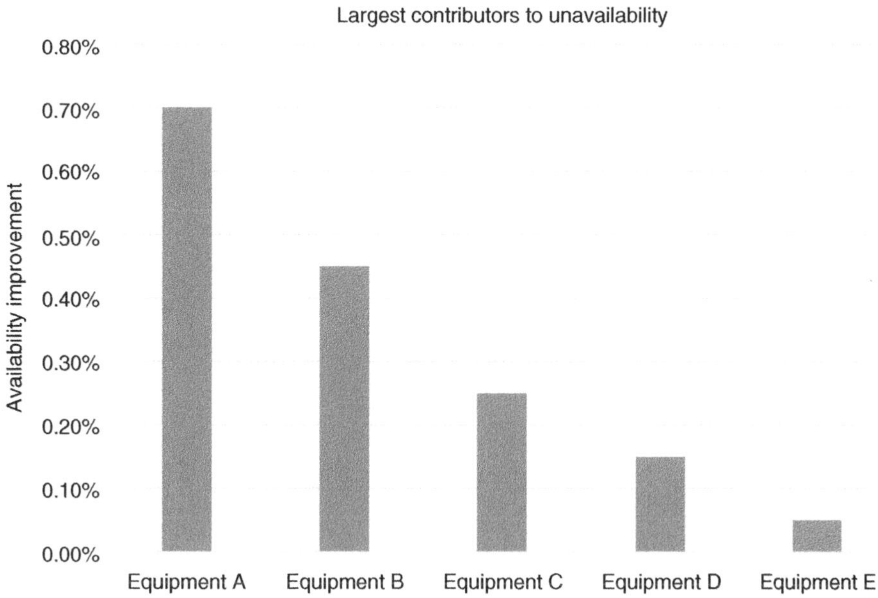

Figure 5.7 Largest Contributors to Unavailability

require a turnaround to improve equipment A's reliability, whereas for equipment B, it could be done immediately.

It's important to note that improving equipment reliability is one such way to increase system availability. In a future chapter entitled "Improving Availability," we will examine other methods of improving system availability.

5.5 Submodel – Detailed Process Unit Model

Up to this point, we have used simple blocks for each of our 11 major areas within the semi-regen unit. While it's certainly possible to perform analyses at this level, sometimes it's necessary to drill down further into a unit. For example, in this RBD, the debutanizer is shown as only one block when in reality, it consists of numerous pieces of equipment. Thus, to adequately analyze the debutanizer, we need to detail it further. However, it's still nice to retain the simpler representation of a single block when analyzing the Reformer.

There's where the concept of a "submodel" becomes helpful. Within the higher-level RBD, the submodel functions essentially like a single block. However, when one clicks on the submodel, it brings up another RBD – this time with the blocks associated with the debutanizer, as shown in Figure 5.8.

This RBD functions exactly similar to the first RBD shown in exactly the same way as already described. For the debutanizer to be "Available," it must be possible to complete a path from the starting point to the ending point.

One important thing to note about this RBD is that there are several instances of "parallel" paths. In this case, it's the spares for each of the four pumps associated with this unit – feed, reflux, distillate, and bottoms. What this means is that if either pump A or pump B is operational, then we can still draw and move from the Starting Point toward the Ending Point as long as the other pieces of equipment are operational.

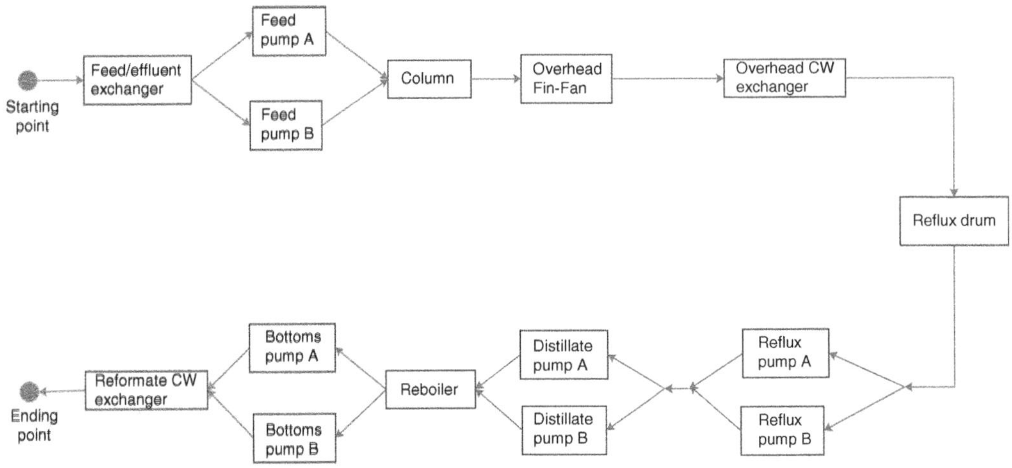

Figure 5.8 Debutanizer Reliability Block Diagram

5.6 What Level of Detail Is Required?

A common question is what level of detail is required for a Monte Carlo simulation, and what functional blocks are required? That's entirely a user decision since there is nothing inherent in Monte Carlo simulation technology that dictates that it must be a certain way. In that sense, Monte Carlo simulation is just a generic simulation technology. It allows the user total flexibility in how one defines the problem. Of course, with flexibility, comes responsibility.

A fairly standard way to proceed is to simply include the major pieces of equipment within the unit. However, there's no question that this is only a starting point. For example, what you frequently discover is that some equipment will almost never have any input on a unit. For example, most towers don't fail of their own accord, so including the tower itself is usually overkill. (It's usually the equipment around the tower that is more likely to fail.) What I tend to do is still include it. If I determine that it isn't affecting anything, then I simply zero it out. That way, it doesn't have any impact. That forces discipline into my methodology.

It's also important to note that we can add other blocks that are not equipment-related at all if it can impact unit availability. For example, we could add a block "Following Procedures," if our unit was staffed with new employees, who occasionally make operational procedure mistakes, and those actions periodically cause units to go down.

In fact, there is no end to the types of blocks that could be added. I would just caution some organizational discipline when adding these other ad-hoc blocks. Normally, an organization will settle on the set of standard blocks that they would add to any model to take into account those "other" factors.

5.7 Definitions

As noted in a previous chapter, Monte Carlo simulation is often referred to within our industry as Reliability, Availability, and Maintainability (RAM) modeling. Each of those letters actually has meaning.

Availability is the normal output of a Monte Carlo simulation and is often expressed as a probability distribution, as discussed previously in this chapter.

On the other hand, **Reliability** and **Maintainability** are the two major inputs for each block in an RBD.

Reliability is the input that describes the likelihood that an asset (a piece of equipment or it could be a collection of assets like in our example) will work as desired. This is often expressed as the probability that it will perform its intended function and is usually in the form of a mathematical distribution. Common distributions are Weibulls and normal distributions, though other forms are also possible.

On the other hand, **Maintainability** is the amount of time necessary to return the system to its operational state. It is also modeled as a probability distribution with the same options just mentioned.

5.8 RAM Software Tools

There are many different commercially Available RAM modeling tools. This book will be agnostic with regard to specific tools and will not advocate one tool over another. At the end of the day, software tools are merely enablers and won't make or break your program. So, don't be lulled into a slick sales pitch. What's more important to your program's success are the various business aspects that will be detailed throughout the remainder of this book's chapters.

While there are free Monte Carlo software programs, I'd stay away from those for a professional organization. At the other end of the spectrum are tools that allow one to write code for any event that occurs. Even though it is not specifically a requirement, it's a nice feature that can be exploited for more difficult issues, and will likely be desired for more advanced organizations. However, many operating companies will be able to start with more mid-range tools that are both professional, full-featured, yet relatively cheap.

5.9 Challenge to Monte Carlo Simulation Vendors

While we are discussing the two different technologies, I would like to issue a challenge to our Monte Carlo simulation vendors.

The 2004 NPRA presentation, described in the previous chapter, used two different technologies to optimize gasoline pool availability – Monte Carlo simulation technology, as well as an optimizer. While I was responsible for the thought process in the presentation, a highly competent modeler who was able to combine the two technologies together to implement the project that I had desired. But it took some work.

To the best of my knowledge, no vendor company has used these technologies paired together in this fashion as part of their standard offering. What I am hoping is that the discussion in this book can spur the major Monte Carlo simulation vendors to seamlessly integrate optimization technology together within their Monte Carlo engines. Or, at the very least, make it easier to use. That way, our industry could more easily implement the ideas contained within this book. I know that it will eventually happen, but the sooner it occurs for our industry, the better.

In the meantime, we can still implement these solutions, as we had to do in APC for many years before the technologies became integrated with one another. However, for now, it will just take some more work to execute each project.

5.10 Conclusions

Monte Carlo simulation technology, often referred to as RAM Modeling (Reliability, Availability, and Maintainability Modeling) is a key component of our asset-based approach to plant optimization. It is a generic, probabilistic approach for difficult mathematical problems that cannot be easily solved deterministically, with RAM being specifically tailored toward process industry problems.

A Reliability Block Diagram (RBD) is a logical representation of the key pieces of equipment (or subsets of equipment) that, if Not Available, could cause the system to be unavailable. It is fundamentally different from a Process Flow Diagram (PFD), even though many of the individual pieces of equipment may be the same. It has a starting and ending point with blocks in between – configured in a series or parallel configuration, or more normally a combination of the two.

Blocks are often a single piece of equipment but could be the subcomponents within a single piece of equipment or could even represent an entire process unit. The system is Available if one can trace a path of blocks that are Available from the starting to the ending point.

Submodels are special blocks on an RBD that if one clicks upon bring up a different RBD with supporting details. These make building more detailed models easier, since they allow the user to hide details, unless necessary.

Monte Carlo simulation determines if a block is Available through a random number process with the likelihood of availability determined through a failure probability distribution based on the historical likelihood of being Available. If the block should fail, the time it remains down is determined through a repair probability distribution based on the historical time to repair.

A Monte Carlo simulation for the system is normally conducted over a multi-year time frame, with 20–30 years being typical. Each run from time zero through ending time is termed a history. Each Monte Carlo simulation consists of hundreds, if not thousands of history runs, performed so that the software can generate a statistically relevant set of data representative of the system.

The most important results are system availability probability distributions, which generate mean, absolute minimum, and absolute maximum availabilities, as well as a Pareto chart that details equipment that is the largest contributor to unavailability.

A more complete description of Monte-Carlo simulation can be found in "Appendix A – Nuts and Bolts of Monte-Carlo Simulation". In addition, a chapter is devoted to how to find high quality data to use in the simulations – "The Consultation Process – The Main Work Process".

6

Optimizer Overview

The other technology required to optimize process plants based on better asset decisions is true optimization technology, which will sit directly above the Monte Carlo engine and manipulate targets to it. To give the reader a better flavor of how to define optimization problems, this chapter will provide a more detailed look into that technology beyond the introductory text and example in Chapter 4. We'll achieve that by proposing optimizer definitions for each of the different optimization problems that will be discussed throughout this book.

As mentioned previously, an optimizer consists of four important elements:

- Independent Variables
- Dependent Variables
- Constraints
- Objective Function

6.1 Independent Variables

Independent variables are those variables that the optimizer will choose, or manipulate, to drive toward a value that maximizes or minimizes the user-defined objective function. What variables are chosen as independent variables is totally up to the user.

In the example presented in Chapter 4, the independent variables were each process unit's availability target. While we decided to choose all the availability targets in that example, there's no explicit requirement to do so. We could have easily included only the crude and vacuum units' availability and forgotten about all the other process unit availability targets. At the end of the day, it's the user's business and technical requirements that drive the decisions of what independent variables to include in the problem definition.

In addition, as we will see in a later chapter, there's no requirement that independent variables be the same type of variable. For example, a problem could include variables for improvement projects, tankage size, equipment reliability design options, and task optimization, among others – all defined within one overall problem. In fact, later in the book, we will include an entire chapter on combined optimization, where we will broaden the problem definitions to include different independent variable types. We'll find that the only thing limiting the choice of independent variables is a user's imagination.

Plant Optimization in the Process Industries: Incorporating Equipment/Assets in the Decision-Making Process, First Edition. Marty Moran.

6.2 Dependent Variables

Dependent variables are any equations, or their results, that are a function of some or all of the independent variables. They can be used as either inputs into the objective function or as constraints.

Numerous dependent variables were used in the previous problem definition in Chapter 4. For example, each process unit had its own equation that converted the process unit availability choice into a cost for its related improvement project(s). In addition, there was a simple equation that added together all the improvement project costs. What dependent variables are chosen can vary widely from one optimization problem to another.

6.3 Constraints

Constraints are an important part of every optimization project. That can be either independent or dependent variable constraints, and virtually every project utilizes both of them.

6.3.1 Independent Variable Constraints

Independent variable constraints are the upper and lower bounds that the optimizer can move a particular independent variable. For example, the maximum/minimum crude unit availability in the previous example problem was set to be 99.75% and 95%, respectively. This allowed the optimizer the ability to move that variable anywhere within that 4.75% range, but not outside of it.

As we mentioned back in Chapter 4, from this point forward, we will assume that every independent variable has a maximum and minimum variable associated with it since all the major optimization packages automatically provide a high/low limit for each of those variables.

6.3.2 Dependent Variable Constraints

Dependent variable constraints are an important part of most optimization projects. They are used extensively. These consist of any constraints the user would like included in the problem definition that are not already independent variable constraints.

In our previous problem, we included dependent variable constraints for both total improvement project spending and gasoline component production limits. While our example problem had several dependent variable constraints, there is no limit to how many can be included.

The one important requirement, from a practical standpoint, is that every dependent variable constraint must be a function of at least one independent variable. Otherwise, it will never be affected by any changes in the independent variable selections.

It's important to note that just because a dependent variable is defined, that doesn't mean that when the optimizer executes, it will run up against that particular constraint. What constraints the optimizer encounters depend on many factors, one of which is the independent variable ranges. For that reason, whenever you are examining results, always ask yourself if you run up against a limit, whether it's fundamentally because the independent variables associated with it don't have enough room to move.

6.4 Objective Function

The objective function is a user-defined equation that includes either only independent variables or a combination of independent and dependent variables. The user will set up the objective function to be either a maximization or a minimization. For the problem in Chapter 4, the objective function was to maximize profit.

Since dependent variables are calculated in an equation with independent variables on the right-hand side of the equation, at the end of the day, the objective function is ultimately a function of only independent variables.

It's important to point out that each independent variable should have some discernable effect on the value of the objective function over at least a portion of its defined range. If not, the optimizer will quickly learn that it doesn't have any effect and won't adjust that variable. In general, that's not a big deal. However, it means that this variable has no effect on the objective function and probably shouldn't have been included as part of the problem definition anyway.

6.5 Optimizer Problem Definitions

The only optimization problem discussed up to this point in time has been the example problem of determining the optimal process unit availability by choosing the most desirable improvement projects. To give the reader a broader understanding of how to define optimization problems, this section will present an overview of problem definitions for all the other optimization problems that will be discussed in this book.

By necessity, this will be a very brief introduction to those problems. However, while reading this section, we'd like readers to think about how they might structure optimization problems for their own plants. In fact, after reading later chapters, the reader might want to return to this section because it represents a rather concise summary of all the different optimizer definitions, arranged neatly in one chapter.

The following are the new optimization problems discussed in this section:

- Turnaround Timing, Duration, and Equipment Selection Optimization
- Maintenance Plan Options Optimization
- Spares Optimization
- Task Optimization
- Storage Tankage Determination/Optimization
- Combined Optimization
  - Optimal Improvement Project and the optimal mix of crude feeds.
  - Turnaround equipment selection combined with the optimal choice of crude feeds.

For the sake of completeness, we will start this section by including the optimization problem from Chapter 4 – Optimal Improvement Project (determination of process unit availabilities).

6.5.1 Optimal Improvement Project (Determination of Process Unit Availability)

The objective of this optimization problem is to choose the best combination of improvement projects for the main refinery process units that optimizes profitability. By virtue of choosing the best improvement projects, this will also determine the optimal availability targets for each process unit.

6.5.1.1 Independent Variables

The independent variables for this optimization are the following:

- Process unit availability targets

6.5.1.2 Dependent Variables

There are four main dependent variables for this problem:

- Each unit's improvement project function
- Total product volume increase

- Individual gasoline component totals
- Total improvement project expenditure

Each time the optimizer chooses a process unit availability target, it will refer to the improvement project graphs for its particular unit to determine how much it must spend to achieve that new availability target.

Then, production values are calculated based on the overall system availability. These are then multiplied by their appropriate wholesale prices to determine the total revenue increase for that choice of new availability targets.

Each major component stream may be included as a constraint since the plant may not be able to sell all the incremental product volumes.

The last major calculation is the total improvement expenditure, which will be used as a constraint, as well as in the profitability calculation.

6.5.1.3 Constraints

Total improvement project expenditure will be included as a constraint, as well as gasoline production.

6.5.1.4 Objective Function

The goal of this optimization problem is to maximize profit, subject to meeting constraints. Fundamentally, the optimizer balances the production increase from money spent against possible improvement project expenditures.

6.5.2 Turnaround Timing, Duration, and Equipment Selection Optimization

While turnaround planning always involves many different facets, when deciding the best time to take a turnaround one consideration is often how declining equipment reliability might affect unit availability. When evaluating this concern, it often makes sense to examine its effect over more than just one turnaround cycle, so that any decisions made concerning this turnaround don't cause equipment to fail in the middle of the next production run.

6.5.2.1 Independent Variables

The typical independent variables for this problem would be:

- What equipment will be restored in a particular turnaround
- Time when each turnaround begins (in some cases, but not always)
- Duration of each turnaround (in some cases, but not always)

A couple of comments. This optimization can still be used even if the turnaround timing and/or duration is being set by other factors. In that sense, this is probably the most flexible optimization problem described in this book.

To clarify, what this means is that if the starting time and/or the duration (critical path) is dictated by something other than equipment reliability, those variables would be fixed, and the optimizer wouldn't adjust them. However, since it is not always obvious what equipment should be restored within a turnaround, we want to give the optimizer the ability to examine a list of equipment and be allowed to choose whether to restore individual equipment or not - and in which turnaround.

In other words, determining what equipment is restored while the unit(s) are down will always be one of the independent variables, though which equipment would be included could vary from one turnaround cycle to another. (Of course, it's almost always guaranteed that this will be multiple pieces of equipment – not just one.)

Let's take an example to clarify those comments. Let's say that the turnaround start date is being set by an improvement project, but its duration is somewhat flexible with a range of possible dates. The optimizer could then be run to evaluate whether certain equipment makes sense to be included, within what turnaround cycle, and what the optimal turnaround duration is.

6.5.2.2 Dependent Variables

The sheer quantity and range of dependent variables for this problem are hard to generalize since they could span a wide spectrum.

For example, there may be constraints on one particular type of labor resource. In that case, we would include a calculation of how much time is required for certain resources as different equipment is chosen to be restored within a turnaround.

6.5.2.3 Constraints

There are many possible constraints, but a standard one is process unit availability. Normally, one would want availability to remain above a certain value at a given future date. This would ensure that any equipment restoration decisions made for this turnaround cycle result in sufficient process unit availability at a future date.

In addition, it's easy to envision including various constraints related to resource limitations, such as labor and/or budget constraints.

6.5.2.4 Objective Function

The goal of this optimization is to determine the maximum profit, subject to meeting all constraints. Profit is production revenue minus turnaround costs, such as labor and material costs.

Incremental profit is driven by decreasing the time that the unit remains down for maintenance and/or reducing what work gets done during the turnaround. However, that is balanced against potentially declining equipment availability.

The constraint highlighted in the previous section is important because it forces the optimizer to restore enough equipment and drive the duration to last long enough to maximize profitability. Otherwise, the optimizer will want to run the unit as long as possible, since each turnaround curtails production and consumes cash in the form of people and equipment repairs.

6.5.3 Maintenance Plan Options Optimization

For some more complicated repair scenarios, we may have options that span the range from those that can return the equipment to service quicker but cost more, all the way to some that are much cheaper but take longer. Of course, we may have options in between those extremes. Which option best balances increased production versus costs for that production isn't always obvious. In those cases, maintenance plan options optimization is an attractive means to evaluate those opportunities.

Unit availability could be constrained if the optimal availability has been determined independent of this problem. If not, the optimizer could evaluate scenarios where a quicker repair, even if considerably more expensive, is the best business decision since the increased production would more than offset the increased costs.

In the detailed chapter on maintenance plan options, we will present three different optimization problems – one set of maintenance plan options for one piece of equipment, several different maintenance plans for different equipment in one process unit, and the more generalized case where we evaluate different maintenance plan options for equipment scattered in different process units. The more generalized the case, the more interesting the optimization result possibilities.

6.5.3.1 Independent Variables

The independent variables would be:

- Maintenance plan options in discrete steps for equipment (or more than one) in one or more process units

Fundamentally, the optimizer would evaluate different choices from a table of maintenance options. These choices would document the repair cost, the time to complete it, and any production impact. This could be a single piece of equipment or numerous equipment in one or more process units(s).

6.5.3.2 Dependent Variables

Dependent variables would be summations of key aspects of the maintenance plan options, such as labor resources, total costs, etc.

6.5.3.3 Constraints

There are many potential constraints. Some of the more obvious are labor resources, contractor availability, total costs, etc.

6.5.3.4 Objective Function

Similar to the previous problems, the goal of this optimization is to maximize profit, subject to meeting constraints.

6.5.4 Spare Parts Optimization

Plants don't want to be caught without spare parts if the equipment should fail – especially if losing that equipment would result in a loss of production. On the other hand, retaining excessive spare parts can result in significant procurement and storage costs. Since the optimum likely lies in between those extremes, spare parts are a great candidate for optimization.

In the detailed chapter that follows, we will present four different optimization problems – the optimal quantity of spare parts for a single piece of equipment, the optimal quantity of several different spare parts for several pieces of equipment within one process unit, and the more generalized case where the optimizer determines the spare parts requirements for different equipment scattered in different process units. The more generalized the case, the more interesting the optimization result possibilities. The last example is for a common spare used for the same equipment type used in multiple process units.

6.5.4.1 Independent Variables

The independent variables for this optimization would be:

- Quantity of spare part holdings for equipment(s) in process unit(s)

6.5.4.2 Dependent Variables

Equations were regressed for each spare associated with a piece of equipment for both an average and maximum production loss. Average production losses for all equipment were summed together and included in the objective function. On the other hand, the maximum production loss for each equipment was included as a constraint.

Another dependent variable is the total cost of incremental spare parts purchases.

6.5.4.3 Constraints

There were constraints included for maximum production loss for each spare. In addition, there was a maximum spare parts budget constraint.

6.5.4.4 Objective Function

The goal of this optimization is to determine the maximum profit, subject to the maximum production loss constraint. This balances potential increased production through holding more spares minus the cost of those increased spares.

6.5.5 Task Optimization

Maintenance task intervals are usually set once and rarely changed. However, plant conditions can change significantly over time, or maintenance technology could improve allowing longer intervals at the same risk profile. In addition, the plant could face resource issues where it becomes necessary to allocate limited resources more effectively.

While this optimization could be done for a single task, it is probably more applicable and appropriate for multiple tasks, or for the same task spread across many units. It's highly effective when there are resource constraints. This optimization could also be combined with spare parts or maintenance plan options, if necessary.

In the detailed chapter that follows later in the book, we will present three different optimization problems – one set of maintenance tasks for one piece of equipment, several different maintenance tasks for different equipment in one process unit, and the more generalized case where we evaluate different maintenance tasks for equipment scattered in different process units.

In the typical case, we would likely tightly constrain process unit(s) availability within rather narrow limits. This would force the optimizer to focus just on minimizing task costs at the same availability targets. Or availability could be allowed to vary on some or all of the units, which would free the optimizer up to increase production on that unit, if additional production more than offset the increased maintenance task costs. The more generalized the case, the more interesting the optimization becomes.

6.5.5.1 Independent Variables

The independent variable for this optimization would be the following:

- Maintenance task(s) frequency

6.5.5.2 Dependent Variables

A typical dependent variable would be resource skill limitations or budgets.

6.5.5.3 Constraints

Usual constraints are labor costs for one or more of the skill pools or total labor time.

In addition, as mentioned earlier, we would probably want to constrain unit availability within a narrow range, since that was likely determined in a previous exercise.

6.5.5.4 Objective Function

Similar to the other optimization problems, the goal of this is to maximize profit, subject to meeting constraints. Profit is incremental production revenue minus maintenance task costs, such as for labor budgets and parts.

More than likely, increasing profit is driven more by reductions in maintenance tasks than increasing production if unit availability is constrained. If availability is allowed to float, then the optimizer could add tasks to improve production (if that actually happens) up to the point of diminishing returns or if it hits a budget or labor constraint.

6.5.6 Storage Tankage Determination/Optimization

Intermediate storage tankage could potentially provide a buffer for a downstream unit when an upstream process unit shuts down. In other words, a downstream unit could be insulated from an upstream unit's "unreliability" when enough intermediate tankage exists.

However, storage tankage is a scarce resource and consumes considerable physical space. In general, storage tankage is rarely added within an existing facility. However, one could consider repurposing tankage, if an analysis shows that is the right economic decision.

As shown later in the book, this determination/optimization can be a natural complement to determining the best improvement projects. That's because there are times when more storage is a better economic decision than trying to incrementally increase process unit availability. In those cases, the best answer may be more storage or a reallocation of current storage. That typically occurs when a processing unit has relatively high availability and incremental equipment improvements are unduly expensive relative to other storage tankage options.

The example in the book solves this problem largely via a series of Monte Carlo simulations to make it easier for the user to understand the fundamental tradeoffs. In many cases, optimization is not required. However, we will describe the problem below as if we were performing a full-blown optimization.

6.5.6.1 Independent Variables

For this optimization problem, the independent variables would be:

- Tankage size

6.5.6.2 Dependent Variables

There are not many dependent variables that spring to mind for this problem.

6.5.6.3 Constraints

The most likely constraints would be capital available for tankage repurposing.

6.5.6.4 Objective Function

It's probably starting to sound like a broken record, but the goal of this optimization is to maximize profit, subject to meeting constraints. Fundamentally, the optimizer balances increased production against the cost of increased tankage. However, the problem could also be to allocate available storage tankage in a more optimal fashion.

6.5.7 Equipment Design Optimization

Plant designers are constantly wrestling with whether they should choose more reliable equipment or whether they could get away with cheaper alternatives. In other words, can a more expensive choice be cost-justified? If so, how do I prove it?

Unfortunately, too many equipment design decisions are made based on initial capital budget constraints and don't consider the implications of how equipment performs over the long haul.

Similar to many of the previous optimization problems, we will present three different problems – one set of equipment reliability options for one piece of equipment, several different reliability options for several different equipment in one process unit, and the more generalized case where we evaluate different reliability choices for different equipment scattered among different process units.

6.5.7.1 Independent Variables

The independent variables for equipment optimization would be:

- Equipment reliability choices shown in discrete steps

Fundamentally, the optimizer would choose an option from a table of design choices that highlights equipment failure times and their associated cost.

6.5.7.2 Dependent Variables

Once the optimizer has made an equipment choice, it would calculate, or look up, its associated cost. Another dependent variable is the total design budget.

6.5.7.3 Constraints

Budget constraints could also be included.

6.5.7.4 Objective Function

This optimization strives to maximize profit by balancing the cost of more reliable equipment against possible incremental production capability gained with more reliable equipment, subject to meeting constraints.

6.5.8 Combination of Improvement Project/Crude Feed Mix Optimization

In this problem, we will combine choosing the best mix of improvement projects along with the best crude mix of two different crude types.

We can think of this problem as the planning department considering running some different crudes in the future but wanted to give us a "heads up" as we contemplate future improvement projects. That way, if we needed to take a different direction with what projects we implement given a different diet of crudes, we could do so. This would allow management to make the best decision given all relevant information.

Including crude A/B in the optimization problem is necessary because each crude has a different feed cost and yield. However, each crude also affects equipment performance in significantly different ways due to each crude's differing sulfur content.

Therefore, the objective of this optimization problem is to choose the best combination of improvement projects along with the best crude mix to optimize profitability. As in the previous optimizations involving optimal improvement projects, by virtue of choosing the best improvement projects, this will also determine the optimal availability targets for each process unit.

6.5.8.1 Independent Variables

The independent variables for this optimization are the following:

- Process unit availability targets
- Percentage of crude A/B being run

6.5.8.2 Dependent Variables

There are many dependent variables for this problem. As already noted for choosing the optimal improvement project, the following would be necessary:

- Each unit's improvement project function
- Total product volume increase
- Individual gasoline component totals
- Total improvement project expenditure

For each new process unit availability target, the optimizer will determine the cost for the improvement project to achieve that new availability target.

Gasoline production would also be included as a constraint since the plant may not be able to sell all the incremental product volume.

Total improvement project expenditures would also be calculated and used as a constraint.

To account for how different crude mixes affect furnace equipment, we would also need to include dependent variable equations for the furnace tube failure rate determined based on the crude mix being run. In other words, different crude choices can dramatically affect equipment performance, which shows up as differing system availability. Equipment failure rates would be based on estimates provided by equipment experts.

Then, after the Monte Carlo engine has run, production values are calculated based on the overall system availability. These are then multiplied by their appropriate wholesale prices to determine the total revenue increase for that choice of new availability.

There will also be a dependent variable calculation for total crude cost, since the price of each crude varies in the marketplace.

6.5.8.3 Constraints

There would be several likely constraints:

- On the cost side, a total improvement project would be included.
- Gasoline production value constraints to ensure we meet demand, but do not overproduce such that the increased volume cannot be sold.
- In addition, crude acquisition costs may be a constraint.

6.5.8.4 Objective Function

The objective of this problem is to maximize profit, which is the difference between incremental production revenue minus incremental improvement project spend and crude feed costs, subject to constraints.

6.5.9 Combining Turnaround and Future Feed Composition

This optimization will share similarities with the previous problem. In this case, we will be considering what equipment to restore for an upcoming turnaround, but we will also make those choices knowing that we will also be choosing a different crude feed mix for the future. Therefore, the optimizer will need to determine both the optimal crude split and the proper equipment to restore.

Note that what equipment is to be restored will be potentially different depending on the crudes selected. That's because the crudes selected may affect equipment availability, hence system availability. That, in turn, directly determines production quantities.

6.5.9.1 Independent Variables

The independent variables for this problem are:

- What equipment will be restored in a particular turnaround
- Percentage of crudes A/B, etc. being run

While we could have also included turnaround duration or start date in the optimization problem definition, we left them out for this particular problem.

6.5.9.2 Dependent Variables

There are seven main dependent variables for this problem:

1) Each unit's improvement function
2) Individual gasoline component totals
3) Total product volume increase
4) Total product revenue
5) Total improvement project expenditure
6) Furnace tube failure rate
7) Total crude feed cost

Each time that the optimizer chooses a process unit availability target, it will use the improvement project graphs to determine how much it must spend to achieve that new availability target. There is one equation for each different process unit.

The refinery material balance in Appendix allows us to calculate gasoline component production for each new availability target. The final gasoline component streams are then added together and multiplied by the wholesale gasoline price to determine the total revenue increase for the new availability targets.

The total values for each of the individual gasoline component streams are also included as constraints since the planning department can estimate the maximum/minimum values that it needs to blend gasoline successfully later.

Another calculation is the total improvement project expenditure. This will be used in two different places. First, it will be one of the major dependent variable constraints. In addition, it will be an important aspect of the objective function calculation.

Then, after the Monte Carlo program has run, production values are calculated based on overall system availability. These are then multiplied by their appropriate wholesale prices to determine the total revenue increase for the new optimizer choices.

To account for how different crude mixes affect our furnace equipment, we need to include dependent variable equations for the furnace tube failure rate determined based on the crude mix being run. In other words, different crude choices can dramatically affect equipment performance, which shows up as differing system availability. Equipment failure rates would be based on estimates provided by equipment experts.

There will also be a dependent variable calculation for total crude cost since the price of each crude varies in the marketplace.

6.5.9.3 Constraints

There will be two constraints defined for the problem.

The first will be the minimum gasoline pool quantity since we are contractually obligated to deliver certain gasoline volumes.

The second will be a minimum crude unit availability, which will be calculated to reflect the poorer long-term availability when running higher quantities of crude A. This is due to its higher sulfur values and related higher failure rates, as shown in the graph earlier in the chapter.

6.5.9.4 Objective Function

The objective of this problem is to maximize profit, which is the difference between incremental production revenue minus crude acquisition costs and the combination of incremental improvement project spend and increased crude feed costs.

6.6 Conclusions

Every optimization problem will require defining independent and dependent variables, constraints, and an objective function.

Independent variables are those that the optimizer will manipulate within user-defined maximum/minimum limits.

Dependent variables are used in equations which are fundamentally a function of the independent variables. They are used within the objective function or as constraints. They allow the user considerable flexibility to tailor the solution to meet their needs.

Initial optimization definitions were proposed for the optimization problems that will be discussed throughout this book. These include:

- Optimal Improvement Project (Determination of Optimal Process Unit Availability)
- Turnaround Timing, Duration, and Equipment Selection Optimization
- Maintenance Plan Options Optimization
- Spares Optimization
- Task Optimization
- Storage Tankage Determination/Optimization
- Equipment Design Optimization

We also included two different Combined Optimizations. These were:

- Optimal Improvement Project along with the optimal mix of crude feeds.
- Turnaround equipment selection combined with the optimal choice of crude feeds.

7

The Consultation Process – The Main Work Process

After reading the previous chapters, you were starting to get excited until it dawned on you that you need data for the analysis. You start to panic.

"Where am I going to get the data? Our plant doesn't have particularly good data. I love this concept, but will I ever be able to use it because we have such poor data."

Is all lost?

It's time to take a deep breath and relax. There is hope, and a solution.

7.1 Nobody Has Excellent Data in the Process Industries

You can join the crowd. Nobody has good data!

Well, almost nobody – at least, nobody in the process industries. On the other hand, there are industries where reliable failure data does exist. Two good examples are the aircraft and automobile industries. However, they have a significant advantage over the process industries.

That's because those in the automobile industry can easily collect a large quantity data of failure data for the major components in a car – such as the quantity of alternator failures and their likely failure modes. However, the automobile industry two significant advantages over oil refineries and petrochemical plants.

First, the number of cars on the road is extremely large. So, they have a huge pool of candidates where they could collect data.

Second, cars largely operate in a stable and "clean" environment. Even if it is not quite perfectly clean, it would be easy to create several different environmental cohorts and still have enough data to be able to accurately determine equipment failure characteristics for each of those segments. The same can largely be said for aircraft turbines, which also operate in "clean" environments.

However, the situation in the process industries is totally different – for several reasons.

The first concern in the process industries is sheer data availability. I can verify that after working as an asset management vendor for many years, I can only think of one refining client that had excellent data. I have also heard rumors of a second, but since I haven't seen it personally, I can't vouch for it. So, maybe there are two process industry companies out of hundreds with good data.

The data issue has been enough of a stumbling block that many clients have balked at moving forward with a program like this, forgoing the benefits that they could have obtained if they implemented such a program. However, engineering has, and always will be, the art of making assumptions. If engineers had waited for perfect data, we'd still be living in a hunter-gatherer society.

Plant Optimization in the Process Industries: Incorporating Equipment/Assets in the Decision-Making Process, First Edition. Marty Moran.

In order to create a Monte Carlo model, one only needs data "accurate enough" to adequately describe the problem so they can make a good business decision. There are certain words in that sentence that are so important that they should be repeated – "accurate enough." One does not need perfect data, and we would likely never be able to obtain it anyway. What one needs is data that is representative enough of the problem at hand to be able to solve it. It's important that we never lose sight of that fact. So, we won't likely have the quantity of data at our disposal like those in other industries, such as the automobile and aircraft industries.

However, we also face another major problem. That is, equipment failure rates are a strong function of operating conditions and, unfortunately, those can vary tremendously. This simple fact complicates our data collection effort tremendously. But there's still hope, and a solution.

7.2 Why Operating Conditions Are so Important in the Process Industries

Operating conditions matter much more in determining failure rates in the process industries than in most other industries. Equipment in a process plant does not exist in isolation. How often it fails, and how that occurs, is often a direct function of its operating conditions, such as feed quality, contaminant levels, pressure and/or temperatures, or some combination of those, as well as how operations run the equipment.

Comparing equipment performance among different locations, even for a large refining company, can be misleading because one refinery may have a processing scheme that preferentially processes one type of crude, whereas another similar refinery, in another part of the country, processes a different type of crude.

In addition, spot crude opportunities can differ substantially from one location to another. For those unfamiliar with the refining industry, there is often the possibility of buying "one-off" crudes, not normally run, usually at low prices – often referred to as "distressed crudes." So, different crude diets can make equipment performance comparisons challenging.

Let's take pumps as an example. Normally, a given pump from a reputable manufacturer can be used in various services. Contaminants can vary widely from one service to another, heavily influencing operability. How much "tender loving care" operations provide a pump can dramatically influence its functional life. In addition, maintenance practices and skills can vary significantly from one plant or application to another. Last, the number of equipment that operates in the same environment type is exceedingly small – often, it's just that single piece of equipment. Unfortunately, you can't develop cohort data based on a single piece of data.

That's why the reader should be weary of industry-standard equipment failure rate databases. They receive a lot of hype, but I would stay away from them. Assigning accurate data is more complicated than just reading a number out of a table.

So, given this reality, how can one proceed?

7.3 Tapping into Your Company's Innate Knowledge

Fundamentally, the heart and soul of any asset optimization program is your company's knowledge of the process and equipment, as well as the particulars of your business. Of these, knowledge of the equipment is paramount. That's the hard part. The Monte Carlo part borders on trivial once you understand the equipment.

The good news is that the information you desire is readily available. It resides in the heads of the experienced people in your organization – often the guys with the gray hair. These are your veteran Operations, Maintenance, Reliability, and Equipment experts.

Don't underestimate them. Once you appreciate that fact, your job becomes how to extract their knowledge and institutionalize it. In the west, this is a time-critical mission since this group is retiring at a rapid pace. The good news is that it's easier than you think.

As noted in the previous section, data at each plant is unique – for that plant. That's why you will need to spend a lot of time with those previously mentioned because they are the ones who will understand how operating conditions in the plant will affect failure rates. You wouldn't be able to obtain that from some standard industry publication. Another benefit of this group is that they will be able to give accurate estimates of the time to repair, which also varies from plant to plant.

To extract that knowledge, it's important that you get the right people in the room to engage in the "conversation." You will need participation from Operations, Maintenance, and subject matter experts in various equipment types as the projects and program proceed.

The conversation that ensues from getting the right individuals in the same room at the same time is invaluable and *far, far* more valuable than any one person looking at failure data in a Computerized Maintenance Management System (CMMS).

In many ways, the only way to accurately estimate a plant's performance will come from those who know. I don't see how it can be any other way. Therefore, a Consulting approach is the best way to obtain the data necessary for your program.

7.4 Golden Opportunity To Test the Approach

While I innately believed that it was possible to obtain the necessary model data through a pure Consulting approach, I hadn't actually done it. However, a golden opportunity presented itself when there really wasn't any other way to obtain the data.

I had been tasked with building a plant-wide RAM model for a consulting company working directly for a brand-new multi-billion dollar coal gasification plant in the United States.[7]

When I was hired to build the model, the plant was brand new and in the final stages of construction. It was a large facility and similar in scale to that of a small to mid-sized refinery. Given its size, we decided to concentrate on building a model with only the major equipment in each process unit.

The constructed model had 19 process areas and consisted of 23 submodels with three main levels. There were roughly 250 main pieces of equipment. Once spares and additional equipment in parallel paths were accounted for, the complete model had 350 pieces of equipment.

My colleague, who had hired me to build the model, had been both a Reliability Department manager, as well as a former Turnaround Manager, at one of the largest oil companies in the world. He knew the equipment side of the business backward and forward. (What's ironic is that our paths had crossed in prior years, yet we didn't know it. I had executed a Monte Carlo modeling project for people who reported to him about eight years prior at a large refining site.) His comments about his experience with the Consulting process are detailed in Textbox 7.1.

On the other hand, my expertise was that I understood how Monte Carlo simulation technology could be structured to solve the more important asset problems in our industry. During our project, I coached and educated him on how we could use that technology to solve the problems his client was facing.

I emphasized that the Monte Carlo simulation is not a "silver" bullet. At the end of the day, Monte Carlo simulation engine is mathematics – nothing more, nothing less. It's what goes into

Textbox 7.1 A Guide to the Consultation Process for Managers and Engineers by Greg Weber, President of Fidelis Associates

Reliability and availability are key performance indicators for any engineering system or process. They measure how well the system or process can deliver its intended function under various conditions and over time. However, reliability and availability are not only determined by technical factors of a single piece of equipment, such as design, maintenance, testing, and monitoring. They are also influenced by configuration and operation of the equipment.

As Marty proclaims, the organizational structure of most companies separates the Process experts from the Equipment experts. The organization develops a 'binary' view of plant operations, as either process or equipment. Rarely is anyone willing to describe the problems which are concurrent across these two realms. Hence, the Consultation Process is used to breakdown these barriers.

Why is a Consultation Process required?

The project we worked on had basic information on a facility design in equipment specifications and configuration drawings, but the knowledge of how the process would work as a system - during start-up, shut-down and other scenarios - was contained within the collective knowledge of Operations, Maintenance, Reliability and Equipment experts.

However, individuals from these departments each took a limited functional view. This required us, as part of the Consultation Process, to ask broader questions which ultimately revealed the key characteristics of the facility.

The consultation process uncovered and exposed the hidden influences on reliability and availability, by:

- Providing an objective and unbiased perspective, that can challenge the assumptions, biases, and blind spots of the participants
- Creating a safe and respectful environment, that can encourage the participants to share their honest opinions, experiences, and concerns
- Using appropriate and effective tools and techniques, that can elicit and capture the tacit and implicit knowledge and insights of the participants
- Facilitating the dialogue and discussion, that can foster the exchange, integration, and synthesis of diverse and conflicting views and information
- Enhancing the learning and innovation, that can stimulate the generation and evaluation of new and creative ideas and solutions
- Building the trust and commitment, that can motivate the participants to take ownership and responsibility for the results and outcomes

mathematics that is important. Software may be a requirement since it facilitates arriving at an answer, but it is not a panacea.

I then tasked my colleague with conducting the consulting meetings by reviewing the major equipment in each unit. More specifically, I asked them to take the simple RBDs that I had produced for each unit, similar to the semi-regen reformer in a previous chapter, and walk through each block on the RBD to get agreement from the team on values for both failures and repair times, as well as some other minor information. To my amazement, they completed this task within a two-week time period.

Perfectionists might cringe at this level of detail, but the reality is the failure rate of equipment is dependent on so many factors that often the best estimates come from those who know the equipment well, the skill level of how operations will operate the equipment, as well as the maintenance people who will maintain the equipment. Given how site-specific this information is, you'll never get a more accurate estimate from some book.

We concluded that one failure rate was sufficient for most equipment. If more than one failure mode was necessary, we added it, but it turned out that was rare. In addition, an analogous "time to repair" was chosen for each equipment. That assumption did not work in every instance, but it was our starting point.

The Consulting approach worked beyond our highest expectations, and our final availability results were ultimately consistent with estimates from other independent sources.

In fact, most organizations should be able to use this simplistic approach for their Asset Optimization program. If one ultimately finds this approach inadequate, one can always add further detail. However, I would only do that, if truly required.

In fact, the golden rule of ALL Monte Carlo programs is that detail should ONLY be added if it is necessary to help answer the question being posed. There are far too many Monte Carlo model corpses in our industry. Often, they were ultimately pronounced dead due to being drowned in detail. I hope you avoid that fate.

7.5 Consulting Meeting Details

This section will describe how to conduct the Consulting meeting, and the information that needs to be collected.

It's better if the process is run by outside consultants, especially the first time you use this approach, because they will have an unbiased opinion. Plus, they are usually good at running the meetings to "extract" information from your experts. You probably want to think that it's a similar process to a HazOp, RCAs, FMEAs, or RCMs.

The Consulting meeting should be run by a Facilitator and attended by those with unit or equipment expertise. Representatives from Operations, Maintenance, Reliability, and experts with special equipment expertise, should be in attendance. The Monte Carlo modeler can attend the meeting, but it is not an absolute requirement.

Normally, this meeting is a separate meeting from other company meetings. However, it may be possible to perform this Consultation process by piggybacking it with other consultation meetings that may be already underway or scheduled in the near future. For example, many of the same skill sets are needed for HazOp, RCAs, FMEAs, or RCMs. Thinking opportunistically, it may be possible, with just a small amount of extra effort to sit in on one of those other meetings and obtain the required information.

In fact, a company may want to consider just adding this activity to one of those other processes as an adjunct, even if there is not an active project. This will help generate the required data for the future. In fact, this should be done whenever there is an opportunity to do so, even if there is no Asset Optimization project currently planned for the unit.

7.5.1 Meeting Preparation

In general, there are two prerequisites for the meeting – creation of the Reliability Block Diagram (RBD) and tracking spreadsheet. These are both created in a separate project step – Initial RBD Development – before conducting the Consulting meeting.

The Facilitator uses knowledge gained from the Kickoff meeting and any follow-up technical detail meetings to create its best guess of the RBD. He may use process understanding, equipment knowledge, process flow diagram (PFD), and equipment details, in addition to previous experience in creating the RBD. For each block in the RBD, he also creates an analogous line in the spreadsheet (to be explained shortly).

The more accurate it is – the better. However, it doesn't have to be perfect since it will be refined as part of this process.

7.5.2 Meeting Flow

The Facilitator will run the meeting and walk the group through every line in the spreadsheet. As the meeting proceeds, the group will decide if the block in the RBD is necessary, or if other blocks not shown on the RBD should be there.

Once everybody agrees that the block is necessary, the Facilitator will lead a discussion to determine how often the equipment would fail over a 5, 10, or 20-year time period. Answers should be in simple, nonmathematical terms. For example, a typical failure rate might be "once in the first year, and once every five years thereafter."

The Facilitator would then walk the group through a similar exercise for Repair Time, how much production would be lost if the block failed, specific and general notes, and how often preventative maintenance (PM) is done and how much equipment restoration occurs when a PM is performed.

7.5.3 Spreadsheet Documentation

Information from the meeting is captured in a spreadsheet such similar to that shown in Table 7.1.

Table 7.1 Example Consultative Meeting Documentation

Equipment	Tag	Failure Time	Repair Time	Throughput Loss	Block Notes
Feed Product Exchanger	HE1			100%	No issues
Feed KO Drum	Drum1	One failure in five years	Demister change out FOFI four days	100%	No major issues.
Pump 101	Pump1 a/b	One failure in two to five years per pump	Five days to repair	50%	two 100% pumps
KO Drum	Drum2			100%	No issues
Chiller	HE2	Failure in first five years	FOFI 7 days	100%	Needed to run the train.
Recycle Compressor	Comp1			0%	Can run without this. No spare.
Pump 102	Pump2 a/b/c/d	One failure in two to five years per pump	Twelve days repair due to access issues	See Notes	Three pumps needed to run; cut rate with two pumps to 70% (four 33% pumps installed)

Each block in the RBD will correspond to a line in the spreadsheet. Information required for each block are the following:

- Equipment Description
- Tag or Asset Number
- Failure Frequency
- Repair Time
- Throughput Loss upon Block Failure
- Block Notes
- Other Notes (not shown)
- Preventative Maintenance Information (not shown)

Each of these are explained below:

7.5.3.1 Equipment Description

This is a simple description of the equipment familiar to unit personnel.

7.5.3.2 Asset Number

This is the asset number in the asset register or could be a less formal identification. What's important is that it have meaning to unit personnel.

7.5.3.3 Failure Frequency

For each block in the RBD, the group must come to a consensus as to the failure frequency likelihood of the block (usually a single piece of equipment) during a given time frame. The consensus does not have to be as specific, as you might believe. For example, it could be something as simple as "once in the first year; once more in years 2–5."

The time frame could be anything that the group believes is long enough to be representative. However, longer is almost always better because longer time frames have more uncertainty associated with it. The minimum time would be five years for most plants. Failure estimates must cover the entire time period.

If failures are not likely to occur during the agreed-upon time frame, then it can be recorded as "No Issue."

7.5.3.4 Repair Time

Another key piece of information is how long it will take to repair the item. This is not just the equipment repair time, but rather the total time from when feed is removed from the unit until the time that the unit is once again producing on-spec product.

The correct value for Repair Time in this context incorporates many different factors, such as:

- Time to cool down the unit
- Time to isolate the equipment
- Time to remove the equipment so that it can be repaired
- Time to repair the equipment
- Reinstallation time
- Time to reintroduce feed, bring the unit up to temperature producing an on-spec product

Since there are so many factors that play a role, it's important to get a perspective from all the appropriate people because many of these are very site or locale dependent.

For example, how much time to remove equipment will depend on the installation area and how easy it is to get equipment in and out of the area. It may also depend on the availability of equipment to be rented. Repair time will depend on the skills of those making the repairs.

Obviously, the time for some of these steps is fixed. For example, the time to cool the unit down and bring it back to temperature producing on-spec products are beyond our control. They are largely fixed by process and equipment constraints.

7.5.3.5 Throughput Lost upon Block Failure

If a block should fail, we need to understand how much production will be lost. For most equipment failures, we will lose all production. However, if it is only a partial lose, we need to understand the approximate percentage of production lost.

7.5.3.6 Block Notes

This field is for capturing any notes specific to the block itself.

7.5.3.7 Other Notes (Not Shown)

This field allows the Facilitator to record any more general-purpose notes that are not necessarily block-specific.

7.5.3.8 Preventative Maintenance Task (Not Shown)

If there are major preventative tasks associated with the block, we would want to understand how often the PM is performed, typical task time (similar in nature to repair time – loss of production back to production back on-spec), and the equipment percentage restored to full functionality.

7.5.4 Analysis Process

Once the Consulting exercise has been completed, you will want to review the information captured in the spreadsheet. *What you will be amazed by is how easily and accurately you'll be able to determine the unit's availability from just perusing the information in the spreadsheet.* For the most part, you can just add up the total failure time from the information in that column. It's a little more complicated than that since redundancy makes certain failures, such as pumps, largely irrelevant. Nonetheless, it provides a good estimate – probably a lot more accurate than you would have ever believed.

7.5.5 Final Thoughts About the Consulting Process

What's so impressive about this consulting process is that these are group decisions utilizing knowledge from across the company. Everybody participates in the decisions. It is the best estimate of how the equipment performs from those who truly have the experience to make those judgments. In many ways, this is the hardest and most important part of an Asset Optimization program.

As I told my colleague when we did our project together, he really did the hard work. He never genuinely believed me, since too many people think that reliability tools, like Monte Carlo, contain some sort of black art that discerns intelligence from thin air. It doesn't.

All I really did later was to translate their knowledge from the spreadsheet and put it into a form so we could perform the mathematics in the Monte Carlo engine. I had the easy job.

There are many side benefits of this discussion over and above the information contained in the spreadsheet. My experience has been that many people in plants become too "siloed," with not nearly enough cross-sharing of information. Therefore, one important outcome of this process is that it has the potential to educate other plant personnel on how the unit actually works from those who truly have the knowledge. Those participating in the meeting may learn more about other aspects of the unit than they never knew beforehand.

7.6 Monte Carlo Modeler Software Inputs

After the meeting has been completed and a first phase analysis has been done by the Facilitator and extended team, the Monte Carlo modeler takes the data from the spreadsheet and converts it into values for use within the Monte Carlo software program. This section describes that process. There are four values that he must review for each block:

- Failure Frequency
- Repair Time
- Throughput Loss upon Block Failure
- Preventative Maintenance Information

7.6.1 Failure Frequency Distribution

Failure frequencies from the Consulting meeting were high level estimates. For example, "one failure in the first year and another sometime during years 2–5." One of the jobs of the Monte Carlo modeler will be to take those high-level estimates and translate them into the best mathematical distribution that captures the estimates made by the extended team.

Typical mathematical distributions might be normal, logarithmic, exponential, Weibull, single value, etc. This book will not go into the details behind those distributions. If one is not familiar with the mathematical details behind those distributions, the best way to come up to speed is through a Reliability Analytics class. The vendor that you select for your Monte Carlo software might also be able to help you find the training class for your experience level.

Honestly, I wouldn't get too hung up on being overly precise. It doesn't matter all that much which distribution gets chosen. (I know that I probably just irritated a lot of Monte Carlo modelers, but sorry it's really true.) The only real requirement is that it match the performance.

For the project with my colleague, what I did was make a parameter estimate for each distribution type and then run a single block Monte Carlo simulation to see if how accurate my initial estimate was. I then tweaked it until I got to a point of diminishing returns. The final value that I settled on was then entered into the spreadsheet.

7.6.2 Repair Time Distribution

The determination of the mathematical distribution for Repair Time is exactly the same process as described in the previous section for Failure Frequency. Therefore, there's no reason to repeat that discussion here. The final value of the Repair Time is also entered into the spreadsheet.

7.6.3 Throughput Loss upon Block Failure

For the Monte Carlo software packages that are normally used for an Asset Optimization program, we will need to estimate exactly how much production is lost when the block fails. For most equipment failures, we will likely lose all production. However, there may be some equipment failures in which we only partially lose production.

The format of how this will be entered might vary from package to package, though it is often entered as a percentage. If we lose all production, we will enter 0%. If we can still process half of normal production, we would enter 50%.

7.6.4 Preventative Maintenance

The general idea behind Preventative Maintenance (PM) is that by performing a certain task, we extend the life of that asset by a certain amount of time. This concept can also be used within the Monte Carlo simulation modeling environment to better reflect an equipment's performance after a PM has been performed. Mathematically speaking, this means that the failure frequency distribution will need to be "reset" in some fashion.

Normally, we don't specifically adjust the failure frequency per se, but rather adjust another factor, called the "Just Like New" factor, that has a scale of 0–100% that indicates how close to "New" the equipment becomes after the PM.

7.7 Data from Asset Management Systems

Earlier in this section, we put forth a strong argument in favor of the Consulting process where experts provided their best estimates for failure frequency, repair time, throughput loss on block failure, and PM. We argued against blindly using asset data from an external source and were also lukewarm about internal databases, as well. However, for more mature organizations, it may be possible to use their asset performance management systems as an aid in obtaining failure or repair data.

As mentioned earlier, most organizations struggle to keep good data. Thus, the most obvious problem with trying to use internal failure data for an asset optimization program is data quality. However, that's not the only problem. The second strike against it is that it is a time-consuming activity because one has to take time to comb through the data to determine its validity.

My suggestion is that these sources should rarely be used without a knowledgeable individual who can review it. However, if these data sources do exist, it's possible that they might be a valuable source of supporting information.

In many ways, I would use the data in any internal system the way a good doctor uses test results. As one doctor once told me, if a doctor orders a test and doesn't have a really good idea of what to expect when the results come back, run fast. He's probably clueless as to what the real issue is. Good doctors only use test results as a "confirmation" of what they believe might be the problem and as a way of specifying the exact spot of the illness.

In fact, a doctor once specifically articulated exactly how he used test information. I had described some numbness that had recently occurred, most likely from bike-riding. He later told me, he was formulating in his head exactly which nerves might have been affected, as I was describing the problem. He said any tests that he would have ordered in my case only served to pinpoint the exact point of damage.

In fact, that's exactly how asset management data should be used within RAM modeling. You don't want some modeler, who has no equipment knowledge blindly using that information. You want somebody knowledgeable about the equipment making the final decision, in the same way that you don't want nonmedical personnel examining the results from a medical test and then making a final diagnostic on your health issues.

The bottom line is that data used blindly is virtually always a bad idea. On the other hand, experience is invaluable in solving many different problems. However, the combination of experience with accurate data using modern tools can be immensely powerful.

7.8 Data Storage/Structure

The consulting process will generate a significant quantity of data that will need to be managed. For the first project, it probably makes sense to store project data within Excel. However, as time moves on and additional projects are executed, the company will need to decide on the best place to store the data, so that pertinent personnel have access to it.

Since each company uses different IT systems, there is no right answer for each company. The best choices are probably the CMMS or higher-level asset management system database. If those are not available, then it might make sense to create a SQL or Access database.

One also needs to understand the required data structure. The following is a fairly complete representation of the data that must be stored. Descriptions are provided for the data where it is not obvious.

7.8.1 Block Information

For each block that can fail, the following information/fields are the most logical:

- Functional Location
  - Only used to track back to the main asset database. Not absolutely required. Just nice for tracking purposes.
- Equipment Name
- Equipment Tag
- Equipment Type
- Failure Description
  - High level description captured during the consulting sessions
- Repair Description
  - High level description captured during the consulting sessions
- Failure Time – Consulting
  - Time value captured during the consulting sessions, but entered as a text field
- Repair Time – Consulting
  - Time value captured during the consulting sessions, but entered as a text field
- Comments
  - General purpose field that the consulting team can use
- Failure Distribution Type
  - Type of mathematical distribution modeler chose from the consulting exercise. Normal values are normal, Weibull, logarithmic, single value, etc.
- Failure Time – Modeling
  - Time value that modeler calculates based on the values from the consulting exercise. At this point, the values are now numeric.
- Repair Distribution Type
  - Type of mathematical distribution modeler chose from the consulting exercise. Normal values are normal, Weibull, logarithmic, single value, etc.
- Repair Time – Modeling
  - Time value that modeler calculates based on the values from the consulting exercise. At this point, the values are now numeric
- Spares used?
  - If the Repair task requires spare parts, then the digital value for this block should be set to "Yes." If not, it should be set to "No."

7.8.2 Condition Monitoring Tasks

Some blocks may have a Condition Monitoring task associated with it. The fields that should be associated with a CM task are the following:

- CM Task Description
- CM Task Frequency
- CM Task Restoration Level
  - This describes the percentage of functionality that the task has restored the block to after completion of the condition monitoring task. For example, if 100%, then it means that the block performs like it is "Like New" after the CM task.
- Spares used?
  - If the Condition Monitoring task requires spare parts, then the digital value for this block should be set to "Yes." If not, it should be set to "No."

7.8.3 Spares Fields

Spares can be consumed by either a repair or condition monitoring task. The fields required are as follows:

- Spare Description
- Cost
- Normal Lead Time
- Expediated Lead Time
- Reorder Quantity

7.9 Conclusion

The Consultation meeting is a critical activity within the asset optimization program. This is a group activity organized by a Facilitator who leads a cross functional group of plant personnel (Operations, Maintenance, Reliability, equipment experts) through an exercise where they discuss each block (usually single equipment) within a preliminary Reliability Block Diagram (RBD) to principally determine failure and repair time estimates. This group will also determine how much production is lost if the equipment should fail, as well as normal conditioning monitoring activities.

The benefit of this activity is that it is the best estimate of what will is likely to happen in this particular plant, as opposed to industry-specific data, which, by its nature, is more generic.

The estimates from the Consulting meeting are handed off to the modeling team, who translate these into the mathematical distributions used within the Monte Carlo simulation software.

The consultation effort is normally a separate activity as part of the Asset Optimization project. However, there may be times when this activity can be combined with other activities, such as a HazOp, where the same group of experts may be assembled.

The information from the Consulting effort is captured in a spreadsheet, but as the program expands, it should be formalized into a database managed by the IT department. The ultimate decision of how to best achieve this goal will be plant-specific.

8

Turnaround Considerations

Turnarounds are the most significant maintenance activity in the oil refining and petrochemical industries, and, in fact, for most process plants. The intense activities surrounding turnarounds consume considerable resources, such as outside labor and material costs. By definition, turnarounds are disruptive to production. However, every plant knows that they must perform them periodically.

The decision of when to take a turnaround, and for how long, involves balancing many different, and often competing factors. Sometimes, a turnaround must be undertaken to replace or restore a catalyst's activity. Other times, plants must shut down to execute periodic inspections dictated by regulations or local laws. Or the plant might decide it needs to shut down to implement a capital improvement project. There might be other occasions when equipment availability becomes limited, thus making it necessary to shut down to repair assets responsible for causing the production decline.

Last, a plant may decide to take a turnaround at a certain date because a key outside supplier or a downstream customer has decided to shut down their facilities. Obviously, a given process unit does not exist in isolation, so turnaround timing on one unit can be driven by factors external to that unit – whether that be within our fence line or outside.

The trend over time has been to stretch the time between turnarounds. While on the surface, this seems to make business sense, it carries risks that must be managed. That is, the longer that a unit is run a unit before being turned around, the higher the likelihood that equipment availability, and hence process unit availability, will decrease below an economically acceptable threshold. Of course, no manager wants to unduly risk an unexpected shutdown due to equipment failure. This could potentially expose the plant to catastrophic economic losses, and be career-limiting.

Once the management team has decided when the unit should be shut down, they must also decide for how long the unit should be down for, and what we should repair/restore during the turnaround.

Anybody who has been involved with turnaround planning knows that turnaround management involves a considerable amount of art. While more science has been injected into the process in recent years, there's little doubt that art still plays a significant role in turnaround planning. Of course, there's plenty of politics involved, also.

I'll never forget a conversation with a senior refinery engineer years ago at a local Houston refinery. He posed an interesting question. "Can you provide a work process where we can use science to better decide if we should replace or repair a crude tank? If not, we are going to sit around a table and pull the answer out of thin air." His question always intrigued me because there are still plenty

Plant Optimization in the Process Industries: Incorporating Equipment/Assets in the Decision-Making Process, First Edition. Marty Moran.

of decisions that we make in the process industries that seem to be driven by thin air, rather than science. Turnaround planning is certainly one of them.

For that reason, we are going to propose ideas in this chapter on how we can use more science to make better turnaround decisions. This methodology will allow us to assess the probability of future equipment failures if we decide to delay a turnaround or how to restore particular equipment during this turnaround and wait until a future turnaround.

That's important because equipment in process plants can often operate for a long period of time before failing. However, if we push that envelope too hard, we may be surprised by what may fail in the middle of the next production cycle. Since decisions made today may affect production years in the future, we may need to look out multiple cycles in the future to fully appreciate that risk.

The methodology presented won't address all of the many reasons why a plant might decide to undertake a turnaround – catalyst degradation, capital improvement project, mandatory inspections, supplier/customer issues. However, it can be used in conjunction with those decisions to better understand the risk of assets failing in the future.

The approach we will describe will allow us to simulate this idea using any level of detail. The optimizer can choose the length and duration of several future turnarounds and what equipment is restored to proper functioning during each turnaround. Appropriate constraints can be included in the calculations that bracket the start and end of the turnaround, and what equipment can, or cannot, be restored during a particular turnaround.

The objective function is to maximize profitability over several cycles. To do that will require calculating process stream availability, hence yields, from the different production runs, and including the cost to make the equipment repairs during each turnaround.

Conceptively, to perform these calculations, the Monte Carlo simulation may need to be run several times with different sets of data within a particular optimizer iteration. This is shown graphically in Figure 8.1, where we consider two turnaround cycles in our example refinery. The

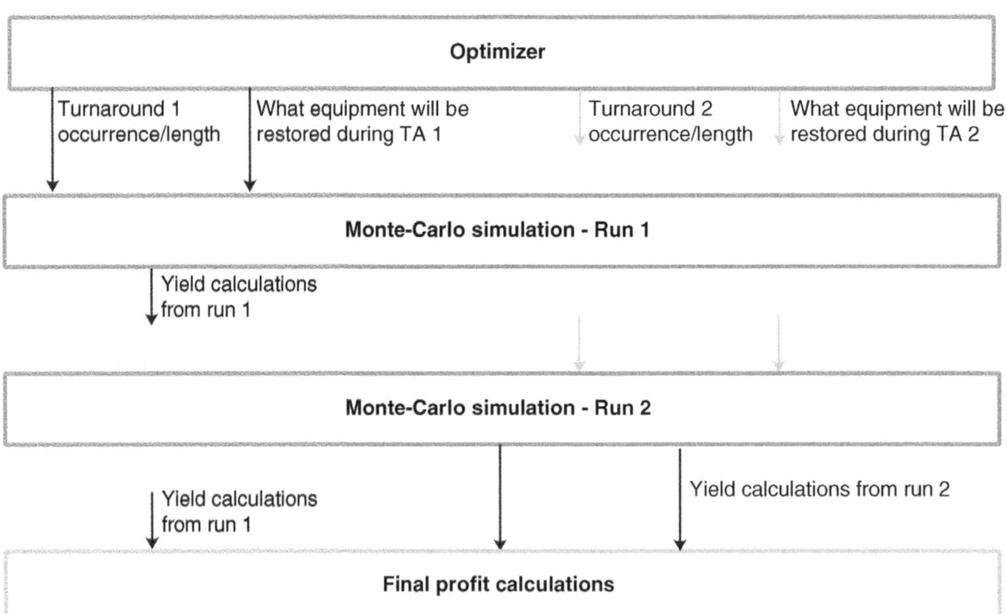

Figure 8.1 Turnaround Optimization Overview

optimizer can be set up to choose when each of the turnarounds occurs, the length, and what equipment should be restored in each turnaround – or any combination of those.

The Monte Carlo program will then be run with the appropriate data for each run. Availabilities are calculated for each run to generate production volumes, and the costs to service the equipment during that turnaround. This provides the optimizer with the revenue and costs to calculate incremental profitability. The optimizer can then iterate to find the optimal values of occurrence, length, and what equipment to restore in each turnaround.

8.1 Example Problem Overview

To clearly explain this concept, let's set up the detailed problem from our fictional refinery that we will use throughout this chapter.

Let's assume that we have just completed year 4 of a 5-year run, and management has decided we will be shutting down the crude unit for a 21-day turnaround exactly 1 year from now. That timing was chosen so an important debottlenecking project could be implemented, which would increase overall crude oil production.

Therefore, the equipment must run another year before it's shut down for repairs. The plant has already decided that the unit will be run for another five years after the completion of this turnaround.

Since we already know we are going to be down, we are trying to decide what equipment should be included in this proposed 21-day turnaround. Some equipment has already made the cut and is not up for discussion. However, there is other equipment where no final decisions have been made yet. The decision to include or not include is not quite as straightforward compared with the equipment decisions already made.

It's important to note that all the equipment in this example is all in series from a reliability perspective. In such an arrangement, the availability of a "system" of this equipment is simply the multiplication of the individual availabilities. Thus, if all 12 equipment in the example we will be considering had 99.75% individually, then the overall "system" availability would be 97.5%. So, even though most of the equipment has exceedingly high availability individually, when multiplied together, it drops significantly. Remember that nuance as we are working through the examples in this chapter.

8.1.1 Undecided Equipment List

The following is the list of equipment that we are still considering to be included in this turnaround. The 12 equipment in the list, A–L, are entirely fictitious, but let's assume that if any of them should fail, it would bring down the entire crude unit. All values in Table 8.1 are in years.

The second column, **Current Age**, lists how long the equipment has been running since last being fully repaired, with the third column, **Mean Time Failure**, being the mean time to failure estimated by our equipment experts. Column 4, **Estimated Failure Date**, represents the approximate time when we expect the equipment to fail based on the mean failure time using simple math.

The next column, **Std Dev Failure**, is the best estimate from our equipment experts of the time when 68% of the likely failures would occur concentrated around the mean. In other words, 32% of the failures will be less than the mean, while the other 32% will be greater. From our perspective, we are only worried about equipment less than the mean since those would occur before the end of the next turnaround.

Table 8.1 Equipment to be Considered for Turnaround

Equipment	Current Age	Mean Time Failure	Estimated Failure Date	Std Dev Failure	Earliest Failure	Latest Failure
A	3.5	10	6.5	1	5.5	7.5
B	3	17	14	3	11	17
C	3.5	14	10.5	1.5	9	12
D	5.5	14	8.5	1.5	7	10
E	0.5	10	9.5	1.5	8	11
F	3	8	5	1.6	3.4	6.6
G	3.5	12	8.5	1.5	7	10
H	3	13	10	2	8	12
I	5	15	10	2	8	12
J	7	16	9	2	7	11
K	4	12	8	1.5	6.5	9.5
L	7.5	18	10.5	3	7.5	13.5

So, if we want to know the range where the equipment would fail, it would be ±1 standard deviation around the estimated failure date. Thus, the last two columns represent the earliest and latest time when each equipment would fail, **Earliest Failure and Latest Failure, Assuming** those failures occurred within one standard deviation around the mean.

So, equipment A could fail somewhere between 5.5 and 7.5 years from now, based on one standard deviation. But we are still one year from the turnaround. Thus, 1 year down the road, those figures would be 4.5 and 6.5 years, assuming it doesn't fail between now and then.

8.1.2 Turnaround Failure Ranges

Let's rearrange our table, dropping out some of the original columns, since we now understand how they were used to develop the data. The rearranged data are now shown in Table 8.2. Equipment rows have also been prioritized to show the equipment with the smallest earliest failure (at turnaround) dates.

In the rearranged table, the first pair of columns is the time to failure using today's date (one year from turnaround), and the other is the analogous pair one year from now when the turnaround begins. It is the second set of ages that we are most concerned with.

We now need to peruse the list to determine which equipment needs to be included in this turnaround and which can wait until the next.

Looking at the estimated failure (at the turnaround) column, we realize that equipment F might only last 2.4 years until failing on the low side and 5.6 years on the high side, so its chances of making it to the next turnaround aren't good. It should be added to our list.

The same can largely be said for equipment A. On the low side, it shows 4.5 years, and 6.5 years on the high side. Probably needs to be added to the list.

The remaining equipment seems to be in fairly good shape. Or is it?

Table 8.2 Listed by Earliest Failures

Equipment	Earliest Failure (Today)	Latest Failure (Today)	Earliest Failure (at TA)	Latest Failure (at TA)
F	3.4	6.6	2.4	5.6
A	5.5	7.5	4.5	6.5
K	6.5	9.5	5.5	8.5
D	7	10	6	9
G	7	10	6	9
J	7	11	6	10
L	7.5	13.5	6.5	12.5
E	8	11	7	10
H	8	12	7	11
I	8	12	7	11
C	9	12	8	11
B	11	17	10	16

8.1.3 Considering All Probabilities of Failure

Up to this point, we've only included one standard deviation, but that only covers 68% of all total failures. We are still exposed for the remaining 32% of failures.

So, let's add the second and third standard deviations to see how exposed we really are. Two standard deviations cover 95% of all possible failures, while three standard deviations take us all the way up to 99.7%.

Since we are only really concerned about those on the low side, increasing our range to two standard deviations increases the number of failures on exposed side by another 13% (95% – 68%/2), while going all the way to three standard deviations covers another 2.35% (99.7% – 95%/2).

At this point, let's review our table with this new standard deviation information after adding some new columns. The first two columns show the earliest possible failures for the third and second standard deviations. (Note that we've only included third and second standard deviation information on the low side, since that is where we are exposed.) The result of these changes is shown in Table 8.3.

We've also dropped out equipment A and F, since we have already decided that they will be added to the equipment being restored in the turnaround. (Therefore, from this point forward, we will no longer be referring to equipment A or F.)

Once you peruse the list, it gives one pause, since many of our equipment are more exposed than when we were only looking at one standard deviation. In fact, almost any of the equipment could easily fail if one looks at their third standard deviation.

Further down the list, equipment E, C, and B are all okay when only considering their second deviation, but looking at their third deviation, they all look potentially vulnerable. The question is, "Do you want to take the chance and not include them?" I guess that's why they pay the managers the big bucks. They get to make the tough decisions.

Table 8.3 Including second and third Standard Deviations

Equipment	Earliest Failure 3 Std Dev	Earliest Failure 2 Std Dev	Earliest Failure (at TA)	Latest Failure (at TA)
K	2.5	4	5.5	8.5
J	2	4	6	10
D	3	4.5	6	9
G	3	4.5	6	9
L	0.5	3.5	6.5	12.5
H	3	5	7	11
I	3	5	7	11
E	4	5.5	7	10
C	5	6.5	8	11
B	4	7	10	16

8.1.4 Considering Production Loss, Turnaround Repair Time, and Costs

However, before making any final decisions, let's add more information that will muddy the water even further.

Table 8.4 includes new information in the form of three new columns on the right side of the graphic. The first new column details the production loss if the equipment should fail, labeled as **"FOFI Time,"** which is shorthand for "Feed Out – Feed In." The second column, **Repair Time**, is the amount of time it would take to repair the equipment if it should fail during the production run. The last column, **Repair Costs**, is a qualitative measure of the cost of repairs if it should fail outside the turnaround.

We will assume for purposes of this discussion that the turnaround repair times will not be a problem. In other words, they are not on the critical path and can be easily restored during the turnaround. Later on, we will make some runs where we will alter that assumption to see how it

Table 8.4 Add Cost and Repair Time Information

Equipment	Std Dev Failure	Earliest Failure 3 Std Dev	Earliest Failure 2 Std Dev	Earliest Failure (at TA)	FOFI Time	Repair Time	Repair Costs
K	1.5	2.5	4	5.5	1.5 days	0.5 days	High
J	2	2	4	6	2 days	0.5 days	Moderate
D	1.5	3	4.5	6	21 days	7 days	Very High
G	1.5	3	4.5	6	10 days	6 days	High
L	3	0.5	3.5	6.5	2 days	0.5 days	High
H	2	3	5	7	7 days	3 days	Moderate
I	2	3	5	7	21 days	10 days	Very High
E	1.5	4	5.5	7	3 days	1.5 days	Moderate
C	1.5	5	6.5	8	2 days	1 days	Moderate
B	3	4	7	10	3 days	1 days	Moderate

affects the final optimization solution. However, for our first set of problems, we will assume that none of this equipment will be on the critical path.

So, let's examine what implications this new information might have in our decision-making process.

If we examine equipment K or J, we see that if either of them should fail, there would be a loss of a couple of days or less to bring it back online at full production. That's not too bad, but it's equipment D where the loss would be extremely painful. If equipment D should fail, we'd lose 21 days of production. It might fail in the second standard deviation, but there's only about a 13% chance of this occurring. Examining the third standard deviation, there's roughly a 15% chance that might occur, but a complicating factor is that it's an expensive repair. The answer to include, or not, isn't obvious. So, the real question is, "Do we want to take that chance?"

If we look back at equipment K and J, we see that the cost to repair for K is high, but J is moderate. So, "Do we include K, but exclude J?"

The situation for equipment G is troublesome. Slightly outside the first standard deviation, but inside the second. If it should fail, we'd be looking at 10 days of lost production. Wow, that would be hard to swallow.

When we consider equipment L, what we worry about is the failure standard deviation. While its first standard deviation is outside our current turnaround limit, one starts to get worried about its second, and especially, its third standard deviation. The good news is that its production loss and turnaround repair time aren't as significant, though its repair costs are still high.

If equipment I failed, it would be very painful since it incurs 21 days of lost production. On the positive side, it's outside our limits for both the first and second standard deviations. It's only the third standard deviation, where the situation becomes more problematic.

With equipment E, there's some more production loss, but its repair costs are less. The odds of this failing aren't high, since it's outside the first and second standard deviations, and it's not much inside its third standard deviation. This might be one of the safer ones to exclude.

Equipment H only has only a small chance of failure in its 3rd standard deviation, but would incur seven days of lost production. Ouch. Its repair time is only a couple of days, and it's not terribly expensive.

Equipment C is another that we might be able to exclude. Less chance of failure, its production impacts are relatively small, and repair costs are moderate.

While equipment B is quite favorable when examining its second standard deviation, it's a bit more problematic when looking at its third standard deviation. Its turnaround repair cost isn't that high, but its production loss would be three days.

Of course, in a world with no constraints, we'd probably repair all of them. However, that's not the world in which we live. Budget constraints are ever present.

8.1.5 Initial Group Recommendations

It's now time to make some final decisions.

Experienced engineers could argue the pros and cons of including or excluding equipment for turnarounds in the above list in different ways. They've been doing this for decades. Some are much safer bets than others.

Near the top of the list, equipment D would likely need to be restored since it carries a high production loss, if there is a failure.

Other equipment can almost certainly be excluded. At the bottom of the list, we might be able to exclude equipment C and maybe E. However, I wonder how many turnaround planners would want to bet their retirement money on it?

In between the top and bottom of the list, it's not an easy call. For many piece of equipment, it's possible to argue it either way. Even worse, we still haven't even included any real-world constraints – such as turnaround budgets, which might drive the decision one way or another. Once we do that, it will almost certainly get even more complicated.

The exercise that we just conducted is similar to how real-world plants have been making these kinds of decisions for years. In our case, we only started with a dozen equipment. Imagine if the list had been hundreds? It just shows the kind of difficult decisions that must be made with imperfect information. While experience will always count for much, the question is, could we use more science and mathematics to give turnaround planners and other decision-makers some firmer ground to stand on when they make these kinds of decisions?

We hope that you agree that the answer is "Yes" after working through the next three optimization examples. Our first optimization problem will be to choose the equipment to restore in several turnarounds, given the start dates and durations of both turnarounds have been set.

8.2 Results Expectations

Before we actually detail the three optimization problems, let's take a look at what results we might be expecting and see how those expectations might change as we work through our examples.

Table 8.5 summarizes the relevant information. We are starting with the following equipment: mean failure times, standard deviations, current ages, turnaround costs per repair, and in a later problem the time to extend the turnaround, if the equipment is repaired.

One of the benefits of the problem setup that we have chosen is that it is easy to play the "What If" analysis to see how various scenarios play out. Let's assume that we choose not to repair any of this equipment during each of the next two turnarounds. How does that affect the availability of the individual equipment and total crude unit availability? Table 8.6 shows those results.

The first two columns show values of 0.01, which indicate for purposes of this example that the equipment will not be repaired during the turnarounds. (There are actually some correlations built within Excel that won't allow a zero value, so this was the approximation to simulate "zero.")

Table 8.5 All Relevant Equipment Information

Equipment	Current Age Before First TA	Mean Time Failure	Std Dev Failure	Estimated Failure Date	TA Repair Costs ($)	Extend TA (days)
B	4	17	3	13	375,000	
C	4.5	14	1.5	9.5	250,000	
D	6.5	14	1.5	7.5	2,500,000	2
E	1.5	10	1.5	8.5	400,000	
G	4.5	12	1.5	7.5	675,000	1
H	4	13	2	9	380,000	
I	6	15	2	9	2,100,000	
J	8	16	2	8	485,000	
K	5	12	1.5	7	900,000	
L	8.5	18	3	9.5	1,400,000	

Table 8.6 No Repairs

Equipment	Average Availability End Run 1 (%)	Average Availability End Run 2 (%)
B	99.940	92.518
C	99.991	72.563
D	99.352	27.211
E	99.925	52.524
G	99.347	35.484
H	99.715	61.025
I	99.544	59.839
J	98.909	50.730
K	98.903	29.669
L	98.327	67.472
Unit	**94.10%**	**0.13%**

What this shows is that if we don't repair any of the equipment during the first turnaround, individual equipment availabilities remain quite high, but the overall crude unit availability is just slightly above 94% due to the multiplication of these together since they are in series from a reliability perspective. While 94% is low, it's not catastrophic for the current turnaround, but it's clearly devastating for the second run. This makes sense once you consider how equipment ages and changes over time if nothing is repaired, assuming the turnaround cycle is five years. Table 8.7 shows the dramatic increase in equipment age at the end of turnaround 2.

Now, let's play "What If" and assume that we will repair everything during the first turnaround. How does that change our results? Table 8.8 shows how it affects individual equipment and also unit availabilities.

Table 8.7 Ages at Turnaround 1/2 End – No Repairs

Equipment	Current Age Before First TA	Real Age After TA1	Real Age End Run 1	Real Age After TA2	Real Age End Run 2
B	4	4	9	9	14
C	4.5	4.5	9.5	9.5	14.5
D	6.5	6.5	11.5	11.5	16.5
E	1.5	1.5	6.5	6.5	11.5
G	4.5	4.5	9.5	9.5	14.5
H	4	4	9	9	14
I	6	6	11	11	16
J	8	8	13	13	18
K	5	5	10	10	15
L	8.5	8.5	13.5	13.5	18.5

Table 8.8 Repair All Equipment First Turnaround – Nothing Second Turnaround

Equipment	Average Availability End Run 1 (%)	Average Availability End Run 2 (%)
B	99.916	99.558
C	100.000	99.516
D	100.000	99.293
E	100.000	75.833
G	100.000	93.753
H	100.000	96.582
I	100.000	99.301
J	100.00	99.733
K	100.000	93.916
L	100.000	99.702
Unit	**99.92%**	**62.64%**

The good news is that we have dramatically improved our results for the first production run with availabilities of essentially 100%. (It's not quite 100% in the actual Monte Carlo results, but it's shown here that way since it was so close.) The majority of the equipment for production run 2 also has high availabilities, but there are several with extremely poor availabilities, especially Equipment E, which drag the overall results for the processing unit down.

As Table 8.9 shows, the ages of all equipment are reset to zero after turnaround 1 has been completed. The equipment has aged to 5 years after the first production run is over, and 10 years after the second production run since nothing was repaired during turnaround 2.

Table 8.9 Ages at Turnaround 1/2 End – All Equipment Repaired Turnaround 1

Equipment	Current Age Before First TA	Real Age After TA1	Real Age End Run 1	Real Age After TA2	Real Age End Run 2
B	4	0	5	5	10
C	4.5	0	5	5	10
D	6.5	0	5	5	10
E	1.5	0	5	5	10
G	4.5	0	5	5	10
H	4	0	5	5	10
I	6	0	5	5	10
J	8	0	5	5	10
K	5	0	5	5	10
L	8.5	0	5	5	10

The reason for equipment E's poor results is quite clear. It has a mean time to failure of 10 years, and that is roughly where it is at the end of turnaround 2. Realistically, it would need to be repaired within each cycle.

Given the turnaround costs presented earlier, repairing all equipment during turnaround 1 would incur a total cost of $9.3 million.

So far, we haven't shown production benefits, but we would expect a credit for achieving such a high availability during the first production run and being severely penalized during production run 2.

So, now, let's try the opposite. Let's pretend that we won't repair anything during the first turnaround and, instead, repair it all during the second turnaround. Table 8.10 shows those results.

What we see is that once again, we have slightly more than 94% availability during production run 1, but we have now achieved a rather high availability of almost 100% for production run 2. Of course, we have shifted all costs to the second turnaround.

So, we have now shown the two extremes. There's also the possible solution of fixing all equipment during both turnarounds. However, that's quite expensive since we are literally doubling our turnaround costs with relatively little upside.

Up to this point in time, we have not considered any constraints. What we will eventually see as we examine the problem results is that constraints can greatly affect the problem solution.

Two of the obvious constraints are total turnaround costs during any of the turnarounds, as well as minimum availability constraints during any of the two production runs. For example, we couldn't possibly tolerate a 0.13% availability for production run 2, as shown previously for the case where we didn't repair any equipment during either turnaround.

During our examples, we will impose a minimum availability constraint of 95%, even if it is unrealistically low for a crude unit since it is the first unit in the overall processing scheme. Nonetheless, we will use that limit value to illustrate the concepts.

Total turnaround costs were slightly more than $9.3 million to repair all equipment in one turnaround. What if there are budget constraints that limit our expenditures to $8,000,000, or even $6,000,000, or $4,000,000? The optimization results would obviously change.

Table 8.10 Repair All Equipment Second Turnaround

Equipment	Average Availability End Run 1 (%)	Average Availability End Run 2 (%)
B	99.940	100.000
C	99.991	100.000
D	99.352	99.976
E	99.925	99.698
G	99.347	100.000
H	99.715	99.810
I	99.544	100.000
J	98.909	100.000
K	98.903	100.000
L	98.327	100.000
	94.10%	**99.48%**

What will also affect the optimization results will be the credits that we receive for production increases. However, we now have a flavor of what to expect as we work through the examples in the remainder of this section.

As mentioned in other chapters, the absolute values in the following examples are not nearly as important as how the combination of values are relative to one another. It's the concepts we are trying to convey.

8.3 Solution Approach

The problems in this chapter were solved using the Solver within Excel as the main optimization algorithm.

The probabilistic nature of Monte Carlo was mimicked through the use of Excel's Data Table features, but Monte Carlo was not used directly. Rather, the ideas behind Monte Carlo analysis were incorporated through the use of regressed equations with the data generated from Data Tables.

The solution approach required modeling equipment on/off switches, and subsequently turnaround costs, as well as estimating production run 1 and 2 availability.

Let's discuss each of those separately.

8.3.1 Discrete vs. Continuous Optimization

The decision to repair a piece of equipment is obviously a discrete decision. You either decide to repair the equipment, or you don't.

This is a classic integer optimization problem. The actual optimization should be to adjust the combinations of repair/not repair for each of the 10 pieces of equipment in each of the 2 turnarounds. That's 20 different decisions which could result in millions of different combinations. That's why integer optimization is so difficult. There are just literally so many different possible combinations.

Normally, these types of problems are handled in different steps. One approach is to reduce the problem size by eliminating as many of the combinations as possible. We'll see in some of the results later that there are quite a few of the decisions that don't change across a wide spectrum and could actually be eliminated.

Since this is such a difficult mathematical problem, we will take some shortcuts in this book to show the concepts knowing that a more well-developed mathematical approach would be needed for a generic commercial application. Therefore, we will treat the problem as a continuous problem, rather than pure discrete, so we can continue to use Excel's Solver since it doesn't handle discrete problems well. Rather, it works much better with smooth, continuous functions.

For that reason, we modeled equipment on/off switches as a linear value from 0 to 1. However, it's clearly not a linear function. To make the function more closely approximate a discrete choice, we used that linear value and modified it within a weighted exponential function, which is referred to as "adjusted age." The exponential function was so steep that it did a reasonably good job of approximating a discrete choice. Adjusted age was then used to calculate turnaround costs.

8.3.2 Modeling Production Run 1 and 2 Availabilities

The solution approach required that we estimate both production run 1 and 2 availabilities for all equipment individually. We accomplished that through a modified Monte Carlo analysis. We didn't incorporate Monte Carlo directly in the optimization setup, but rather included those concepts based on regressions built using Excel's Data Tables functionality. We were forced to use that approach since the Monte Carlo analysis results are quite noisy, and the Solver requires repeatable derivatives to a number of significant digits, which isn't possible when using Monte Carlo directly.

There are significant differences and difficulty levels in calculating production run 1 and 2 availabilities. Production run 1 availability is solely a function of the decision to repair that equipment, or not, during the turnaround proceeding production run 1.

For example, equipment D's age before production 1 run begins is 6.5 years old. It has a mean failure time of 14 years with a standard deviation of 1.5 years. If we don't repair it, then the equipment will be 11.5 years old at the end of the production run. It would have a decent chance of failing sometime during that particular production run. However, if we did repair it in the turnaround process, its age at the end would only be five years old, and it's not likely it would fail.

Compared with production run 1 availability, estimating production run 2 availability is considerably more complex since it involves two different repair decisions for each piece of equipment. That is, the decision to repair the equipment before the first production run and the same decision before the second production run. There are three distinct possibilities.

The first is relatively straightforward and is exactly similar to that described for turnaround 1. If we chose to repair that equipment during turnaround 2, then determining its availability would be similar to what was described for turnaround 1. That's because the equipment will just continue to age until turnaround 2 and then be repaired.

However, there are two other possibilities related to turnaround 1 equipment repair decisions.

The first is the case when we repair during turnaround 1, but not turnaround 2. That's OK, but not as good as when repairing all equipment before turnaround 2. The other case is when we don't repair the equipment in either turnaround 1 or 2, which would have the poorest availability of all the cases.

What we learn is that estimating production run 2 availabilities is a function of two different variable choices and encompasses three completely different cases as just described. Modeling those accurately requires a lot of data since the regressions use "adjusted ages." In addition, it became obvious that we needed to include many more terms to adequately describe the production run 2 availabilty dataset compared with the simple correlation for production run 1.

While we had mentioned earlier that the independent variables were a linear value from 0 to 1, we should note that we ended up modifying that to be from 0.01 to 1. That's because, the log terms in the regressed equations for production run 2 could not directly accept a value of 0 for numerical purposes. However, that tradeoff did not materially affect solution quality.

8.3.3 Trading Off Increased Production vs. Turnaround Costs

At a high level, this optimization problem features a tradeoff between two important and opposite aspects of the objective function – increased production against increased turnaround costs. Obviously, increased production increases revenue and is desired, but that may require increased turnaround costs to achieve that goal.

In our problems here, we will impose a 95% availability constraint limit for both production runs. In fact, that's probably too liberal, but we will use it for illustration purposes. However, since we are using a 95% limit, that implies that we only have a 5% range maximum (100–95%) where we could possibly increase production. That's a limited area to work within. Plus, there's a point of diminishing returns such that once we get above a certain availability, for example 99%, it may take a considerable amount of turnaround spending to increase equipment availability much, if at all, beyond this figure.

In cases where there are higher production margins, it's logical to assume that we may want to increase turnaround spending, but that runs up against the point of diminishing availability returns. On the other hand, if turnaround costs are higher, we may back off slightly on production.

8.4 First Problem – Fixed Start Date and Duration

Let's recap our example problem and accompanying data up to this point in time.

The crude unit is four years into its run, and we have tentatively scheduled a turnaround one year from today. The duration will be 21 days, and the critical path is a debottlenecking project to increase crude capacity. There will be another turnaround of five years from the time that this next turnaround is completed.

We were unsure whether to include 12 equipment, labeled equipment A–L, with failure, repair, and production loss characteristics as described in the previous section. Our job is to perform an optimization to determine which equipment we recommend be included, and which we do not.

At this point, let's define the independent and dependent variables, as well as any constraints and the objective function.

8.4.1 Independent Variables

The independent variables for this problem are:

- Equipment A–L → Should they be restored in the upcoming turnaround?
- Equipment A–L → Should they be restored in the following turnaround?

While we mention equipment A through L as our choices, we have already eliminated two pieces of equipment from our list – equipment A and F, so there will realistically be only 10 pieces of equipment participating in the optimization.

If the equipment is chosen to be included in one of the turnarounds, their age will be reset to 0 at the time of their restoration.

8.4.2 Dependent Variables

There will be a running total of budget costs.

Profit will be the incremental revenue of the next two run cycles added together minus total turnaround costs.

8.4.3 Constraints

The most important constraint are minimum process unit availabilities for both production runs averaged over their cycle. In addition, we will include a maximum turnaround budget constraint.

8.4.4 Objective Function

The goal of this optimization is to determine the maximum profit, subject to meeting all constraints.

8.4.5 Results

Many optimization runs were made to examine different aspects of the problem. In general, we used a two-step process.

First, we ran the Solver model explicitly and examined the results. In many cases, the optimizer pegged the results as either a decisive "Repair" or "Do not Repair." In some cases, the answer was quite close to one of the two decisions, so it was fairly obvious what to do. However, in other cases, it wasn't quite as straightforward. In those instances, we manually changed values and observed the effect on the objective function. There was not perfect alignment between the two approaches, but it seemed to work decently well when used in the sequence just described.

8.4.5.1 Base Case

Table 8.11 shows optimal base case results where there are no active constraints using a margin of $100,000 per day.

What the results seem to indicate is that it's better to spend more money in the future turnaround, which will drive the availability slightly higher in the second production run. It decides to give up somewhat on the first production run and allow a slightly lower availability. Of course, these results are based on no active constraints.

Incremental profitability is a negative number because we are spending more money repairing equipment than any possible incremental production. However, that is to be expected during a turnaround. What the optimizer will try to for all examples in this chapter is to try and minimize how negative profitability is.

Table 8.12 shows the decisions for each equipment for both of the turnarounds.

Table 8.11 Base Case – Results

Availability Production Run 1	98.78%
Availability Production Run 2	99.28%
Turnaround 1 Costs	$4,264,124
Turnaround 2 Costs	$7,982,897
Profitability	−$4,828,938

Table 8.12 Base Case – Equipment Repair Decisions

	TA1	TA2
B	No	Repair
C	Repair	Repair
D	No	Repair
E	No	Repair
G	Repair	Repair
H	Repair	Repair
I	No	Repair
J	Repair	Repair
K	Repair	Repair
L	Repair	No

What we noticed as we ran through all the different possibilities was that the decision for certain equipment didn't change for any of the cases. That was the case for equipment B, C, and J. Therefore, we won't show that equipment in any future tables.

In the case of equipment B, since it had a long mean failure time, the optimizer always decided not to repair during the first turnaround but was forced to do it during the second turnaround. Otherwise, it's availability would be unacceptably poor during the second production run. In essence, it traded off equipment B availability between the two production runs.

For both equipment C and J the best decision was to repair during both turnarounds. For J, it was always necessary to repair during the first turnaround.

However, the decisions for J during the second turnaround, and both equipment C decisions were more marginal. In fact, one could argue it either way. In essence, the production gains were very nearly the same as the increased turnaround costs. However, if the margin had decreased or turnaround costs for that equipment had increased, the decision would have gone the other direction.

8.4.5.2 Constraining Turnaround Costs

So, let's start constraining turnaround costs and see how the results change. Table 8.13 shows the results if we limit turnaround 2 costs to under $4,000,000.

Unsurprisingly, production 1 availability increased because we repaired more equipment in the first turnaround as indicated by its higher spend. On the other hand, production 2 availability decreased. The profitability isn't dramatically different from the base case. However, it did change some important equipment decisions as shown in Table 8.14. That is, both equipment D and I were moved from being repaired in the second turnaround and brought forward to the first turnaround. All other decisions remained the same.

Table 8.13 Turnaround 2 Cost Constraint – Results

Availability Production Run 1	99.87%
Availability Production Run 2	97.88%
Turnaround 1 Costs	$8,564,288
Turnaround 2 Costs	$3,682,734
Profitability	−$5,413,252

Table 8.14 Turnaround 2 Cost Constraint – Equipment Repair Decisions

	TA1	TA2
D	Repair	No
E	No	Repair
G	Repair	Repair
H	Repair	Repair
I	Repair	No
K	Repair	Repair
L	Repair	No

Now, let's try moving the other direction from the base case by further constraining turnaround 1 spending. Let's set a limit of remaining below $3,000,000. Table 8.15 shows those results.

The same general trends continue. That is, production 1 availability decreased, while production run 2 availability increased very slightly. In this case, it's obvious that we are moving further away from the real optimum as shown by the significant decrease in profitability. The only equipment decision that changed as shown in Table 8.16 was moving equipment L from being repaired in turnaround 1 to turnaround 2. What we now realize is that all equipment are being repaired during turnaround 2.

If we continue by further decreasing the spend during turnaround 1 to below $2,500,000, we get the following results as shown in Table 8.17.

In this case, we decide to forgo Equipment G during turnaround 1 as shown in Table 8.18 to get back below the limit. Profitability is getting worse.

If we continue by further decreasing the spend during turnaround 1 to below $1,500,000, we get the following results as shown in Table 8.19.

Table 8.20 shows that in order to achieve this cost reduction, we were forced to remove Equipment K being repaired in turnaround 1.

At this point, we can't further reduce turnaround 1 costs more since if we did, we'd violate the production run 1 availability constraint.

8.4.5.3 Margin

Now, let's examine how differing margins may change turnaround decisions. In our base case mentioned earlier, the margin was set at $100,000. At a high level, we might imagine that as margin increases, we would want to add turnaround activities in either of the two planned events. On the

Table 8.15 $4.0 MM Turnaround 1 Cost Constraint – Results

Availability Production Run 1	97.15%
Availability Production Run 2	99.59%
Turnaround 1 Costs	$2,955,379
Turnaround 2 Costs	$9,291,642
Profitability	–$7,258,076

Table 8.16 $4.0 MM Turnaround 1 Cost Constraint – Equipment Repair Decisions

	TA1	TA2
D	No	Repair
E	No	Repair
G	Repair	Repair
H	Repair	Repair
I	No	Repair
K	Repair	Repair
L	No	Repair

Table 8.17 $2.5 MM Turnaround 1 Cost Constraint – Results

Availability Production Run 1	96.52%
Availability Production Run 2	99.60%
Turnaround 1 Costs	$2,324,376
Turnaround 2 Costs	$9,291,642
Profitability	−$7,738,981

Table 8.18 $2.5 MM Turnaround 1 Cost Constraint – Equipment Repair Decisions

	TA1	TA2
D	No	Repair
E	No	Repair
G	No	Repair
H	Repair	Repair
I	No	Repair
K	Repair	Repair
L	No	Repair

Table 8.19 $1.0 MM Turnaround 1 Cost Constraint – Results

Availability Production Run 1	95.48%
Availability Production Run 2	99.64%
Turnaround 1 Costs	$1,483,040
Turnaround 2 Costs	$9,291,642
Profitability	−$8,741,542

Table 8.20 $1.0 MM Turnaround 1 Cost Constraint – Equipment Repair Decisions

	TA1	TA2
D	No	Repair
E	No	Repair
G	No	Repair
H	Repair	Repair
I	No	Repair
K	No	Repair
L	No	Repair

Table 8.21 Base Case – Equipment Repair Decisions

	TA1	TA2
B	No	Repair
C	Repair	Repair
D	No	Repair
E	No	Repair
G	Repair	Repair
H	Repair	Repair
I	No	Repair
J	Repair	Repair
K	Repair	Repair
L	Repair	No

other hand, if margin decreased, such that production was no longer valued as highly, we might remove activities. In fact, the results obtained verify those intuitions.

Table 8.21 shows the base case decisions.

When the margin decreases to $75,000, the optimizer no longer chooses to repair equipment J in turnaround 2. In addition, when the margin decreases further to $25,000, the optimizer will no longer choose to restore equipment C during turnaround 2.

On the other hand, if margin increases to $200,000, it makes business sense to repair E during turnaround 1. If it increases further to $300,000, then we should add E and I to turnaround 1, as well as L to turnaround 2.

8.5 Second Problem – Fixed Start Date, but Flexible Duration

In this optimization problem, we fix the start date of the first turnaround, but allow the duration to float. Our first turnaround will commence one year from today, with another turnaround beginning exactly five years afterwards.

We will use the same example problem as in the first optimization problem with one change in the data set. That is, the repair of the equipment D and G can't be done within the original 21 day window for these two equipment. Therefore, if we decide to restore them in the turnaround, it will require the turnaround be extended by the amount shown in Table 8.22.

Thus, if we decide to service equipment G in the turnaround, it will require the turnaround be extended by 1 day to 22 days. If we decide to include equipment D, it will require two more days, extending the entire turnaround duration to 23 days. Let's assume that if we decide to include both equipment that the duration is independent of one another. Thus, if both are included, the duration would be 23 days.

Let's recap our example problem and accompanying data.

Table 8.22 Fixed Start Date – Flexible Duration

Equipment	Std Dev Failure	Earliest Failure 3 Std Dev	Earliest Failure 2 Std Dev	Earliest Failure (at TA)	FOFI Time	TA Repair Time	TA Repair Costs	Extend TA (days)
B	3	4	7	10	3 days	1 days	Moderate	
C	1.5	5	6.5	8	2 days	1 days	Moderate	
D	1.5	3	4.5	6	21 days	7 days	Very High	2
E	1.5	4	5.5	7	3 days	1.5 days	Moderate	
G	1.5	3	4.5	6	10 days	6 days	High	1
H	2	3	5	7	7 days	3 days	Moderate	
I	2	3	5	7	21 days	10 days	Very High	
J	2	2	4	6	2 days	0.5 days	Moderate	
K	1.5	2.5	4	5.5	1.5 days	0.5 days	High	
L	3	0.5	3.5	6.5	2 days	0.5 days	High	

The crude unit are four years into their run, and we have tentatively scheduled a turnaround one year from today. The duration will be 21 days, unless we decide to include either equipment D or G, where it would be extended by the amount per the discussion above. The turnaround critical path is a debottlenecking project to increase crude feedrate. There will be another turnaround five years from the time that this next turnaround begins.

We were unsure whether to include 12 equipment, labeled equipment A–L, with failure, repair, and production loss characteristics as described earlier in the chapter. Our job is to perform an optimization to determine which equipment we recommend be included – and which we do not.

At this point, let's define the independent and dependent variables, as well as any constraints and the objective function.

8.5.1 Independent Variables

The independent variables for this problem are:

- Equipment A–L → Should they be restored in the upcoming turnaround?
- Equipment A–L → Should they be restored in the following turnaround?

Similar to the previous problem, even though we state that our equipment choices are A through L, we already eliminated two of those equipment choices from our list. That is, equipment A and F, so there will realistically be only 10 equipment participating in the optimization.

8.5.2 Dependent Variables

We will not explicitly be choosing the duration. However, it will be done implicitly through equipment choice. If we choose equipment G, then overall run time until the next turnaround will be reduced by either one day, or two days if we should choose equipment D. This will have the effect of reducing total production for that time period.

Profit will be the revenue of the next two run cycles added together minus total turnaround costs.

8.5.3 Constraints

As with the previous problem, the most important constraints is minimum process unit availabilities for both production runs. In addition, there will be a maximum turnaround budget constraint.

8.5.4 Objective Function

The goal of this optimization is to determine the maximum profit, subject to meeting all constraints.

8.5.5 Results

On the surface, this problem would seem to be one in which there would be extreme interest since it is debated so often among turnaround planners. In essence, we are trading off lost production from extending the turnaround, but gaining the increase in equipment availability, which should improve process unit availability, once production resumes, if we decide to restore the equipment.

However, that analysis is only accurate if we are examining one production run/turnaround cycle. Once we extend our thoughts to include future production/turnaround cycles, our thought process changes completely.

That is, if we delay restoring equipment to start back up quicker, saving the production loss, we don't really save the turnaround cost over the long haul. We merely delay when we spend it (which may be important in certain circumstances). However, not restoring equipment may have a devasting effect on future reliability.

Let's take a look at equipment D as our example case. Its current age before turnaround 1 is 6.5 years, and it has a mean failure time of 14 years.

There are fundamentally four different cases we can consider since we can choose to restore it or not for each of the two turnarounds. These are shown in Table 8.23 with the appropriate ages at important points depending on whether we choose to restore or not.

The least risky option is to restore during both turnarounds. This incurs the most cost because we have both turnaround and lost production costs.

From an equipment risk standpoint, the next best option is to restore it during turnaround 1 and not to restore it during the subsequent turnaround. However, once again, we incur the lost production before turnaround 2 if we choose this option.

The next choice would be the opposite. That is, restore during the second turnaround, but not the first. At least in this case, we save the lost production after the first turnaround (which usually impresses management, but only pushes it to the second turnaround).

Table 8.23 Possible Equipment D Repair Decisions – Flexible Duration

	TA 1	TA 2	Age Entering Production Run 1	Age Finishing Production Run 1	Age Entering Production Run 2	Age Finishing Production Run 2	Risk
Case 1	Restore	Restore	0	5	0	5	Least
Case 2	Restore	No	0	5	5	10	Next best
Case 3	No	Restore	6.5	11.5	0	5	Getting worse
Case 4	No	No	6.5	11.5	11.5	16.5	Unacceptable

The last option is the worse since its age would be 16.5 years, which is beyond its mean failure time. No plant would take that risk.

What this analysis illustrates is that initial age makes a huge difference in how we analyze our options. Case 3 is worse than case 2 only because its initial age of 6.5 years is greater than the turnaround duration. If it were less, then case 3 would be less risky than case 2.

That's because in each of these cases, one of the production runs had an initial age of 0 and an ending age of 5. So, those production runs are exactly equal.

It's the other production run where the difference occurs. Since the age was greater than the turnaround duration, its ending age, by definition, was greater than the other case. If its initial age had been 3 years, then its ending age would have been 8. In that case, it would have been less risky.

So, what this entire analysis shows us is that we are ultimately trading off the lost production in front of a production run, but there's an inherent availability difference that is entirely dependent on equipment age. In fact, that age difference will ultimately drive the results more than any other factor.

For that reason, there was not really a need to make any model runs. The results would have been obvious and only dependent on the exact numbers used in the previous discussion.

8.6 Last Problem – Flexible Start Date

In this problem, the start date of the upcoming turnaround will remain fixed, but the start date of the next turnaround can float, up to ½ year in either direction. Therefore, the duration of production run 1 could extend anywhere from 4.5 to 5.5 years. The same is true for production run 2. However, the total duration of both production runs together will remain fixed at 10 years. Thus, if one is extended by a given amount, the other is reduced by that same amount.

8.6.1 Independent Variables

The independent variables for this problem are:

- Equipment A–L → Should they be restored in the upcoming turnaround?
- Equipment A–L → Should they be restored in the following turnaround?
- Start Date of Next Turnaround

As with both of the two previous problems, even though we state that our equipment choices are A through L, we have already eliminated two of those choices from our list. That is, equipment A and F, will not participate, leaving 10 equipment in the optimization.

8.6.2 Dependent Variables

Profit will be the revenue of the next two run cycles added together minus total turnaround costs. The production revenue from each cycle is adjusted based on the length.

In addition, there is a slight availability adjustment based on the production run length. For example, if production run 1 is extended by 25 days, its corresponding availability decreases by a certain amount. Obviously, if the production run length increases by 100 days, the availability decreases even more. The amount of the increase/decrease was based on a correlation using data from Monte Carlo estimates determined using Excel's Data Tables functionality.

Of course, the same is true in the other direction. As production run decreases, the availability of that time period will increase. In essence, one is trading off the difference in production run availabilities based on the selection of turnaround repair decisions for each turnaround.

8.6.3 Constraints

As with the previous problem, the most important constraints are minimum process unit availabilities for both production runs. In addition, there will be a maximum turnaround budget constraint.

8.6.4 Objective Function

The goal of this optimization is to determine the maximum profit, subject to meeting all constraints.

8.6.5 Results

Eight model runs were conducted where the unit margin was adjusted from \$5,000 to \$300,000. Interestingly, the effect on profitability was relatively mild and didn't affect any equipment restoration decisions compared with the cases where the turnaround start date wasn't adjusted. Figure 8.2 shows the optimization decision on increasing the length of turnaround one.

As the graph shows, as the margin decreases, the optimizer chooses to lengthen production run 1. This has little to do with the production run extension decision, but rather the equipment restoration decisions that were covered in the first problem in this chapter. If we recall from that

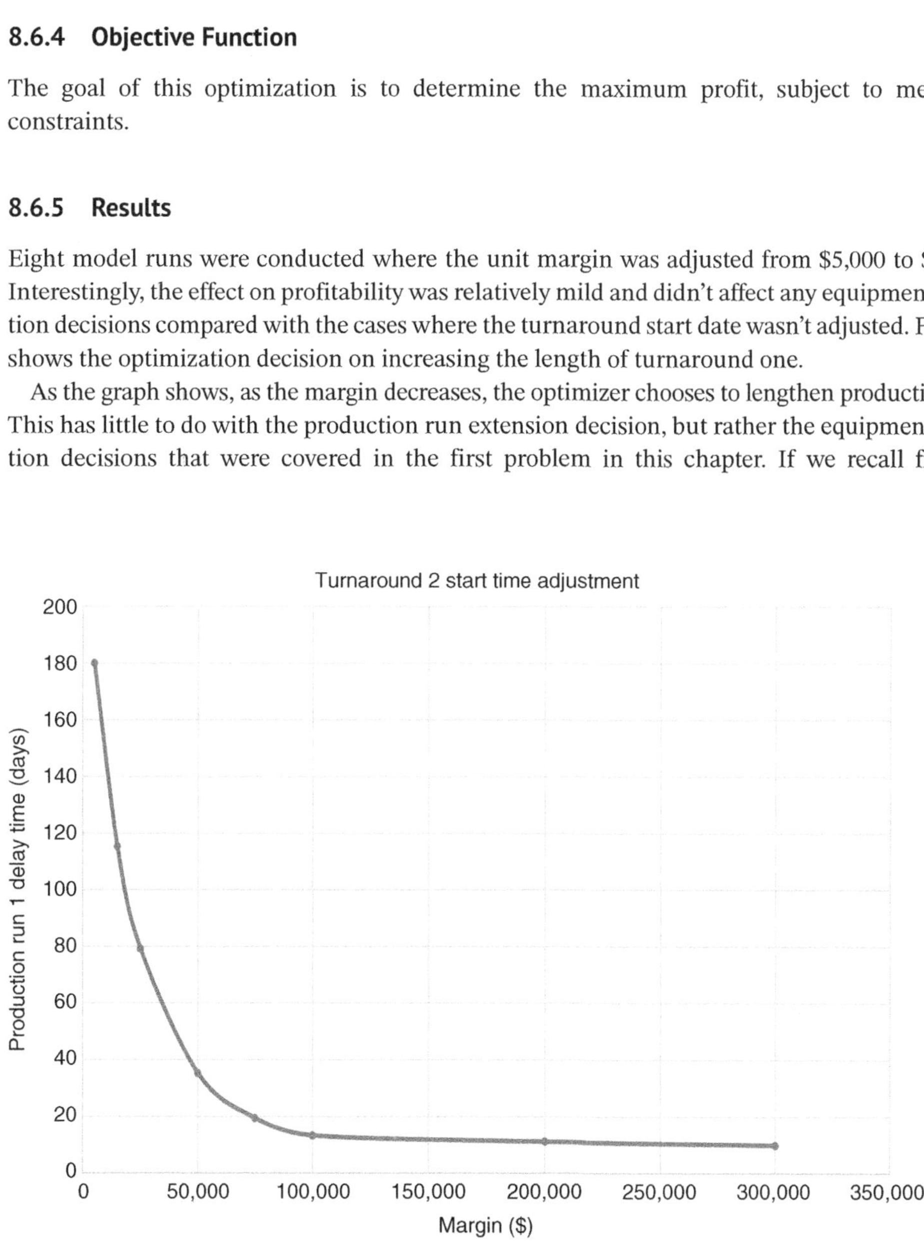

Figure 8.2 Turnaround 2 Start Time Adjustment vs. Margin

discussion, as the margin was decreased, the optimizer chose to repair fewer equipment during turnaround 2 up until the point where it drove production run availability down to 95%.

Thus, at lower margins, there is a large difference between production run 1 and 2 availabilities with production 1 availability more than 4.5% greater than production run 2 at the lowest margins. However, that difference decreases dramatically as the margin increases, essentially disappearing, since the optimizer is choosing to repair more equipment during production run 2.

What this discussion illustrates is that once the optimizer has an additional handle to manipulate, in this case, production run 1 end date (production run 2 start date), then it takes advantage of it to the degree that it can influence profitability. So, it moves it considerably at low margins, when there is a large availability difference between the two production runs, and not much, when there isn't.

Obviously, turnaround start dates are driven by many factors, and it's not likely that equipment reliability, as shown in this example, will be the primary determination. However, it does show what one might do if there is some flexibility.

8.7 Conclusion

Turnarounds are the most significant maintenance activity that a plant undertakes. They are expensive and disruptive.

The decision of when to come down for a turnaround, and for how long, are not trivial because many different factors need to be considered. One of those factors is the risk of equipment failure before meeting the next turnaround date.

In addition to the decision of when to come down, another major decision is what equipment should be serviced during the turnaround given turnaround duration and budget.

The methodologies presented in this chapter gives plant management another tool to help in making those decisions. It's limited in the sense that it only considers equipment reliability, but it is still one of the more important factors in the ultimate decision.

9

What About Process Conditions?

As noted in an earlier chapter, it's my impression that there often appears to be an invisible wall separating those who work on the "process" side of the business compared with those from the "asset" side. While the first half of my career was concentrated on using Advanced Process Control and Optimization to optimize real-time process operation, the second half of my career has been involved in reliability and asset management.

While I have seen numerous asset models being built in recent years, I have virtually never seen them incorporate important process-side information, such as feed quality or other operational variables. It's as if changing process conditions have negligible effects on what asset decisions we might make.

In some "clean" processes, especially those in the chemical industry, that might be a reasonable assumption. However, in other industries, such as the oil refining industry, in which I have spent most of my career, process conditions often have an incredibly significant effect on future equipment performance. In fact, since those effects can be so dominant and important, we need to find a way to include those within our definition of the optimization problem. That is the subject of this chapter. How we resolve that issue will ultimately be the "glue" that allows us to marry the process and asset sides of the business together.

In this chapter, we will create an example where we will determine the optimal feedstock selection based on equipment limitations. While this will be a direct use of how process conditions can be used to make more intelligent asset decisions, this idea can also play a supporting role in other types of asset optimization decisions. For example, we might decide to incorporate the idea of different feedstock options when deciding what should be the proper reliability required for new equipment or the proper mix of improvement projects.

While I will be using oil refining examples in this section, the general idea extends to all sectors of the process industry.

9.1 Examples Where Feed Quality and Process Conditions Play a Major Role

One of the biggest variables in all refining operations is feedstock quality. Some oil refineries, such as those in many Middle East countries, are blessed to be able to process the same crude oil all the time. However, in many other parts of the world, such as in many Western countries, a typical oil refinery relies on an entire diet of different crude oils to be economically viable.

Plant Optimization in the Process Industries: Incorporating Equipment/Assets in the Decision-Making Process, First Edition. Marty Moran.

In addition, many refineries in those countries also take advantage of "opportunity crudes" on the spot market. Thus, they will run unique crudes sometimes only once, or possibly occasionally, especially where there is an economic reason to do so.

Each crude oil is different and consists of literally thousands of different chemical constituents. One of the differences between crudes are "contaminant" levels, which might vary dramatically from one crude oil to another. As can be imagined, all oil refineries have processing steps to remove the more common contaminants. For example, the function of the desalter is to remove water, salt, and sludge-type materials from the crude feed. However, even after doing so, there may be components remaining in the crude that can significantly affect the operating life of downstream equipment. This is especially true if there happen to be upsets in the contaminant removal portion of the process.

While we have only talked so far about feed contaminant levels, equipment life can also be adversely affected by local operating conditions. For example, catalyst disengagement within the Fluid Catalytical Cracking reactor can be critical to what happens to downstream equipment. If disengagement is done properly, then the equipment will perform as designed. However, if there are issues, then downstream equipment can denigrate at a rapid rate.

The same thing happens in an alkylation unit, which contains a highly acidic liquid catalyst. These units are designed to handle highly corrosive acids, but if they end up in places that they were not designed for, the result could be rapid equipment corrosion.

In fact, corrosion often results from a combination of different effects occurring simultaneously. For example, in situations where a contaminant is present above a certain concentration, and there are elevated temperatures, you could see a rapid deterioration in equipment life. On the other hand, take the same exact situation, but if temperatures were normal, corrosion might not occur at all. The techniques introduced in this chapter will allow us to model similar effects, assuming we have a mathematical equation that can properly simulate the phenomena.

While these examples are specific to the oil refining industry, the general ideas have broad applicability across all process industry segments.

9.2 Operating Condition Effect on Failure Data

What the previous discussion illustrates is that for the process industries, in general, and oil refining, in particular, operating conditions and feed quality play outsized roles in future equipment performance. As we span the other industries in which Monte Carlo simulation has been used, such as in automobiles or aviation, we can see that operating conditions don't play nearly as significant a role as compared to the process industries.

Take the airline industry. The weather conditions that a plane operates within are quite well defined. There's always some variation, but those are well known, and there is plenty of data for those other cases for us to include in our simulations.

That's not the case in the process industries. Let's take the example of a pump that could be used in a clean service, such as a water pump, and the same pump used in another service processing erosive material. Sure, there are some precautions that we can take, but the bottom line is that the performance of that same pump in those services could vary significantly. You wouldn't have seen that nearly as often in other industries that have used Monte Carlo simulation.

For that reason, using industry-standard failure rate data, as you might find in a book, usually doesn't make any sense. True failure rates are *far* too dependent on actual operating conditions. In fact, that's the reason why the Consulting process outlined earlier in the book is necessary.

We need those who are familiar with the equipment, and the particulars in a specific plant, to weigh in when estimating failure rates.

9.3 Example Incorporating Process Conditions into Our Problem Definition

Given the realities that feed quality, process conditions, and other operational variables could play such a significant and variable role in our problem definition, how does one proceed?

To answer that question, let's work through an example of how we can incorporate process conditions into our optimization definition.

9.3.1 Problem Overview

Let's use an example where the planning department in an oil refinery is considering a long-term contract to source two different crudes, which we will refer to as crude A and crude B, to help demonstrate this concept. They have stated that they would likely alternate the crudes being run roughly every week due to crude tankage limitations. In addition, they have indicated that for planning purposes, the percentage of each crude would remain roughly the same each year over a multi-year time frame. To make the time frame long enough to realistically model the effect on equipment performance, we've decided that we will simulate this potential crude purchase over a 15-year time frame.

9.3.2 Crude A/B Characteristics

The major difference between crude A and B is that crude B is a lighter crude, as indicated by its API number. Thus, crude B will produce somewhat lighter products and less crude/vacuum residues when initially distilled. The good news is that our example refinery is a complex refinery with considerable "conversion" capacity to convert those resid streams into more valuable light products. In other words, the fact that crude A is heavier will have less of an impact that it could in other refineries, given our operational flexibility. In fact, after processing each crude through all of our 11 process units, the final product distribution for each crude is shown in Table 9.1 based on a feed rate of 102,000 barrels per day (BPD). Please refer to Appendix B for a complete discussion of Crude A/B characteristics, as well as how the material balance changes as the crudes are processed in different process units.

What this clearly indicates is that from a pure material balance standpoint, it's more desirable to process Crude B since it produces lighter products. However, the effect of price also needs to be included since crudes are priced based on several factors, with the distribution of lighter products being one of the more important factors. Another factor is sulfur content.

In that sense, another downside is that crude A has significantly more sulfur than crude B. Crude A is what we refer to as a "sour" crude with a sulfur content of 2% – which is significant. On the other hand, crude B is considered a "sweet" crude since it only has a sulfur content of 0.25%.

Table 9.1 Crude A/B Material Balances

	Gas	Naphtha	Kerosine	Diesel	AGO	Crude Resid	LVGO HVGO	Vacuum Resid
Crude A	1300	18,000	12,900	11,900	13,400	42,300	18,697	23,519
Crude B	1600	25,000	11,000	11,933	12,400	38,000	19,000	19,000

Most people in the refining industry are aware that higher sulfur levels can wreak havoc with refinery operations, so as sulfur content increases, they start to become concerned it might negatively, and significantly, affect equipment life. In this instance, the planning department has reached out to our team, since they have become aware of our program, to ask if they could evaluate what might be the upper limit for running crude A before they make any final decisions.

9.3.3 Equipment Impact

While there are likely many different pieces of equipment that could potentially be impacted by the difference in crude types, let's demonstrate the concept by just concentrating on furnace tubes as our example.

If there is only one crude in our problem definition, we would have assigned one, and only one, failure rate to furnace tubes during the Consulting exercise. However, now that we will be using two crudes, we need to reevaluate how changing the percentage of different crudes being run can potentially affect the furnace tube failure rate. To achieve that goal, we will need a dynamic calculation across a range of feed compositions. That's where our equipment experts will be a significant help.

First, equipment experts will need to estimate how many failures would occur if only crude B were running. They've told us that they expect the furnace tubes to last 30 years based on solely running crude B with a standard deviation of 5 years, so there's plenty of variability when the tubes could fail. (BTW, these figures are totally made up for illustration purposes to make the math easy for the problem. They are not intended to be linked to any realistic equipment estimates observed in refinery operation.)

Now, they will also need to do the same for crude A. For this case, they have decided that crude A would increase the failure rate by 50% over and above crude B. Those two bookends define the different ends of the spectrum.

Now, our experts would need to estimate how the failure rate would change as the percentage of different crude changes. In other words, they would need to estimate the failure rate across the entire range from 100% crude A and 0% crude B, all the way up to the other opposite end – 0% crude A/100% crude B.

While a linear extrapolation would seem to be the obvious answer and might be one possibility, the equipment experts may conclude that there are nonlinear characteristics, as shown in Figure 9.1.

9.3.4 Optimizer Definition

Now, let's define the optimization problem.

The problem will be performed on overall refinery operation. Fundamentally, the optimizer will choose the percentage of crude A being run and calculate the adjusted equipment failure rate, which allows us to calculate the availability over the entire cycle, as well as the availability in any one particular year. Then, using the material balance and cycle availability, we can calculate product distribution, and hence revenue. Profit is calculated as the difference between the product revenue and the total cost of crude feed.

9.3.5 Independent Variables

The independent variable for this problem is:

- Percentage of crude A being run

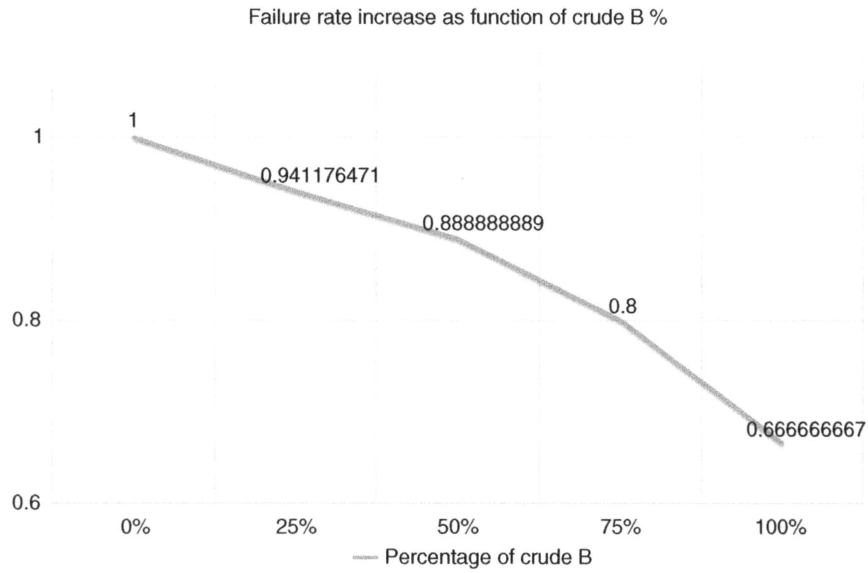

Figure 9.1 Failure Rate Increase vs. Crude B%

Once the percentage of crude A has been determined, we know the percentage of crude B since they must sum to 100%. Therefore, there is only one independent variable for this problem.

9.3.6 Dependent Variables

There could be many dependent variable calculations as part of this optimization problem.

The most obvious one is that of equipment failure rates based on the percentages of crude being run per the failure distribution graph presented earlier.

In previous problems, there was only one overall refinery material balance. However, in this situation, there are multiple crudes, so there will need to be an additional material balance for crude B. Appendix B shows the material balance for both crudes being run, as well as a comparison and contrast between their product distributions as they move through different refinery process units.

After the Monte Carlo simulation is run, there are new values for the availabilities for each process unit, allowing total product values, as well as total product revenue, to be calculated.

There will also be a dependent variable calculation for total crude cost since the price of each crude varies in the marketplace based on the yield structure and sulfur content.

9.3.7 Constraints

There could be many different constraints as part of this optimization problem definition. Two of the more common would be minimum average availabilities over the cycle or minimum availability in the last year of the time period under consideration. In addition, we may need to include some product volume constraints, since the planning department might not be able to sell all the products that we might produce.

In our example, we will include two main constraints. The first will be a minimum availability of 97% in the last year of our 15-year evaluation period. That's to ensure that we don't drop below an unprofitable availability in that year. As is probably obvious, the lowest availability of the

15-year time period is always the last year because we assume for the sake of this problem that the furnace coils are brand new at the start of year one.

Our second constraint would be a maximum gasoline production limit of 67,000 BPD due to storage limitations.

9.3.8 Objective Function

The goal of this optimization is to determine the maximum profit, subject to meeting the two constraints, which is calculated as the sum of all products manufactured for both crudes minus total crude feedstock costs.

9.3.9 Solution Approach

The optimization itself was modeled in Excel using the Solver, which is technically called GRG2, a proven, robust, nonlinear, reduced gradient algorithm. We simulated Monte Carlo using Data Tables in the "What If" analysis under the Data tab, but didn't use it directly.

Unfortunately, it's not possible to have GRG2 call Monte Carlo directly because GRG2 requires precise and repeatable derivatives as it evaluates search directions. Being a statistical, mathematical algorithm, Monte Carlo can produce a small change in the mean value from one run to another. That makes it incompatible directly with GRG2 since GRG2 requires absolute repeatability. However, we can easily circumvent that problem.

In this case, I created some calculations that predicated the availability in each year based on a random age number during that year, along with calculations of the "adjusted" failure rate based on crude selection. Using Excel's Data Tables, year 15 availability and overall availability were simulated based on 2000 iterations.

Since availability is solely a function of failure rates, I was able to create a table of each availability vs. crude A percentage. I then regressed an equation that was used directly in the Solver problem formulation to avoid the Monte Carlo imprecision problem.

9.3.10 Results

Assuming a price of $60 per barrel for crude B, a series of optimization runs were performed, as shown in Table 9.2, varying crude A price to determine the optimal crude split. "A" and "B" are the prices of each crude, while "Optimal" is the percentage of crude A.

The two constraint prediction values are "Availability Constraint" and "Gasoline," whose limits are 0.97 and 67,000, respectively.

After observation, the results make total sense.

When the crude prices are in a similar range, such as $57/$58 for crude A, the optimizer wants to run as high a percentage of crude B, since its material balance is more favorable. The increased light material production offsets the small crude cost differential. However, since it produces a higher percentage of gasoline, it can only run so much because it bumps up against the maximum gasoline production limit. In this case, the minimal split is 30.51%. The Availability constraint is a nonissue since it isn't running enough crude A to cause the equipment degradation to increase enough to become problematic.

As crude A becomes somewhat cheaper, then there is truly an optimum split between the two crudes due to the higher crude price differential. As crude A's price gets cheaper, the optimizer tries to increase crude A to minimize total crude costs since the cost differential offsets the increase

Table 9.2 Optimization Results

A ($)	B ($)	Optimal	Availability Constraint	Gasoline	Profit ($)
58	60	30.51	0.9915	67,000	541,335
57	60	30.51	0.9915	67,000	572,015
56	60	36.02	0.9904	66,683	604,779
55	60	43.39	0.9880	66,192	644,701
54	60	50.82	0.9843	65,622	692,066
53	60	58.32	0.9795	64,971	746,935
52	60	65.87	0.9734	64,239	809,372
51	60	69.55	0.9700	63,855	878,422
50	60	69.55	0.9700	63,855	948,370
45	60	69.55	0.9700	63,855	1,298,113
40	60	69.55	0.9700	63,855	1,647,855

in production revenue. However, eventually, crude A's price drops enough such that the minimum year 15 availability constraint is hit as it runs up against the effect of increased equipment degradation due to the higher sulfur crude.

This set of runs is for one combination of price differences between the two crudes, assuming an assumed base mean failure time of 30 years at a standard deviation of 5 years. One could easily repeat this entire optimization exercise for another value of mean failure time. What one will learn is that the Availability Constraint is quite sensitive, as could be imagined, by reductions in the mean failure rate.

9.4 Conclusion

The process industries are unique in that feed rate quality, operating conditions, and contaminant levels can greatly affect equipment performance and reliability. For that reason, it's important to include their effects in the optimizer problem definition as we consider the best asset decisions to optimize plant profitability.

This chapter showed a generalized methodology whereby we can utilize equipment experts to create estimates of how equipment performance varies as feed quality, process conditions, and/or contaminant levels change. However, this approach is not limited to the aforementioned items. It can be used for any situation where we can estimate a property using mathematical equations that show how equipment performance varies as a function of measurable variables.

Using this methodology, the optimizer can choose the value of feed quality or other relevant process conditions and utilize an intermediate calculation created by our equipment experts that translates it into a variable that can be used within the Monte Carlo simulation engine. Production volumes are then calculated based on the resulting system availability. Since the optimizer knows the production volume increase (or decrease) and the feed cost or other process conditions, it can then easily iterate to the value of feed quality or process condition that optimizes profitability.

10

Opportunistic Maintenance Optimization

While we might be tempted to think that the only way that we can improve a process unit or plant's availability is better equipment reliability, we soon realize that maintenance can also play a significant role.

The "M" in RAM modeling stands for Maintainability. At first glance, this might sound like an overly complicated concept. It's not. In fact, the average Joe would simply call this – "Repair time." In technical terms, Maintainability is the total time it takes to bring a piece of equipment back online after it has failed.

Within the industry, you might hear it referred to as "Feed Out" to "Feed In" – FOFI. Decreasing FOFI can be helpful in improving availability, since anything that can be done to reduce total repair time by decreasing the time for one of these individual steps will allow the equipment to be brought back online to produce the products sooner. FOFI consists of five major steps:

1) Cool the plant down
2) Isolate the equipment
3) Repair the equipment
4) Return it to service
5) Bring it back up to temperature

Some of these steps cannot be changed much, if at all. For example, cooling the plant down and bringing it back up to temperature is largely driven by process or metallurgical considerations. In addition, there's not much we can do to change the time needed for isolating the equipment and returning it back to service once repaired.

However, we do have options for the repair step itself. For example, there might be a choice of repairing equipment "in situ" vs. pulling the equipment and repairing it in the shop. This type of decision is a classic optimization problem. Plus, we might have options for prepositioning equipment or have plans in place ready to execute if we know that there is some likelihood that the equipment may fail soon.

This chapter will explore how we can use optimization to evaluate those situations where we have several maintenance options, with the assumption that the shorter repair time options will be more costly. There are times when the answer to that question is obvious, and we won't need optimization to make the right decision. However, there may be other more complicated situations, when we have multiple options to choose from where optimization might be attractive. This is especially true when there are numerous constraints.

Plant Optimization in the Process Industries: Incorporating Equipment/Assets in the Decision-Making Process, First Edition. Marty Moran.

While these ideas generally apply to all situations, in this chapter, we will use examples where a plant is planning opportunistic maintenance for one or more equipment in one or more process units when/if the opportunity presents itself. In all these examples, the equipment has different time-to-repair options with associated cost implications. As repair time decreases, cost increases. The central optimization question will be how long should the plant be down to complete these repairs given the throughput margin and the repair margins present?

The standard answer is to get the repairs done as quickly as possible. In many situations, that is the right answer. However, one always needs to evaluate the true economics behind possible scenarios before making final decisions.

The ideas presented in this chapter could also be easily incorporated and combined with the turnaround ideas presented in an earlier chapter.

The example problems will be artificially broken down into three main categories:

1) Single Equipment Opportunistic Maintenance
2) Intra-Unit Opportunistic Maintenance
3) Inter Unit Opportunistic Maintenance

Once the reader understands the general ideas behind those examples, there will be no need to force fit their own problems into one of these categories. Rather, you would be able to just use your imagination and technical skills to formulate the correct set of independent and dependent variables, constraints, and objective functions to solve the problem at hand.

The first category is simple and only concerns evaluating maintenance plan options for a single piece of equipment within a single process unit. Honestly, this problem rarely needs full optimization since it can usually be solved via other means. However, we include it here to seed our discussion. On the other hand, the two problems that follow on its heels are much better suited to a full-blown optimization.

Intra-Unit Options Optimization is a fancy way of saying that we are going to evaluate opportunistic maintenance plan options that affect numerous pieces of equipment within a single process unit. On the other hand, Iinter-Uunit Ooptimization evaluates opportunistic maintenance plan options across different process units.

Ironically, the examples in this chapter do not require Monte Carlo simulation – only the optimizer.

10.1 Modeling Maintenance Plan Options

From a modeling perspective, we'll need to transform the discrete maintenance plan options under evaluation into a form that can be easily used by the optimizer. In essence, we'll eventually need to regress those options into one equation with repair cost as a function of repair time.

We'll also need to assume that all possible maintenance plan options are ordered into a list starting with the project with the smallest repair time and progressing to projects of increasing repair time. Of course, costs could, and probably will, vary significantly from one project to another. That's OK and quite normal and should be expected. What is important is that the list is in decreasing cost order as the repair times increase. I don't anticipate that will be an issue since it would be unusual to consider a project with a longer repair time and a higher cost associated with it.

What is possible is that there are significant cost nonlinearities as we progress from one project to another. From the optimizer's perspective, that's not an issue since our optimizer is inherently nonlinear. In those cases, the optimizer quickly figures out that there are non-linearities present and adjusts accordingly.

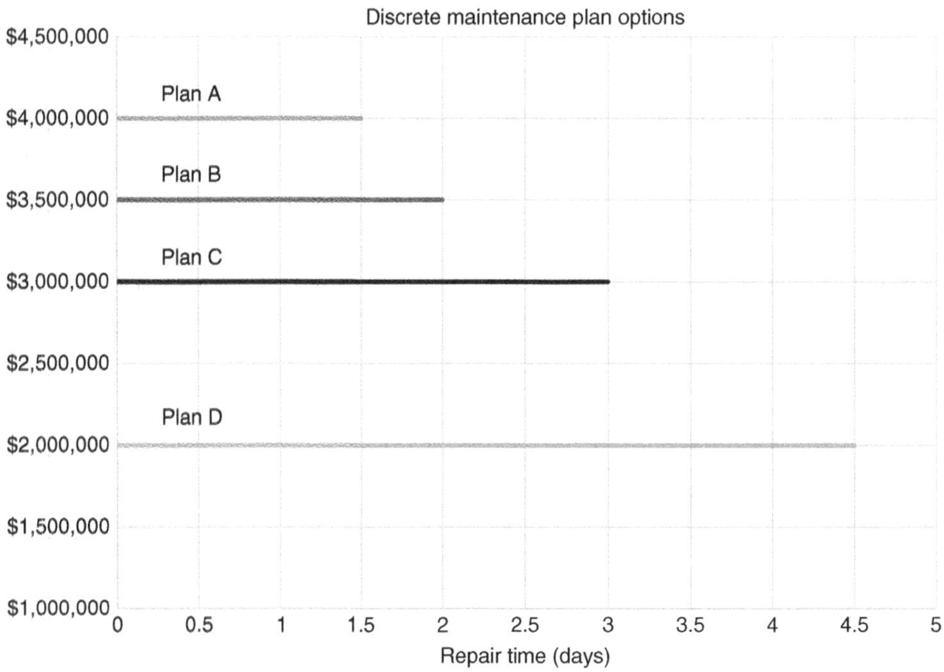

Figure 10.1 Discrete Maintenance Plan Options

Figure 10.1 shows an example of different maintenance plan options that a plant may be considering. As you'll see, the more expensive options take the least time to complete. For example, plan A in the graph takes 1.5 days to complete and costs $4,000,000. On the other hand, plan D requires 4.5 days, but only costs $2,000,000. In this example, the optimizer has four alternatives to consider where it can evaluate more expensive options, but have a quicker return.

If your optimizer doesn't handle mixed integer problems well (discrete problems), then we will need to convert the discrete maintenance plans into a smooth curve, as shown in Figure 10.2. There's some loss of precision when creating continuous curves similar to this, but it's usually tolerable. That's the approach we will take with the examples in this chapter – using smooth functions for use with GRG2 within Excel's Solver.

10.2 Example Problem Data

To illustrate the concepts, let's create a fictitious example using the data presented in this section.

The examples will consider equipment A–G, which are scattered in the Naphtha Splitter, Isom unit, and Reformer. These units typically are associated with one another, since the Naphtha Splitter separates out feed to the two conversion units, so in theory taking these units down as a whole is possible. Practically speaking that would be hard since the Reformer is the main source of hydrogen for the plant. However, let's proceed on the basis that it's possible.

Table 10.1 shows the total cost of maintenance plan options for each equipment (shown in columns) as a function of the days to execute (shown in rows) the project. Some equipment, like equipment A, will have three different options to choose from, whereas equipment B will only have two. On the other hand, equipment C has four, etc.

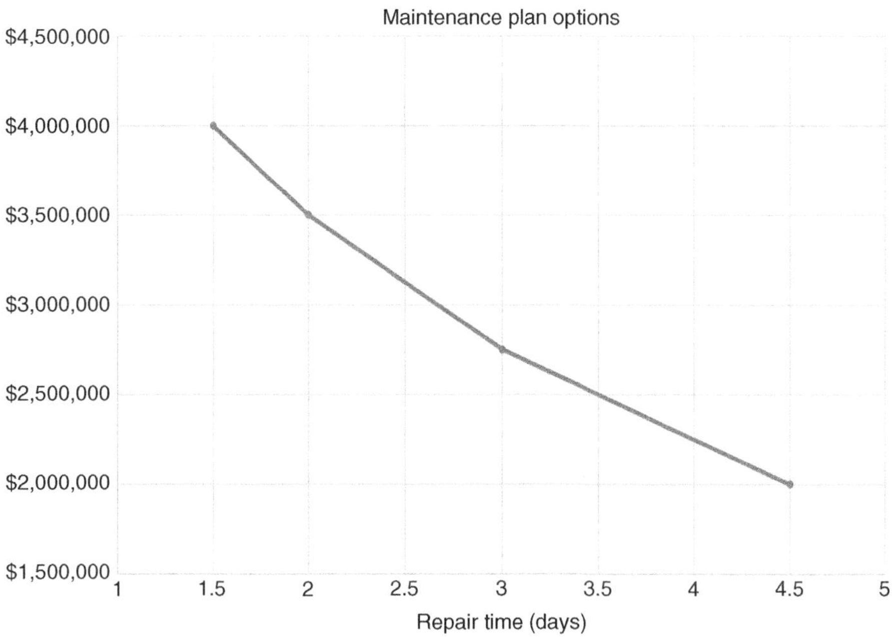

Figure 10.2 Continuous Maintenance Plan Options

Table 10.1 Example Problem Data

	Naphtha Splitter	Isom	Isom	Reformer	Reformer	Reformer	Reformer
Days to Execute	**A ($)**	**B ($)**	**C ($)**	**D ($)**	**E ($)**	**F ($)**	**G ($)**
3.5			500,000		100,000		
4	650,000	500,000	350,000	500,000	95,000	900,000	350,000
4.5			200,000		90,000		
5	500,000	225,000	150,000	400,000		700,000	225,000
6							
7	250,000			100,000		400,000	
8							

The number for the rows is the value for the associated nominal repair time. Therefore, the three values in the first column of numbers in the table are for equipment A and show the values for the repair times of 4, 5, and 7 days.

For equipment A's three options, the cheapest is $250,000 and would take 7 days to execute.

The most expensive option is $650,000 and would require 4 days. There's also an intermediate possibility that costs $500,000 and takes 5 days to complete. The values for all the other equipment can be read in a similar manner.

Figure 10.3 shows the same information except shown in a graphical form as continuous functions.

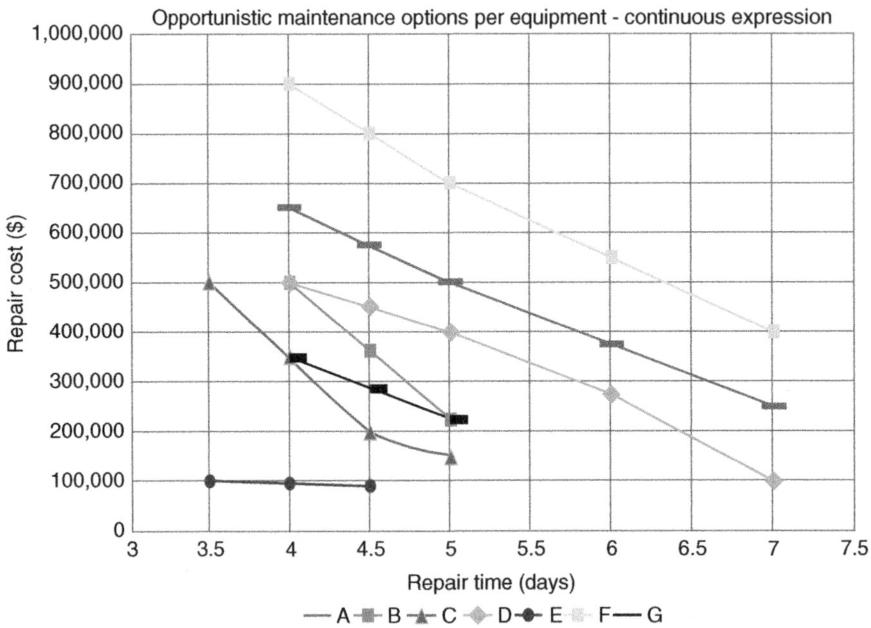

Figure 10.3 Maintenance Options per Equipment – Continuous

One thing to keep in mind as you review the examples in this chapter is that all the figures are relative. I wouldn't worry all that much about the absolute values of any one figure. What's more important is the concepts behind the numbers.

Let's now proceed to discuss each of our three different examples – Single Equipment, Intra-Unit, and Inter-Unit optimization.

10.3 Single Equipment Opportunistic Maintenance Optimization

In this case, we will evaluate the three maintenance plan options for Equipment "A," which is located in the Naphtha Splitter. Taking this unit down by itself might be painful, depending on how long it was down. In a case like this, Operations would likely bypass the unit and send all the feed to the Reformer to keep it going, since it's the refinery's main hydrogen source.

10.3.1 Independent Variables

The independent variable for this problem is:

- Repair time for equipment A

10.3.2 Dependent Variable Equations

The major dependent variable equation for this example converts the three maintenance plan options for equipment A into an equation used internally by the optimizer. This smooth equation (not discrete) allows the prediction of maintenance cost as a function of repair time between the values of 4 and 7 days.

10.3.3 Constraints

Since this is such a simple problem, there won't be any constraints defined.

10.3.4 Objective Function

The goal of this optimization is to determine the maximum profit given the Naphtha Splitter margin.

10.3.5 Results

This problem was simulated in eight different model runs using the GRG2 option within Excel's Solver. What repair option the optimizer chooses is solely a function of the unit margin. As expected, if the margin is relatively low, the optimizer will opt for a longer repair time. On the other hand, as the margin increases the optimizer progressively chooses shorter repair times. Figure 10.4 is a graph of unit margin vs. repair time, with all margins on a per-day basis.

What you notice is that the Optimizer chooses 7 days for Repair Time for values of unit margin less than $100,000. On the other hand, for margin values greater than $175,000, the Optimizer chooses a Repair Time value of 4 days.

There's a transition zone between $100,00 and $175,000, where the Optimizer flips from 7 to 4 days, respectively. It should be noted that this is somewhat deceiving because there's only one maintenance plan choice in this region – specifically at 5 days repair time.

In reality, it's not smooth in this region. It only shows up this way due to the way we are modeling the problem as a continuous function. That's required because GRG2 doesn't work well with discrete functions. However, even though it is not strictly correct, the smooth representation does give users a clear indication of the transition zone allowing them to make an informed choice.

The 5-day repair time in the transition zone is closest to a margin value of $130,000.

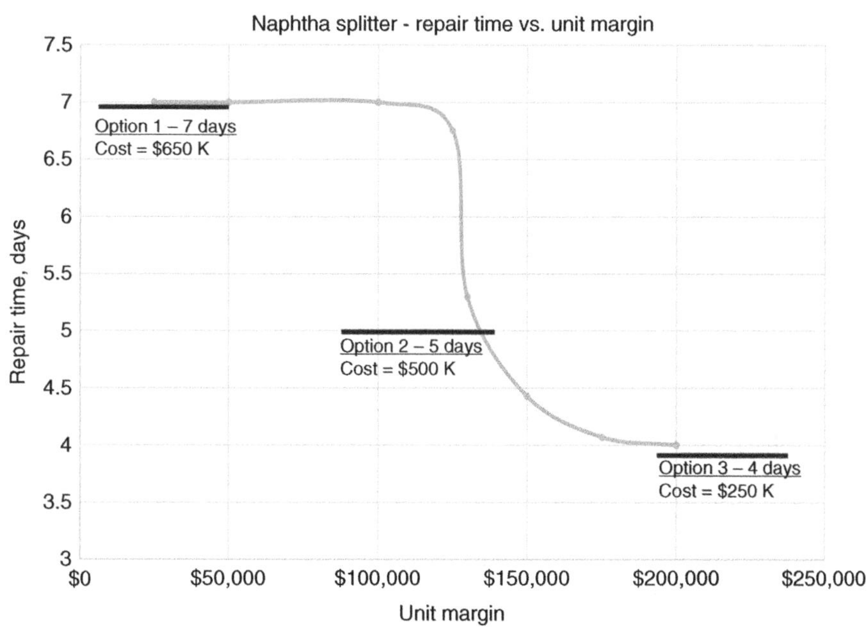

Figure 10.4 Naphtha Splitter Repair Time vs. Margin

10.4 Intra Unit Opportunistic Maintenance Optimization

Intra-Unit Options Optimization evaluates maintenance that touches multiple pieces of equipment within the same process unit. We will evaluate this for two different process units. The first will be for two pieces of equipment in the Isom unit – equipment B/C. This will help illustrate the concepts. Then, we will expand the problem by considering four equipment in the Reformer, equipment D–G.

10.4.1 Isom Unit Problem

In this section, we will detail the Optimization problem solution for the two pieces of equipment associated with the Isom unit. This might be the plant experience when we shut down for opportunistic maintenance in just the Isom unit but keep the rest of the plant operational. This is a fairly representative example, since many refineries don't have Isom units anyway, and the rest of the plant is not typically dependent on this unit.

To understand the problem better, let's review the options for equipment B/C, as shown in Table 10.2.

We have two options for equipment B, while we have four for equipment C. Since both repairs must be completed before the unit can be restarted, certain things become obvious. There are only two realistic options – 4 or 5 days, since those are the only choices for equipment B.

It would be pointless to consider the $500,000 option for equipment C since we could always choose the $350,000 choice at 4 days since there are no options less than 4 days for equipment B. Therefore, for practical purposes, the $500,000 option for equipment C doesn't exist.

On the other hand, even though it lies between the two equipment B options, the $200,000 option for equipment C is possible, since we could choose the $500,000 equipment B option and the $200,000 equipment C option.

10.4.1.1 Independent Variables

The independent variables for this problem are:

- Repair time for equipment B/C

Ironically, there is only one repair time as far as the Optimizer is concerned since the unit can only be started up after all repairs have been completed. While there are discrete repair time choices for equipment B/C, these choices are not truly independent of one another per the discussion in the previous section. The only final options possible are 4, 4.5, and 5 days.

Table 10.2 Isom Problem Data

Days to Execute	B ($)	C ($)
3.5		500,000
4	500,000	350,000
4.5		200,000
5	225,000	150,000
6		

10.4.1.2 Dependent Variables

The major dependent variable equations for this example will convert the maintenance plan options for equipment B/C into two separate equations, which will be used internally by the optimizer. This smooth equation (not discrete) allows the prediction of maintenance cost as a function of repair time between the values of 4 and 5 days for equipment B and 3.5 to 5 days for equipment C.

10.4.1.3 Constraints

There are no constraints for this example.

10.4.1.4 Objective Function

The goal of this optimization is to determine the maximum profit given the Isom margin.

10.4.1.5 Results

This problem was simulated in 14 different model runs using the GRG2 option within Excel's Solver with the results shown in Figure 10.5. What repair option the optimizer chooses is solely a function of the unit margin. Once again, if the margin is relatively low, the optimizer opts for a longer repair time, and vice-versa as the margin increases. All margins on a per day basis.

In this case, the Optimizer opts for the 5-day repair time for margins under $300,000. There's a longer, more gradual change toward 4.5 days, for margins between $350,000 and $525,000. Above $6000,000, the optimizer chooses the quickest repair times.

It should be noted that one needs to be careful in interpreting the gradual decline. That's because there are only two options for equipment B, and there is a fairly significant gap between those two values. On the other hand, there is another value between those options, which muddies the water. As always, engineers must interpret the data to understand all the problem's nuances and limitations.

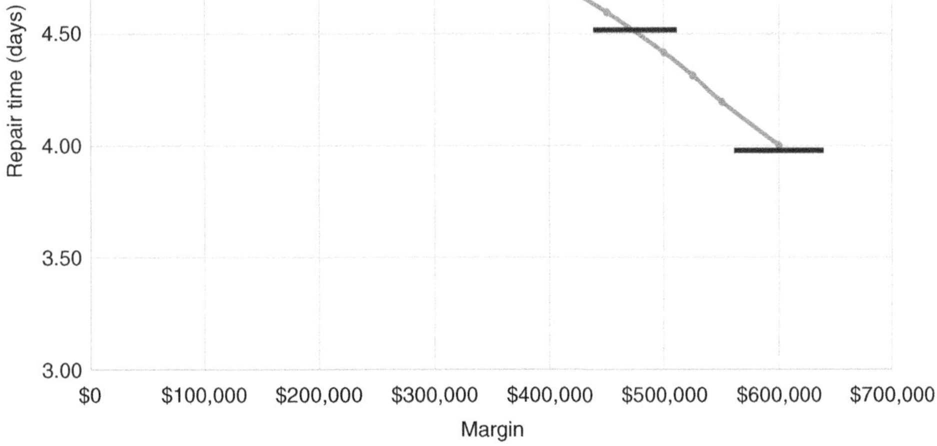

Figure 10.5 Isom Repair Time vs. Margin

10.4.2 Reformer Unit Problem

In this section, the problem will become more difficult since this Optimization problem solution will now consider four pieces of equipment associated with the Reformer. This might be the plant experience when we shut down for opportunistic maintenance for just the Reformer, but keep the rest of the plant operational. Per our earlier comments that's difficult because the Reformer is the source of hydrogen, but let's assume that's not an issue.

Let's dig into the problem by understanding the options for equipment D–G shown in Table 10.3.

We can draw certain conclusions by looking at this data. First, the $100,000 option for equipment E doesn't make sense because it's impossible to repair any other equipment at this time or previously. Thus, if we choose the minimum time when we could repair all equipment that would be 4 days. Thus, we would choose equipment E's $95,000 option. Therefore, there are no occasions when we would take the $100,000 option.

The bottom line is that looking at the table, there are only four realistic options – 4, 4.5, 5, and 7 days.

10.4.2.1 Independent Variables

The independent variables for this problem are:

- Repair times for equipment D–G

Similar to the previous discussion, the repair times for equipment D–G are not really independent since all repairs must have been completed before the unit can be started back up. Realistic possibilities are 4, 4.5, 5, and 7 days.

10.4.2.2 Dependent Variables

As previously mentioned, the major dependent variable equations for this example will convert the maintenance plan options for equipment D–G into equations used internally by the optimizer. This smooth equation (not discrete) allows the prediction of maintenance cost as a function of repair time between the values of their shortest and longest repair times.

10.4.2.3 Constraints

We will include one model run where the total repair cost is constrained.

10.4.2.4 Objective Function

The goal of this optimization is to determine the maximum profit given the Reformer margin.

Table 10.3 Reformer Problem Data

Days to Execute	D ($)	E ($)	F ($)	G ($)
3.5		100,000		
4	500,000	95,000	900,000	350,000
4.5		90,000		
5	400,000		700,000	225,000
6				
7	100,000		400,000	

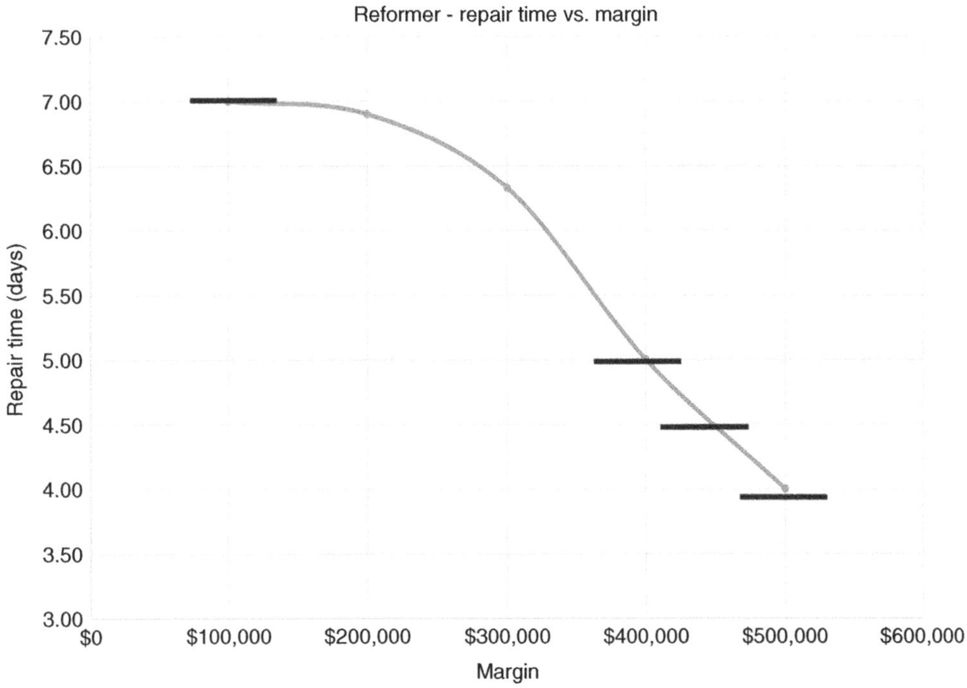

Figure 10.6 Reformer Repair Time vs. Margin

10.4.2.5 Results

This problem was simulated in six different model runs using the GRG2 choice within Excel's Solver. The first five model runs were unconstrained, whereas in the last run, we constrained the total repair cost.

Figure 10.6 is a graph of unit margin vs. repair time, with margins on a per-day basis.

Below a $300,000 margin, the optimizer would choose the longest repair time option – 7 days. The reason it desires the longer times is that the two repair options for 7 days are considerably cheaper than their 5-day options.

Then, there's a quick change to 5 days at about $400,000 and onto 4 days at $500,000 margin.

One model run was also conducted where we constrained the total repair cost to $1,400,000. In that case, the optimizer settled at a repair time value of approximately 5 days.

10.5 Inter Unit Opportunistic Maintenance Optimization

Inter-Unit Opportunistic Maintenance Optimization evaluates maintenance options that touch multiple pieces of equipment across more than one process unit. In this example, we will consider the three different process units already discussed Naphtha Splitter, Isom, and Reformer.

We will assume, for purposes of this example, that all repairs across all units will need to have been completed before the units can be restarted as a group. That makes sense since these process units are related to one another. Instead of different individual process unit margins, in this example, there will be one overall margin.

Let's review the data for those units once again, as shown in Table 10.4.

As discussed previously, the $500,000 equipment C option is not realistic for the same reasons that we presented during the Isom discussion.

Table 10.4 Example Problem Data

	Naphtha Splitter	Isom	Isom	Reformer	Reformer	Reformer	Reformer
Days to Execute	**A ($)**	**B ($)**	**C ($)**	**D ($)**	**E ($)**	**F ($)**	**G ($)**
3.5			500,000		100,000		
4	650,000	500,000	350,000	500,000	95,000	900,000	350,000
4.5			200,000		90,000		
5	500,000	225,000	150,000	400,000		700,000	225,000
6							
7	250,000			100,000		400,000	
8							

The quickest the units could be restarted is 4 days since that the first when all units could have had their repairs completed. Thus, there are only four different viable times – 4, 4.5, 5, and 7 days.

One thing that becomes obvious as more units are added is that there tends to be a large difference between 5 and 7 days, since waiting until day 7 requires an additional 2 days of downtime. Margins would have to be quite low for that option to make financial sense. However, if the repair time differences between 5 and 7 days are incredibly significant, it still might be.

10.5.1 Independent Variables

The independent variables for this problem are:

- Repair time for equipment A–G for the Naphtha Splitter, Isom, and Reformer units

As noted earlier, the optimizer will choose one repair time for all seven pieces of equipment. Since there are only four discrete repair day values possible, the optimization is bracketed between 4 and 7 days of repair time.

10.5.2 Dependent Variables

As in the previous problem definition, the major dependent variables are the equations that convert a series of maintenance plan options into the equations used internally by the optimizer.

10.5.3 Constraints

This problem will include a total budget constraint.

10.5.4 Objective Function

The goal of this optimization is to determine the maximum profit, subject to constraints.

10.5.5 Results

This problem was simulated in six different model runs using the GRG2 choice within Excel's Solver. The first five model runs were unconstrained, whereas, in the last run, we limited the total repair cost. Figure 10.7 is a graph of unit margin vs. repair time, with margins on a per-day basis.

As can be expected, as more equipment needs to be repaired, higher margins are required for the optimizer to choose shorter repair times.

A series of five model runs were also conducted to test the sensitivity of a Total Repair Cost constraint at a very high margin of $1,200,000 across the three units. Those results are shown in Figure 10.8. Above a total repair cost constraint of $3,250,000, the model doesn't encounter the

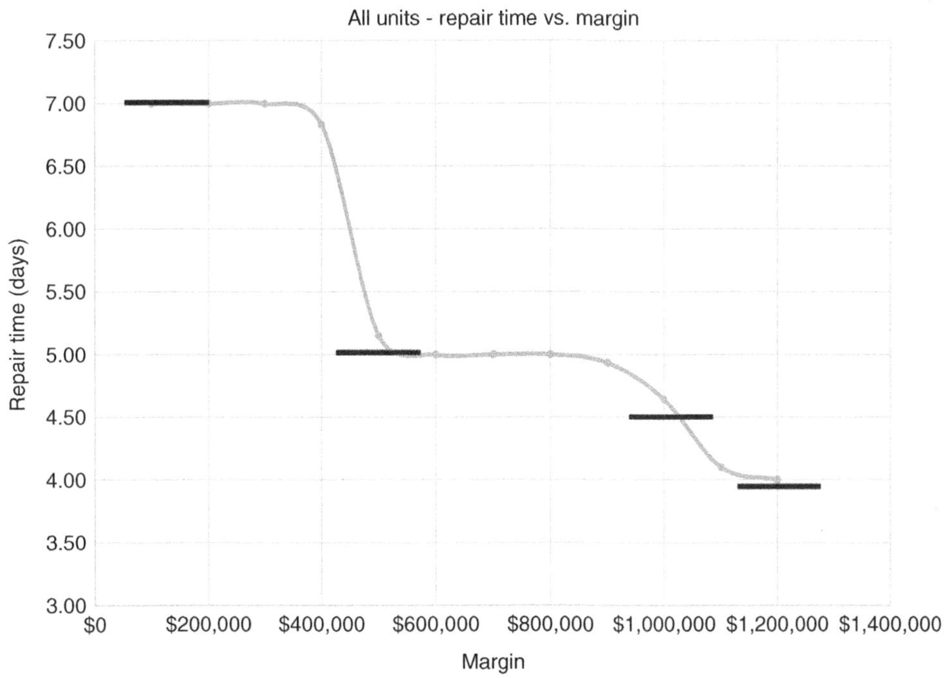

Figure 10.7 All Units Repair Time vs. Margin

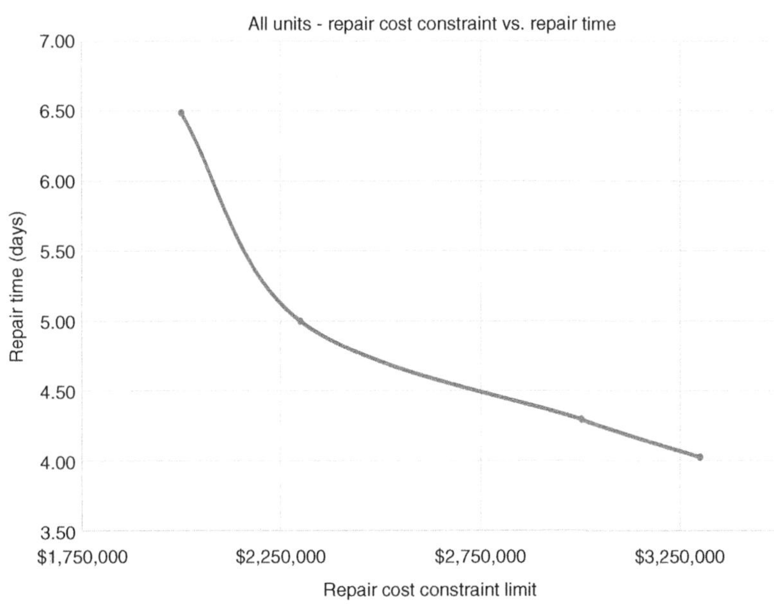

Figure 10.8 All Units – Cost Constraint

constraint limit, and thus, the optimizer chooses the minimum repair time due to the high margin. As the cost constraint becomes more limiting, the optimizer selects longer repair times since their associated costs are smaller.

10.6 Conclusion

This chapter explored different options on how optimization technology can be utilized to improve plant operation by better evaluating different maintenance plan options. The examples considered opportunistic maintenance for several pieces of equipment in three related process units.

Three cases were examined – maintenance plan options for a single piece of equipment, two examples with multiple pieces of equipment within a single process unit, and a comprehensive example considering multiple pieces of equipment across three different process units.

What we learned is that which option the optimizer selected depended heavily on the unit margin. In many real-world problems, the margin may be high enough that we always choose the quickest overall repair time. However, that may not necessarily always be the case.

Budget was also included as a typical constraint for the more complicated cases.

11

Spare Parts Optimization

Typical Replacement Asset Value (RAV) for a process plant could easily be a few billion dollars, with the largest plants potentially reaching $20 billion. Given those significant investments, one can assume that maintaining such a large installed asset base requires an enormous expenditure on spare parts.

It's safe to say that nobody wants to be caught short where when there is a critical need for a spare part that could potentially curtail production. On the other hand, management sends mixed signals about spare parts spending since they often incentivize the organization to keep spare parts holdings to a minimum.

It's the classic catch-22, similar to Goldilocks. Not too hot, not too cold. Just get it right. Much easier said than done.

Optimizing spare parts is a nontrivial, multifaceted topic that could easily expand to cover an entire book. Our discussion, by necessity, will only address certain aspects of the problem.

Before we get to the optimization examples, let's provide some background by covering how spare parts often camouflage other more important equipment issues. In addition, we'll discuss logistical challenges sourcing spare parts in different parts of the world, how spare parts are often initially estimated, some common spare parts issues faced by all organizations, and ideas on how to prioritize spare parts optimization opportunities.

First, let's start with how spare parts often mask underlying equipment issues.

11.1 Spares Parts Dependence Often Masks Other Equipment Issues

All plants need a large quantity of spare parts to run a typical refining or petrochemical facility. However, it's easy to use spare parts as a crutch, and many plants do.

Spare parts can be a buffer against unreliability, just like intermediate tankage. In fact, having buffers against unforeseen reliability issues is a positive. However, most plants retain significantly more spare parts than absolutely necessary. That's because spare parts, similar to intermediate tankage, often mask more fundamental equipment issues.

Whenever a plant identifies excessive spare parts usage as an issue, they need to ask themselves why that is occurring. One would like to think that happens on a routine basis, but the reality is that it is often the exception. If a plant could address the underlying root causes for repeated failures, it's possible we could eliminate the need for spare parts in the first place.

Plant Optimization in the Process Industries: Incorporating Equipment/Assets in the Decision-Making Process, First Edition. Marty Moran.

That's where spare parts optimization can play a positive role. It has the potential to shine a light on existing practices, exposing where plant improvements could address some of the more glaring problem areas. One of those is how we have historically estimated spare parts – vendor recommendations.

11.2 Typical Methods for Estimating Spare Parts

Spare parts issues often start with an initial recommendation of how many spare parts to retain. It's rarely an analytic exercise. In fact, the default is often a vendor's recommendation or a pure guesstimate. I know that because in my first professional position as an analyzer engineer working for a major analyzer vendor in the refining industry, I often put together that recommended list of required spare parts for clients. It was art, that's for sure.

How we estimate spare parts hasn't changed much in the last century. For the most part, we start with a vendor's recommended list, and maybe, in some unique situations, tailor it based on an internal discussion. It's rarely been based on a thorough cost/benefit methodology. However, vendor's have a vested interest in selling you more spare parts since it's one of their more profitable business areas. Unfortunately, they often achieve that goal. It's the classic case of the fox guarding the chicken coop.

Furthermore, if my experience working as an analyzer engineer early in my career is indicative of most plants, there's also a fair amount of spare parts hoarding that takes place. In that position, I learned to keep my personal stock, as did virtually every engineer that I knew. We knew which parts were problematic, and we were determined never to be caught short.

In fact, it wasn't uncommon to walk by a nonfunctioning analyzer and see it stripped bare. It always reminded me of cars in a bad neighborhood, which you often see sitting on crates devoid of tires.

I think that most of us can agree that spare parts being optimized based on a cost/benefit analysis is the exception – certainly not the rule. I have not personally ever encountered it. However, that's not the only problem with regard to spare parts that companies face. Geography often works against many companies.

11.3 Logistical Challenges

For the refinery or petrochemical industry located on the American Gulf Coast, obtaining a difficult spare part may require nothing more than getting into a truck and driving a few miles down I-10 or pulling spare parts from one of several warehouses located nearby. However, that experience can be quite different for other plants – even in certain parts of America.

I will always remember a sales inquiry that landed on my desk when I was sales manager for a major RAM modeling company. The company that called was located on the north shore of Alaska. They were wondering if we could recommend what they should hold in their one warehouse, given its physical location, size, and plant realities.

Talk about an optimization problem! That would have been a fun problem to solve, but unfortunately, the company never moved forward with the project.

However, the question they posed served to underscore the real-world logistical problems that many companies face. Their plant was located a significant distance from major parts centers. They often faced unusually long lead times and limited storage space, and the weather could be a real challenge – especially during the wintertime.

While their situation was clearly daunting, it was easier than what I encountered when working outside the US. I was always amazed at how difficult it was to get spare parts in certain parts of the world. One would think that with plants seemingly everywhere, especially in areas like the Middle East, that spare parts would be plentiful. The reality was just the opposite.

It was relatively rare to find lead times under three months, and for some equipment, lead time could easily stretch to a couple of years. The increase in lead time was the extra shipping time to get the part to its destination, as well as having to move the parts through customs. And areas like the Middle East were certainly better positioned, and in a more favorable position, than many other parts of the world, where plants are even more remote.

In fact, as an analyzer vendor, we spent a considerable amount of time discussing spare parts strategies with our international clients. It was always top of mind for them since they faced a challenging situation.

The bottom line is that managing spare parts outside the western world can be a daunting challenge. But suffice to say that everybody's situation is different.

While that highlights some of the major differences between plants, there are certain challenges that everybody faces.

11.4 Lead Times/Price/Vendor Issues

No matter the plant, every procurement department sourcing spare parts must consider the following issues as they make decide to purchase spare parts.

I was first introduced to these issues at the analyzer manufacturing company since our own internal purchasing people had worn out a path to my desk. They stopped by often inquiring about purchasing recommendations for portions of our analyzer bill of materials required to construct an analyzer. It was baptism by fire. These folks were under serious pressure to always maintain adequate inventory but yet not spend too much. It was a tough way to make a living. Their problems became mine, so I got to know the following issues all too well.

11.4.1 Lead Times

All spare parts have lead times, normally quoted first as a standard lead time. Fortunately, these parts can often be expedited quickly. Of course, that comes at a higher price. Total lead time consists of the time to manufacture the item, ship it to its destination, and move it through customs – if the part is shipped to a foreign country.

11.4.2 Price

Price usually varies by quantity. The more purchased, the cheaper.

Price increases if the part is expedited, and the amount of the increase is driven by how quickly the plant needs the part.

11.4.3 Vendor

Every company has preferred vendors – those that it trusts. In some cases, the organization enters into special contractual relationships with those vendors. At the other end of the scale, every company has vendors that it will only deal with as a last resort. Most vendors fall between those extremes.

11.4.4 Production Impact

While not a procurement issue, every plant is concerned about how production might be impacted and by how much if the equipment should fail and a spare is not available.

11.5 Prioritization

Given that a company typically stocks thousands of spare parts, a reasonable question is which spare parts should we perform an optimization on?

Unfortunately, that question is harder to answer than one might think. While we will present some important guidelines on how one might approach that issue, this book will not detail a true work process for making that determination, since it is actually a very difficult problem to solve and we can only provide some guidelines. Rather, we will provide enough information to help an organization modify its own internal processes to help make progress toward that goal.

The reason we can't provide a specific work process is that companies need to incorporate other aspects of their business into their unique situation. In addition, each company will want to meld spare parts optimization into their normal procurement process. For those reasons, it makes more sense for a company to graft a spare parts optimization process onto their existing work processes than for us to suggest a generic work process.

At this point, let's talk about some key issues an organization might want to consider as they consider modifying their internal processes.

11.5.1 Critical Equipment

A knee-jerk reaction for spare parts optimization is to start with critical equipment. While that might seem like a reasonable starting point, spares for critical equipment are rarely a bottleneck. That's because virtually all plants spend a considerable quantity of time and energy estimating spare parts needed for critical equipment. This equipment garners a considerable amount of attention, and it's not unusual for a committee to be formed to help make those decisions.

Plus, good luck trying to convince the management team to stock fewer spares than what the group decides for that equipment. (I'd bet that nobody on the management team is going to stick their neck out and take that risk.) Therefore, even though it seems counterintuitive, critical equipment is not where you will be spending most of your time.

11.5.2 Focus Areas

However, there are two areas that probably do make sense to focus on. The first is equipment, which would cause a production loss if they should fail. The second focus area is spares that consume a considerable amount of money and touch multiple places in different parts of the plant. These will both be the subject of example problems later in this chapter.

Since this is a wide area with many potential choices, it makes sense to introduce a process to help focus and prioritize our efforts. We will use a ranking methodology to achieve this goal.

11.5.3 Ranking Process

For difficult problems, consultants are fond of using ranking processes to help prioritize the decision-making process. These approaches could include multiple factors, which might influence the ultimate choice. Usually, this consists of identifying the factors that influence a potential decision, assigning a value from 1 to 10 or something similar, ranking the item on that factor, weighing each factor appropriately, and then performing an overall multiplication to determine the final ranking. Once initial results from a smaller example set are examined, the process can be honed.

I've successfully used this process on other problems many times in my career. In fact, you'll see this same idea used later in the book, in coonsiderably more detail, as we consider how to prioritize what projects make the most sense to kickstart an optimization program.

The spare optimization priority list will include factors for Standard Lead Time and Price, Expedited Lead Time and Price, Vendor Trustworthiness, and Availability of Alternate Suppliers. In addition, we would probably want to include the likelihood of the equipment causing a shutdown and how long the plant would be down if a failure should occur. Finally, we should also include a factor that ranks total inventory value. Here's a first pass of what might be included in the list:

- Standard Lead Time
- Standard Price
- Expedited Lead Time
- Expedited Price
- Vendor Trust
- Alternative Suppliers
- Alternative Suppliers Trust
- Production Loss Time
- Total Inventory Value (minimum inventory value multiplied by price)

Each of these categories would need to be further delineated.

It's likely that we would break down **Standard Lead Time** into logical time groupings, as shown in Table 11.1.

Standard Price would also follow a similar format as shown in Table 11.2.

Expedited Lead Time and **Expedited Price** would be exactly similar to the two tables already presented – only with different values.

Vendor Trust would be a simple value, though it could also be a table, similar to the others already presented.

Alternative Suppliers could be a table that identifies the quantity of available suppliers. Of course, there would be another table for **Alternative Supplier Trust.**

The most difficult part of this process is how to combine the various rankings to make a final decision. Should we multiply them together or weigh them before the multiplication to arrive at a

Table 11.1 Standard Lead Time Groups

Lead Time (weeks)	Ranking
0–4	2
4–8	4
8–12	6
12–20	8
20+	10

Table 11.2 Standard Pricing

Price ($)	Ranking
<1,000	2
<5,000	4
<10,000	6
<25,000	8
>25,000	10

final answer? That's where different people and different companies will come up with different and defensible answers on how to perform the final ranking.

In addition, some companies might want to perform the ranking in two different segments – a subset of the factors listed above ranked in one ranking process and then a second ranking using the remaining factors. This would be analogous to using the sorting mechanism in Excel by ranking two different columns. For example, one company might choose to sort the list first by production loss before proceeding with other factors. In addition, the list could be sorted twice – once by production loss and then another time by inventory value.

As you can see from this discussion, this process involves quite a few factors, which can be messy. However, these are the concerns we need to be thinking about.

While rank processing is one way of performing the prioritization, it's hardly the only way. A company might start with some of these ideas and end up with an entirely different process.

While this section could never provide an exact template of how to proceed since that is organization-dependent, it did propose some ideas which is better than beginning with a blank sheet of paper. Now that we have chosen some actual equipment to optimize spare part's holdings for, let's work through some practice problems to gain proficiency.

11.6 Example Problem Data

Let's use a fictitious example to illustrate the concepts. The example problems later in the chapter will allow us to examine the tradeoffs that one must make – as well as the dilemmas we may potentially face.

11.6.1 Basic Problem Data – Multiple Spares Within Process Unit(s) (Problems 1–3)

For these problems, we will have multiple spares, denoted as A–I, as shown in Table 11.3. We will assume each spare is used only for one piece of equipment within one or more process units in the plant.

The column **"Production Loss Time,"** shows how much production will be lost if there is a failure and there is no spare part available, measured in days. **"Age"** is how old the equipment is, denoted in years. **"Standard Lead Time"** is how long it will take to procure and obtain the spare part, measured in weeks. **"Standard Price"** is the cost to buy the part, in thousands of dollars. **"Failure Frequency"** is our expert's estimate of when the part would likely fail. **"Failure Std Deviation"** is the expert's best guess of the failure frequency standard deviation. The last column, **"Normal Inventory,"** is how many spares we are currently holding.

Table 11.3 Example Problem Basic Data

Part	Production Loss Time	Age	Standard Lead Time	Standard Price ($)	Failure Frequency	Failure Std Deviation	Normal Inventory
A	1	1	36	40	5	2.5	1
B	2	2	26	25	4	2	1
C	4	2	30	30	3	1	1
D	3	3	27	25	10	3	2
E	1	1	24	30	6	2.5	1
F	7	3	18	40	8	3	3
G	3	2	48	12	5	1	1
H	5	1	65	27	3	1	1
I	4	5	40	21	12	4	2

There is a lot of valuable information in this table. We'll talk about failure frequency and age in the next section, but for now, let's concentrate on Standard Lead Time and Inventory. You'll see that the Lead Times are often painfully long. The current inventory shows we are covered if a part should fail now, but obtaining a replacement might take a while. In fact, some parts have a lead time of half a year. More about this risk will be discussed in a future section.

11.6.2 First Failure in the Future

For now, let's focus on the age and the likelihood of failure. Table 11.4 removes some of the previous columns since we now have a better understanding of the problem while adding a few new ones. The table has also been sorted, with the part that has the largest production loss listed first and subsequent rows showing less lost production as the rows proceed.

The last two columns are new. "**Early Failure 1st Std Dev**" takes the Failure Frequency and subtracts the Std Deviation Failure. This value represents the earliest estimated failure time for 34% of the values within the standard deviation, based purely on statistics. The next column,

Table 11.4 First Failure in Future

Part	Production Loss Time	Age	Failure Frequency	Std Deviation Failure	Early Failure 1st Std Dev	Early Failure Based on Age
F	7	3	8	1	7	4
H	5	1	3	0.5	2.5	1.5
C	4	2	3	0.25	2.75	0.75
I	4	5	12	2	10	5
D	3	3	10	2	8	5
G	3	2	5	0.5	4.5	2.5
B	2	2	4	2	2	0
A	1	1	5	1	4	3
E	1	1	4	1	3	2

"**Early Failure Based on Age**," adjusts the current age by subtracting the Early Failure 1st Dev column minus the age. The last column is what we are really concerned about.

As we peruse the values in the last column, we note a reason for concern among many parts. Specifically, parts C and B have values less than one year, so they might not make it through to the next year.

Since we have only accounted for 34% of the failures on the low side of the mean, and we are not really concerned about the high side of the mean, implies, in total, we have now accounted for 34% + 50% of the failures or 84% in total.

However, that suggests that there is still a 16% chance of failure on the low side, less than the age we just calculated. Those would be part of the second and third standard deviations that haven't been fully accounted for just yet. When we do consider the remaining 16%, we start to become concerned about parts H and E, which have estimated failure values between one and two years.

11.6.3 Failure Immediately After Part Replacement

It's great to look at the long-term mean and standard deviation of a part failing, but that's not the entire story because whenever you break into equipment and replace various parts, there's always a small probability that after you start up the equipment that part will fail soon thereafter.

It's the classic infant mortality issue, but it could also be that the repair was not done properly, and the part subsequently failing due to human issues. We need to account for those types of effects as we work through the problem. This is especially important since if a failure like that should occur, it would be exceedingly difficult to source a part because lead times are often painfully long.

Table 11.5 shows one way of dealing with that possibility.

The last four columns provide a mechanism for evaluating how exposed we might be. "**Failure Upon Startup**" is a manual entry provided by our equipment experts as to the likelihood that a part fails within the standard lead time. In other words, it's how exposed we would be if we used our last part before we were able to replenish the stock.

The next column, "**Production Loss Startup**," is the multiplication of the Production Loss Time by Failure upon Startup values. It's an indication of how much production, on average, we might lose, if this situation should occur. Of course, in real life, it's not an "average" phenomenon. You either don't lose anything at all or you are down until you get a new part. There's no in-between.

Table 11.5 Start Up Failures

Part	Production Loss Time	Standard Lead Time	Normal Inventory	Failure Upon Startup (%)	Production Loss Startup	Production Loss Dollars ($)	Failure Loss Total ($)
F	7	18	3	5.00	0.35	70,000	(30,000)
H	5	6	1	5.00	0.25	50,000	(23,000)
C	4	9	1	1.00	0.04	8,000	22,000
I	4	18	2	2.50	0.1	20,000	1,000
D	3	27	2	5.00	0.15	30,000	(5,000)
G	3	12	1	5.00	0.15	30,000	(18,000)
B	2	21	1	2.50	0.05	10,000	15,000
A	1	30	1	5.00	0.05	10,000	30,000
E	1	24	1	5.00	0.05	10,000	20,000

However, this column gives us a starting point for an evaluation process. For the first part in the list, the value is 0.35 days. If we assume that our refinery's margin is $200,000 per day, that would net out to a $70,000 loss, as shown in the next column, "**Production Loss Dollars**." That's how financially exposed we are on an average basis. If we subtract the price for that part (not shown in this table, but rather in the first table on this topic), then from a business perspective, we would be losing $30,000, as shown in column "**Failure Loss Total**."

What this implies is that it would probably be a good business decision to have another part in inventory to account for this possibility. The amount of the value isn't as important, but rather, the fact that it is negative suggests that it might be a good investment. The more negative, the more we should consider making that purchase.

11.6.4 Basic Problem Data – Common Spare Across Multipole Process Units (Problem 4)

Up to this point in time, we have been concentrating on multiple spares within one process unit or possibly multiple process units. However, we are going to introduce another problem that we will work through in the next section. In this example, we will have one spare used in multiple locations across the plant. For that situation, we must adequately stock enough parts to address the possibility of multiple failures in different process units at the same time. Therefore, it adds a different level of complexity to the problem. However, the data is much simpler, as shown in Table 11.6.

Now, let's examine where the part is installed, as well as the age of those parts, as shown in Table 11.7.

There are several nuances to this problem. First, the production loss dollar value will depend on what unit the part fails in. In other words, the FCC production loss value will be different and higher than the production loss in Alky due to different marginal values. Table 11.8 shows the value of the daily margin for each of the four units.

Table 11.6 Common Spares Example Data

Standard Lead Time	Standard Price ($)	Failure Frequency	Std Deviation Failure
20	20,000	5	1

Table 11.7 Where Spare Installed and Age

Installation	Unit	Age
A	Reformer	1
B	Alky	2
C	FCC	2
D	Reformer	3
E	Isom	1
F	FCC	3
G	Alky	2

Table 11.8 Unit Margins

	Margin ($)
Reformer	35,000
Isom	7,500
Alky	10,000
FCC	75,000

One thing starts to become apparent as we ponder the data in these tables. That is, optimal results will change each year. Why?

One reason is that age is always changing. Another is that the spare parts price may change. Therefore, the decisions we make on how we manage inventory today will need to change over time. If there's one overarching lesson we learned from this exercise, it's that the optimization of spare parts holding is not a static exercise done once and forgotten about. It will need to be repeated on some common frequency, such as once a year, which might be a good starting point.

11.7 Effect of Failure Standard Deviation

Before we get to the actual problems, it's important to discuss one aspect that will significantly influence optimal spare holdings values for all our examples. That is, the failure time standard deviation. Figure 11.1 shows the average production loss values for the same equipment with a mean failure time of five years and four different failure standard deviations measured in years: 1,

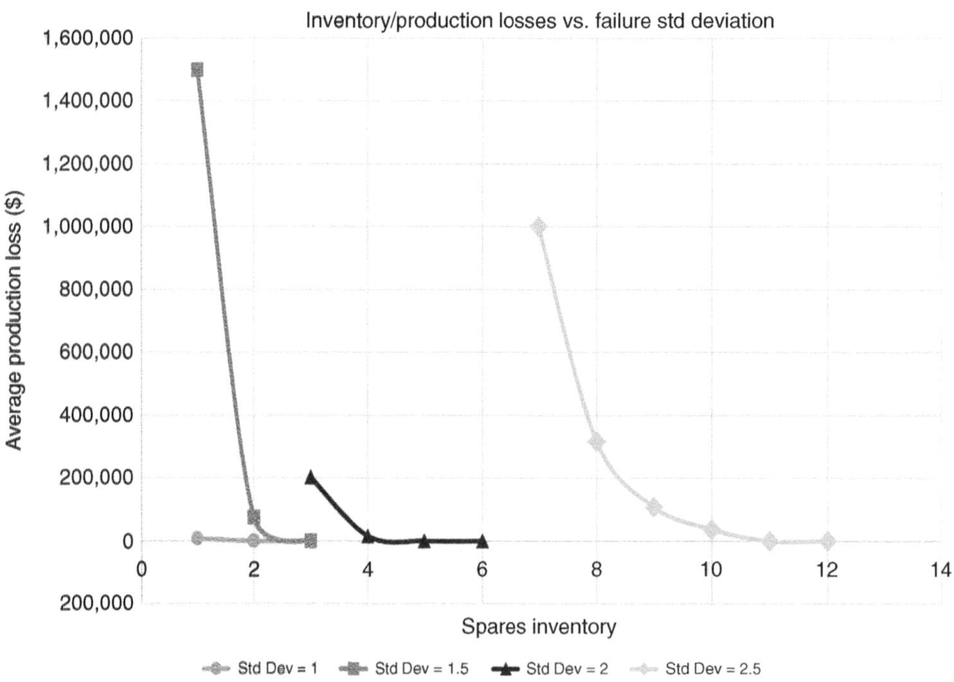

Figure 11.1 Inventory/Production Loss vs. Failure Std. Deviation

1.5, 2, and 2.5. What one sees is that the larger the failure standard deviation, the higher the quantity of spares required to mitigate average production losses.

For a standard deviation of 1, there are almost negligible average production losses, even with incredibly low spare holdings. As we increase the failure standard deviation to 1.5, average production losses skyrocket for the same spare holdings that we had for a failure standard deviation of 1. Then, as we increase the failure standard deviation further to 2, we need to increase spare holding to at least a minimum of 3. When we further increase the failure standard deviation to 2.5, we need to increase our holdings significantly, this time to an absolute minimum of 7.

When we step back and think about these results, they are not surprising. No matter the mean failure time, if we knew exactly within a week or two when a part would fail, we could order the part to be there just before it failed. It's only as we become uncertain as to when the part will fail that we have to be more prepared. If it can fail literally any time in that five-year time frame, we need to have the part in the warehouse right now. If that part is used in multiple locations, then we would really need to increase our holdings significantly because we could be hit with a rash of failures at one time. On the other hand, if we had more predictability, we could order the part to ensure it had arrived just before it failed, even if a lot of the same equipment was installed in the plant.

Given that background, let's move on to the example problems.

11.8 Optimization Problems Overview

The lengthy introduction in this chapter presented the major concepts and data we will be considering as we work through four spare parts optimization problems in this section.

11.8.1 Problems 1–3

Three problems will be similar in concept to those in the last chapter. They are:

1) Single Equipment Spares Parts Optimization
2) Intra-Unit Spares Parts Optimization
3) Inter-Unit Spares Parts Optimization

Like the previous chapter, where we talked about Maintenance Plan Options, we introduce a complicated topic with an easy example.

The first category will only evaluate options for a single piece of equipment within a single process unit. Honestly, this problem rarely needs full optimization since it can usually be solved via other means. However, we include it here to seed our discussion. It will give us some background before we move into the other two problems, which are better suited to a more detailed optimization.

Intra-Unit Options Optimization is a fancy way of saying that we are going to evaluate spare parts that affect numerous pieces of equipment within a single process unit. On the other hand, Inter-Unit Options Optimization will evaluate spare parts holding across different process units.

The distinction between the two isn't especially important from a spare parts perspective but does affect how we solve that problem using our twin technologies of Optimization and Monte Carlo. In the Intra-Unit case, we only have to call one of our Monte Carlo process models, whereas with multiple units, we need to call each unit involved in the optimization. It's more work than the Inter-Unit case, but it is not a big deal. Once the reader understands the concepts in these examples, they can easily formulate the problem to meet their own individual business and technical objectives.

For all three problems in this section, we used Data Tables in Excel, our substitute for Monte Carlo, to determine production loss as a function of inventory for each particular set of failure data – mean and standard deviation. That set of data was then regressed into one equation that predicated production loss as a function of spares inventory. Production loss was regressed both for "average production loss" and "maximum production loss," which were then subsequently used in the optimizer. Similar to most problems in this book, GRG2 within Excel's Solver was then used as the optimization engine.

Maximum production loss was quite important and was used as a constraint in the optimization. What happens is that when you are using Monte Carlo analysis, many iterations result in zero production loss. However, you will get isolated cases where you obtain an exceptionally large production loss. Thus, "average production loss" can be quite low, even though you have isolated large production losses. These effects are shown in Figure 11.2, where Average and Maximum Production Loss are plotted against spare inventory, but the values are shown on distinct axes since their values are so dramatically different from one another.

It's important to somehow incorporate both of these effects into the problem definition. What was decided was that the objective function would be the total cost of all "average production" losses plus any increased spares holding price. On the other hand, maximum production loss was included as an individual constraint, valued at $1,000,000 – for most of the model runs.

11.8.2 Common Spare Problems

In the last problem, we will work through a case where we have one spare part used for the same equipment installed in multiple locations across the plant. For this situation, we will determine the optimal overall number of spares to reduce risk to the point where we can safely mitigate against the effect of multiple failures in different process units at the same time.

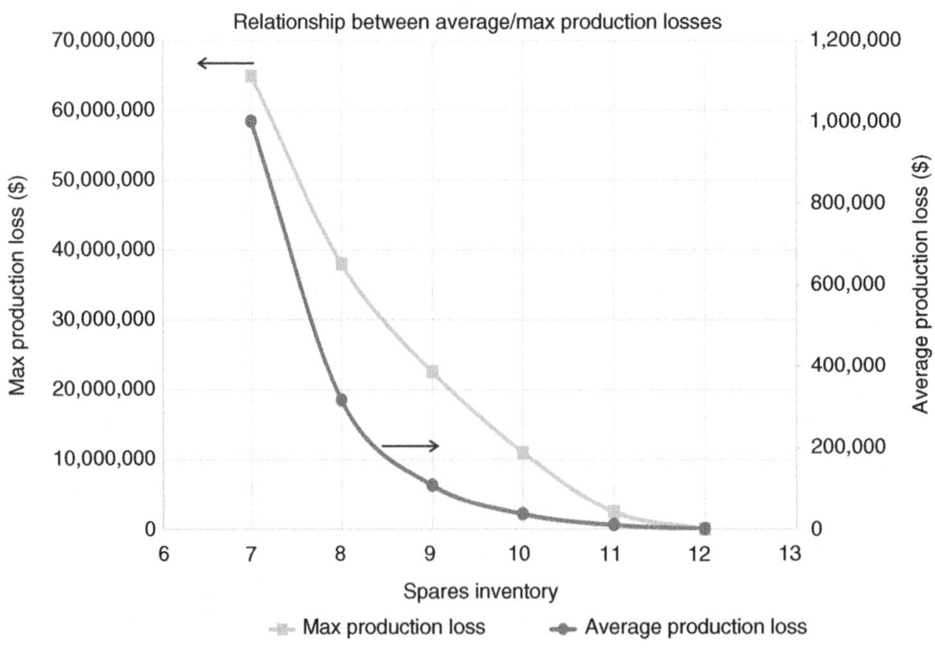

Figure 11.2 Relationship Between Average/Max Production Losses

11.9 Single Equipment Spares Optimization

In this particular example, we would like to optimize one type of spare for one piece of equipment.

11.9.1 Independent Variables

The independent variable for this problem is:

- Quantity of spares to hold for Equipment A

11.9.2 Dependent Variables

There were two dependent variable production loss equations that were regressed from "Monte Carlo like" iterations using Data Tables in Excel. "Average production loss" was incorporated into the objective function, while "maximum production loss" was used as a constraint.

11.9.3 Constraints

The only constraint for this problem was maximum production loss, which was valued at $1,000,000.

11.9.4 Objective Function

The goal of this optimization is to determine the maximum profit, subject to the maximum production loss constraint.

The objective function for this example was the sum of "average production loss" and incremental spare holding costs.

There's actually a subtlety to this problem that's important to understand. That is, average production loss was included directly in the objective function, while maximum production loss was a constraint. Those were somewhat arbitrary decisions. That's because we could have included maximum production loss in the objective function, and what we would have observed is probably a more conservative spares holding optimization result.

However, the problem formulation with maximum production loss as a constraint also considers the same information in a slightly different form. At the end of the day, reasonable people could approach it in different ways.

11.9.5 Results

Equipment A had a mean failure time of 5 years, a standard deviation of 2.5 years, and 1-year age. When running the Monte Carlo analysis over the 20-year time horizon, it was obvious that we would need to maintain an inventory of somewhere between 7 and 12 spares.

Roughly above 10 spares, there would never be any production losses. In other words, 10+ spares would cover the plant in all situations. However, as you drop below 10, you start to encounter some production losses, and those losses become worse as one holds fewer spares. Maintaining only seven spares would probably be a stretch since we would have started to encounter exceptionally large production losses. However, for purposes of the problem solution, we regressed both average and maximum production losses over the range of 7–12 spares.

What's important to note is that standard deviation plays an outsized role in spares holdings. As the standard deviation narrows, equipment failures become much more predictable. However,

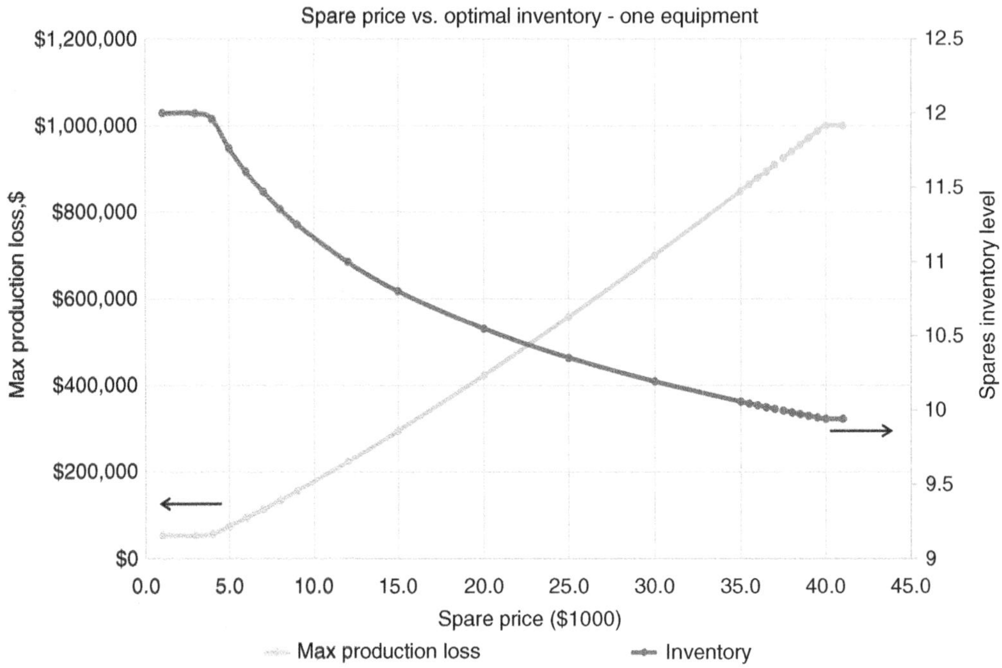

Figure 11.3 Spare Price vs. Optimal Inventory – Single Equipment

the opposite is also true. As failure standard deviation grows, failures seem much more random. In order to cover such random failures, one has to maintain large spare holdings. This comment applies to all problems discussed in this chapter.

Since there is only one equipment in this problem definition, the optimizer determines the optimal spare holding for a given spare part price. Therefore, we varied spare parts prices over a large range, as shown in the graph, to understand how sensitive the optimal spare holding was to spare price.

Figure 11.3 shows two curves. One is the optimal spare holding displayed on the right vertical axis. On the other hand, the maximum production loss is shown on the left vertical axis.

What one notices is that when the spare price is low, the optimizer chooses to hold more spares. Of course, more spares reduce the chance of a production loss. As spares price increases, the optimizer chooses to retain fewer spares. However, maximum production loss keeps increasing until the constraint limit is hit at about $40,000 per spare.

Of course, spares holding is not a continuous function but rather a discrete choice. Nonetheless, the continuous display shows the tradeoff quite nicely.

11.10 Intra-Unit Spares Optimization

Intra-Unit Spares Optimization evaluates more than one set of spare holdings for one or more pieces of equipment within the same process unit. This could be the same spare for multiple pieces of equipment, multiple spares for one piece of equipment, or multiple spares for multiple pieces of equipment. The only caveat for all three scenarios is that this all exists within one process unit. For this problem, we will consider single spares for all three equipment in the Reformer.

11.10.1 Independent Variables

The independent variables for this problem are:

- Quantity of spares to be held for equipment A, D, and I (equipment in the Reformer).

11.10.2 Dependent Variables

Similar to the single equipment example, the main dependent variables will be the regression of the average and maximum production losses. Average production loss will be summed together, along with incremental spare cost, and included as part of the objective function. On the other hand, maximum production loss will be used as a constraint for each equipment.

11.10.3 Constraints

This problem will also include the same maximum production loss constraints for all three equipment as in the previous example.

On the surface, an availability constraint would seem to be a good idea. However, the maximum production loss constraint effectively provides the same protection and is *far* more sensitive to the final result. For that reason, availability constraints were not included.

A total spare cost constraint is also part of the problem definition.

11.10.4 Objective Function

The goal of this optimization is to determine the maximum profit, subject to meeting all constraints.

11.10.5 Results

The results from the Intra-Unit case were less remarkable than one would have originally thought. A couple of things jump out from running this optimization case. First, each of the three equipment acts, more or less, independently of one another.

The only thing that might have tied them together would have been the total spare cost constraint. However, while a total spare cost constraint originally seemed like a good idea, it turns out that it really has no bearing on the final result. That's because the optimizer always rode the maximum production value constraint. In addition, trying to reduce spare holdings runs counter to the maximum production value constraint. So, by trying to enforce both constraints at the same time, the problem becomes infeasible.

We didn't vary the spare price for any of the three pieces of equipment since those results would have exactly mimicked the results in the previous single equipment example. So, there's no new news there.

What really influenced spare holdings results was the Reformer margin, which makes sense when you think about it. As the margin decreases, the maximum production loss value declines for the same spare holdings, so the optimizer can drive it back to the limit by reducing spare holdings.

The seven model runs for this example where the Reformer margin were varied, are shown in Figure 11.4.

For all equipment above $15,000, the optimizer rode the related $1,000,000 maximum production constraint limit. In those cases, the lost production is more painful than the increase in spares cost.

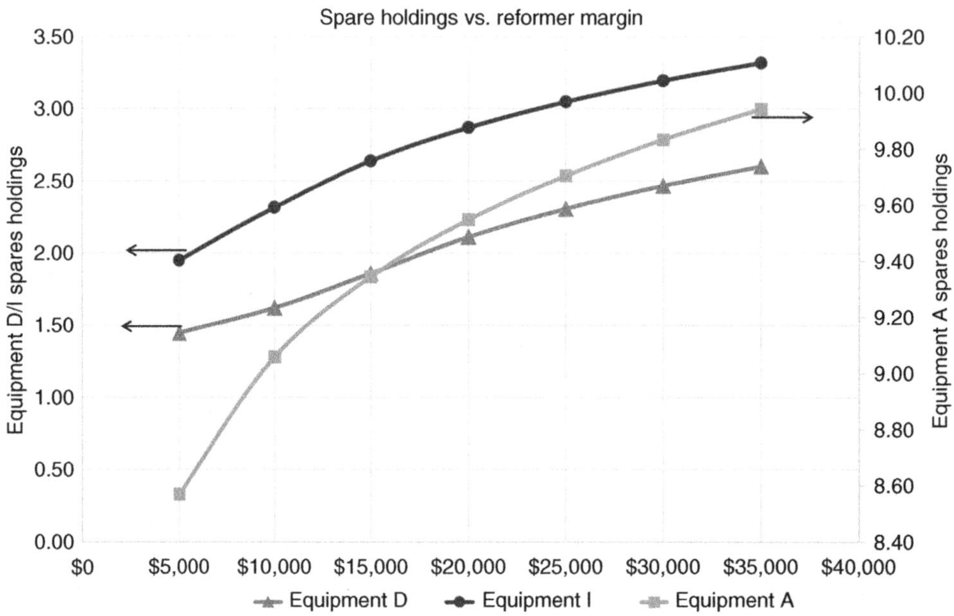

Figure 11.4 Spare Inventory vs. Reformer Margin

However, at $10,000, the optimizer doesn't quite ride the equipment D maximum production constraint limit anymore. At $5000, the optimizer didn't ride either the equipment D/I maximum production constraint limit. For those lower dollar value amounts, the optimizer starts to value the extra spares cost more than the lost production as compared to the higher dollar value margin cases.

11.11 Inter-Unit Spares Optimization

This is the most general spares optimization problem.

Inter-Unit Spares Optimization evaluates more than one set of spare holdings for one or more pieces of equipment across multiple process units. This is similar to the previous problem, except that it occurs across multiple process units. From the Optimization and Monte Carlo problem setup perspective, it's a more complicated problem (though not from the user's perspective) because we have to call multiple process unit Monte Carlo simulations – instead of just one.

11.11.1 Independent Variables

The independent variables for this problem are:

- Quantity of spares to be held for equipment A–I.

11.11.2 Dependent Variables

Similar to the Intra-Unit example, there will be dependent variable equations for each average and maximum production loss regression. In addition, the objective function will sum the total of all average production loss and incremental spare holdings.

11.11.3 Constraints

The same comments apply here for constraints as for the Intra-Unit example. That is, there will be maximum production loss constraints for each equipment. In addition, an incremental spare holdings constraint will be included.

11.11.4 Objective Function

The goal of this optimization is to determine the maximum profit, subject to meeting all maximum production loss and total spare cost constraints.

11.11.5 Results

There were nine model runs that involved all equipment. However, as will be described shortly, there were an additional nine model runs – one for every piece of equipment.

11.11.5.1 Maximum Production Constraint Limit

During the first model run, the maximum production loss constraint was set to $1,000,000, and the optimization performed. Not surprisingly, the optimizer drove each constraint to its limit. So, the results didn't seem very eye-opening at first glance.

However, out of curiosity, the constraint limit was widened to $2,000,000, and the optimization rerun. What happened is that the optimizer drove most of the equipment to the new limit, but there were a couple equipment that settled at values between $1,000,000 and $2,000,000.

So, this prompted an investigation as to whether there was a unique value for each piece of equipment and, if so, what that value was. So, after rerunning the optimization individually for each piece of equipment, the optimizer settled at the following values, as shown in Table 11.9, with values denoted in millions:

Table 11.9 Maximum Production Loss per Equipment

A	1.033
B	10.45
C	5.005
D	2.134
E	4.569
F	2.096
G	1.128
H	1.252
I	3.107

This is not to suggest that we would consider widening the production loss limit and tolerating losses at such high levels, if ever. However, what it does illustrate was that there were unique limits for each piece of equipment. It also explained why, when the constraint limit was set at $2,000,000, the optimizer behaved the way that it did.

The results for the Inter-Unit case had many similarities to the intra-unit example. That is, the optimization treated the equipment as largely independent of one another. In fact, the results were so obvious that only three model runs were completed.

11.11.5.2 Spares Constraint Limit

While the normal process would be to tightly constrain the maximum production limit, in the following model runs, that constraint limit was widened from $1,000,000 to $6,000,000. While not very practical, it was done to test how the optimization would respond if spare parts spending was constrained, as well as to understand how changing margins might affect the results.

Figure 11.5 shows model run results for two different Reformer margins – $25,000 and $35,000 per day, where there were various spare parts limits imposed. Obviously, above certain spare parts limit values, there's enough spare parts budget so that it is no longer constraining and total average production loss plateaus.

The chart clearly illustrates that at lower Reformer margins, there are fewer total average production losses. That makes sense since Reformer production isn't valued as highly in those cases.

Another observation from the graph is that the point where spare cost is no longer limiting is different for different margins. For example, in the $25,000 margin case, that plateau point is about $370,000, whereas for the $35,000 margin, that doesn't occur until approximately $420,000.

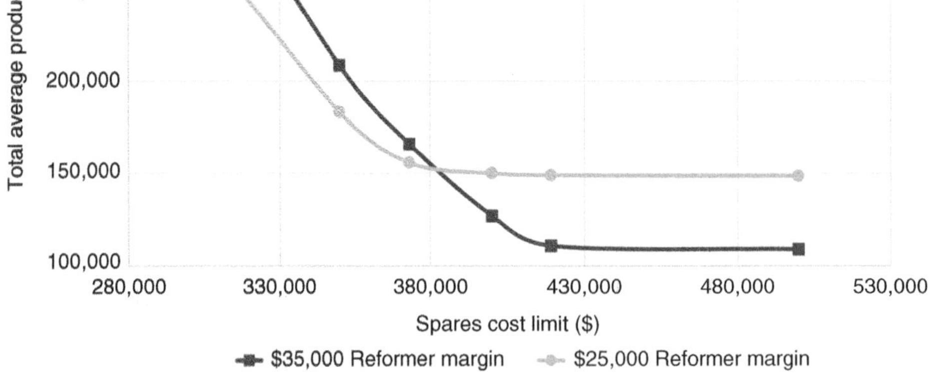

Figure 11.5 Production Loss vs. Reformer Margin/Total Spares Cost Limit

11.12 Common Spare Across Multiple Units

In this example, we will determine the optimal spare holdings for one spare part used in multiple locations within the plant based on data presented earlier in this chapter.

Fundamentally, one would like to reduce the number of holdings across the plant until one reaches a point of diminishing returns where production losses start. Normally, those production losses vastly exceed the cost of spares, but that becomes less an issue as spare part cost increases.

11.12.1 Independent Variables

The independent variables for this problem are:

- Quantity of spares to be held to support equipment A–G across multiple process units

11.12.2 Dependent Variables

There are no dependent variables for this problem.

11.12.3 Constraints

There are no explicit constraints for this problem.

11.12.4 Objective Function

The goal of this optimization is to determine maximum profit.

11.12.5 Results

This problem was modeled in Excel. Unlike other problems, this was not a true optimization. Rather, the model was inherently a complex Monte Carlo depiction of the problem run using Excel's Data Tables. Optimization wasn't performed directly since the multiple Monte Carlo runs clearly showed what would happen if an optimization had been performed.

Twenty-two different model runs were performed where the effects of varying failure time, failure standard deviation, and spare lead time were evaluated.

11.12.5.1 Base Case – Failure Time = Nine Years, Standard Deviation = Three Years

The first case run was the base case. It consisted of a series of six model runs where spare inventory was reduced from 12 down to 7. It assumed that the equipment had a mean failure time of nine years with a standard deviation of three years.

Over the six cases, the average number of failures is 62.5. As would be expected, the number of spares held in inventory doesn't affect the number of failures. That's solely a function of equipment failure characteristics.

What one sees in all the examples in this section is that as one reduces spare inventory, there's a point where dramatic production losses occur, literally falling off a cliff. In fact, the production losses in the graph dramatically underestimate this effect. Figure 11.6 shows the "average" production loss, but individual production losses are often an order or two magnitudes higher. For example, at a spare inventory of 7, the "average" loss is just over $150,000. However, that's deceptive because

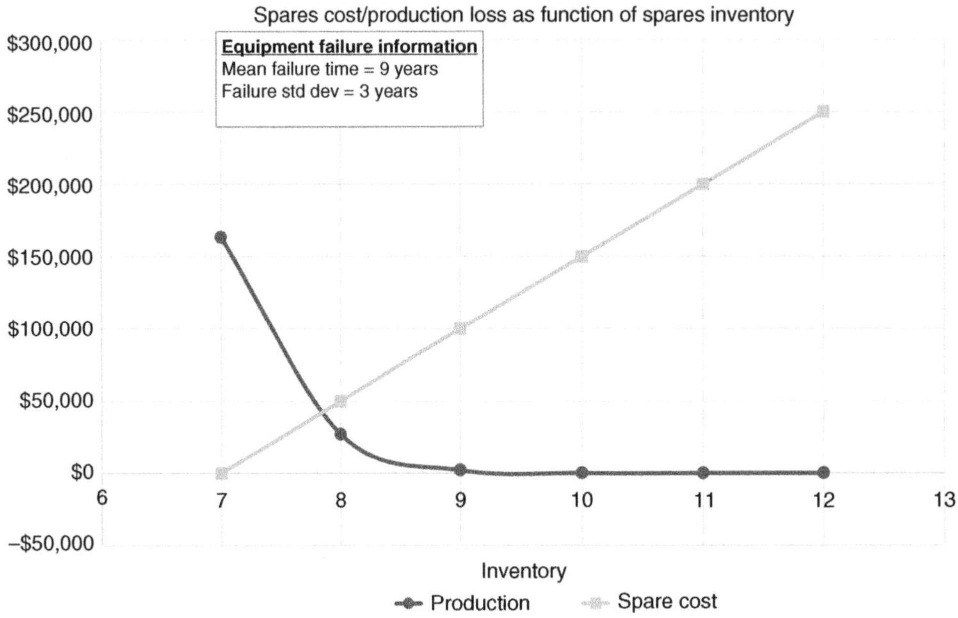

Figure 11.6 Spares Cost/Production Loss vs. Spares Inventory

there could be individual losses as high as $5,600,000. The average is only that low because, in the majority of iterations, the production loss is 0. However, add in a few very large losses during a few iterations, and one obtains an "average" of $150,000. In other words, one has to be very careful as one approaches the cliff.

For the same reasons, the spares cost shown in the graph is deceptively high relative to the "average" production losses. The costs in the graph are based on incremental spares holding above 7.

What this means is that the user will want to err on the high inventory side to prevent a potential catastrophic loss. In other words, the minimum selected could be as low as 9, but more likely, 10 to stay far enough away from the cliff for all potential situations. Moving from 9 to 10 only incurs an additional cost of $50,000, but that could be a minuscule incremental cost if there was an actual production loss event.

11.12.5.2 Increase Standard Deviation to Four Years

In this case, we will maintain the same mean failure time of nine years. However, we will increase failure standard deviation, which will have the net effect of creating some failures earlier than in the previous example. In fact, the average number of failures increased quite dramatically from 62.5 to 152.3, as shown in shown in Figure 11.7.

The same comments apply as in the previous example. That is, "average" production losses dramatically understate your exposure. For example, in the previous example, carrying a spare inventory of 8 could result in losses as high as $5,600,000. In this example, holding only 8 spares could increase your one-time production losses to a maximum of $16,800,000, though "average" losses might only be slightly less than $200,000. The bottom line is that the failure standard deviation

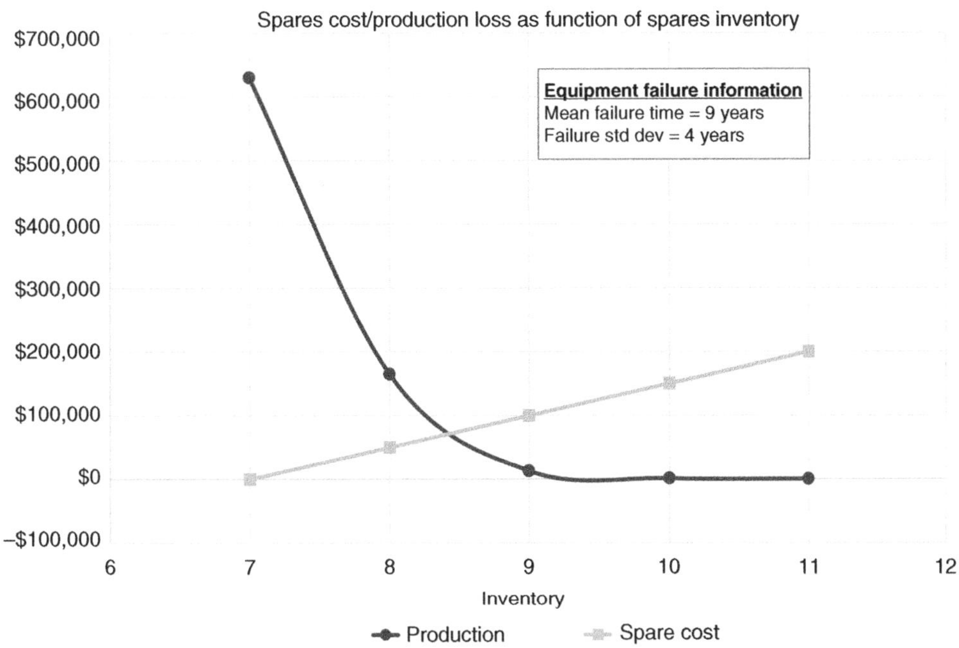

Figure 11.7 Increased Failure Time Standard Deviation

increase would likely lead a prudent organization to increase spares inventory by 1 from 9 to 10 to maintain the same risk exposure as in the previous example.

11.12.5.3 Increase Mean Failure Time to 12 Years

In this example, we will maintain the failure standard deviation at 4 years but increase the mean failure time to 12 years. The average number of failures decreased to 52.8 – the lowest of the three examples.

As one might expect, the longer mean failure time allowed the organization to reduce the spare holding requirement, as shown in shown in Figure 11.8. In this example, one could comfortably reduce holdings to 8 without any real increase in risk exposure and perhaps to 7 if one is open to more risk.

11.12.5.4 Effect of Lead Time

In our last example, we explore how changing lead times affect optimal spare holdings. We return to our original example, where the mean failure time was nine years with a standard deviation of three years. However, we vary the lead time to see where the breaking point is while maintaining a spare inventory level of 8. As our first example showed, that's roughly where we start to experience production losses.

Figure 11.9 clearly shows there's little effect until we hit a lead time of 19 weeks. Then, as we progress to 20, there's a dramatic change when we suddenly start to experience catastrophic production losses.

In thinking about the results from these four examples, there's nothing revolutionary. Our required inventory needs to be adjusted based on lead times, mean failure time, and standard deviations.

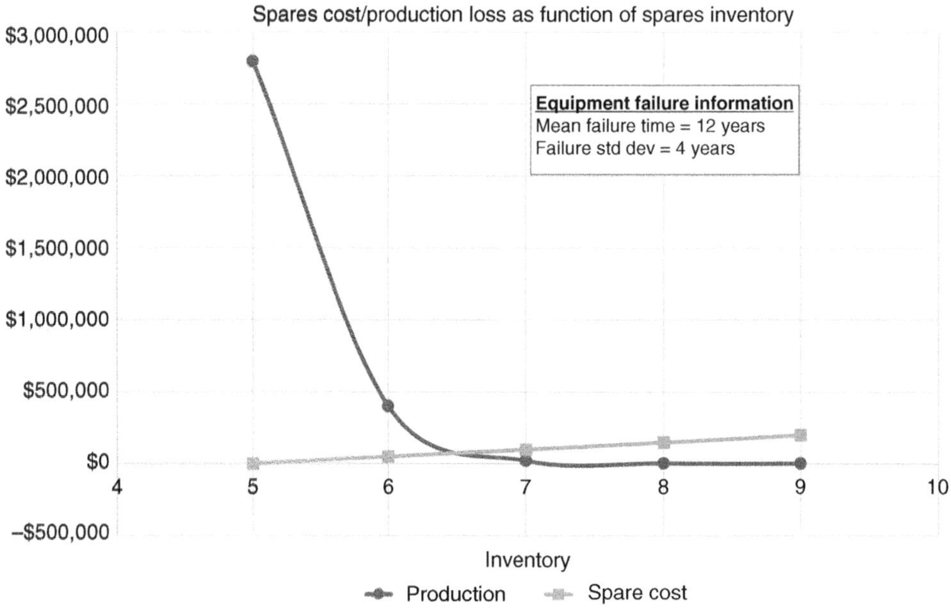

Figure 11.8 Increase Mean Failure Time

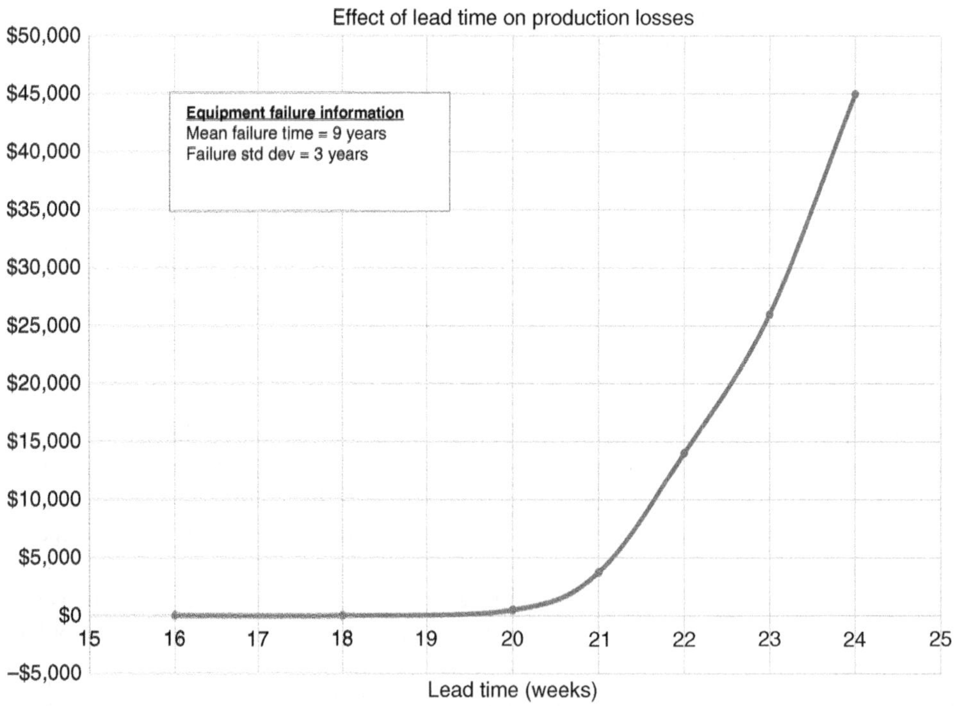

Figure 11.9 Production Losses vs. Lead Time

What the analysis showed was that the actual holding is quite sensitive to the "tipping" point. Therefore, it does make sense to perform this optimization to find the "sweet" spot.

11.13 Full-time Spare Parts Engineer Position

Now that we have a better understanding of how to use spare parts optimization in our organization, this is an important organizational note. Given that there is so much opportunity to streamline spare parts expenditures in most plants, it makes sense for many organizations to consider a full-time spare parts optimization position. This person would be in charge of performing the detailed prioritization of spare parts opportunities, as well as working with the optimization team, or even better, being directly part of the team performing the optimizations. In addition, this new team member would work closely with the warehouse to identify any constraints that should be added to the optimization problem definitions. As an organization gains more experience with its asset optimization program, it should strongly consider this idea.

11.14 Conclusion

In most organizations, spare part holdings are initially set by vendor recommendations or internal meetings. While a decent starting point, they are rarely updated as time progresses. This leads to either excessive spare parts holding costs or the real possibility that a spare may not be available when a unit goes down. This is suboptimal.

This chapter detailed a ranking process that organizations should consider adopting, in part or whole, to help them determine what spare parts should be optimized. This ranking system could then be melded with an organization's own internal spare parts work process.

The chapter then examined three general cases of Spare Parts Optimization – spare optimization for one type of spare for a single piece of equipment, as well as Intra- and Inter-Unit Spare Optimization. These last two optimizations were similar in that they both looked at multiple spare types for one or more equipment. The only difference is that Intra-Unit Optimization is concentrated within one process unit, whereas Inter-Unit Optimization is multiunit.

There was also an example presented where we determined the optimal spare part holding for a common spare used to support the same piece of equipment in multiple process units.

Whereas we might have thought that the Intra- and Inter-Unit examples would have yielded interesting and unexpected findings, the reality was that the single equipment optimization was the most enlightening. We learned that failure standard deviation is super important in determining the optimal spare holding levels. In addition, it was obvious that maximum production loss constraints were a much more sensitive measure than availability constraints. Unit margin also plays a significant role.

Companies should consider the addition of a full-time spare parts optimization engineer as their program matures. This person would be responsible for managing the spare parts prioritization process, identifying additional problem constraints, as well as working with the warehouse and procurement teams.

12

Task/Resource Optimization

Each process plant spends an enormous sum of money on maintenance tasks – many of them routine. Of course, these tasks consume considerable resources, which are often available in limited quantities.

The determination of task frequency is normally set once, years in the past, and rarely changed after that initial setting. However, plants are dynamic. They age, potentially needing more attention, or the demand on maintenance resource pool time allocations could change over time. In addition, technologies could become available that are better able to pinpoint issues at longer task intervals. In short, determining the optimal task intervals and the resources required to execute those tasks is an optimization problem that should be revisited periodically.

Thus, task/resource optimization attempts to determine the proper maintenance task timing that balances restoring equipment to acceptable levels and thus, protecting future availability against the resource costs to perform those activities.

Task timing especially becomes important when there are significant constraints, either budget or labor. For example, if there is a limited supply of a given skillset at the plant, how do I deploy those resources, and where do I get the best bang for my buck? On the other hand, optimization could also determine what resources need to be added to maintain equipment at an optimal level effectively.

Task optimization addresses these challenges. It can be applied locally to single equipment or to as much equipment within a processing unit(s) as needed.

To make the problem easier to understand, we will artificially break down Task Optimization into the same three general categorizations as we did with Spares and Opportunistic Maintenance Optimization. Those are shown below:

- Single Equipment Task Optimization
- Intra-Unit Task Optimization
- Inter-Unit Task Optimization

The first category is simple and only concerns optimizing tasks for a single piece of equipment within a single process unit. Similar to the discussions in previous chapters, this problem rarely needs full optimization since it can usually be solved via other means. However, as before, we include it to seed our discussion.

Intra-Unit Task Optimization is a fancy way of saying that we are going to evaluate maintenance tasks/resources that affect numerous pieces of equipment within a single process unit. On the

Plant Optimization in the Process Industries: Incorporating Equipment/Assets in the Decision-Making Process, First Edition. Marty Moran.

other hand, inter-unit optimization is the more general problem where we optimize tasks/resources that touch many different process units.

As discussed with previous optimization problems, the distinction between the Intra- and Inter-Optimization problems has more to do with our twin technologies of Optimization and Monte Carlo. In the Intra-Unit case, we only have to call one Monte Carlo process model, whereas with multiple units, we need to call each unit where a task is being done. It's more complicated than the Intra-Unit case but is easily performed.

12.1 Example Problem Data

To illustrate the concepts, let's use an example where we have a group of rotating equipment experts who care for the largest compressors in our refinery and alert the plant to any significant issues. The major equipment exists in four different process units in the plant: FCC, Hydrocracker, Reformer, and Isomerization.

Resource skill levels are broken into three different categories: resource 1, 2, and 3. Each of these have different cost burdens to the plant, as well as the sheer quantity of resources available to the organization. Resources 1, 2, and 3 could be Mechanical, Electrical, and Instrumentation. On the other hand, it could be different skill set levels within any one of those particular categories. Or it could be defined in any way that the user wants. The bottom line is that we can include resource constraints in any fashion that makes sense for the problem at hand.

12.1.1 Basic Problem Data

The current task frequency, task time, and resource allocations for equipment A–H are shown in Table 12.1. Task frequency times are in months, whereas the time to execute that particular task are in days.

While the data presented below shows the percentage of Resource 1/2/3 required for each task, it doesn't have to be that way. For a given task, it could be 100% for one task for one discipline, yet 0% for another task. It just needs to accurately portray the resource requirements for that particular task.

Table 12.1 Example Problem Basic Data

Equipment	Unit	Task Frequency (months)	Task Time (days)	Resource 1 (%)	Resource 2 (%)	Resource 3 (%)
A	FCC	6.72	2	10	30	60
B	FCC	6.98	1	10	40	50
C	Isom	11.21	2	5	35	60
D	Reformer	13.30	1.5	10	30	60
E	Reformer	4.80	0.5	20	40	40
F	Hydrocracker	8.39	1	15	30	55
G	Hydrocracker	5.92	1	15	30	55
H	Hydrocracker	11.54	2	20	40	40

Table 12.2 Resource Pricing

Experience Level	Rate ($/day)
Resource 1	4000
Resource 2	2500
Resource 3	1200

Table 12.3 Example Problem Task Data

Equipment	Unit	Failure Time (years)	Failure Std Dev (years)	Repair Time (days)	Material Costs/ Repair ($)	Labor Costs/ Repair ($)
A	FCC	8	2	3	330,000	35,000
B	FCC	10	2.5	3	150,000	30,000
C	Isom	14	3.5	4	340,000	40,000
D	Reformer	9	2	3	500,000	30,000
E	Reformer	12	3	4	450,000	40,000
F	Hydrocracker	16	4	3	400,000	30,000
G	Hydrocracker	15	4	5	200,000	50,000
H	Hydrocracker	13	3	4	1,100,000	40,000

The rates for each resource level on a daily basis are shown in Table 12.2. Given these values, the optimizer will naturally want to minimize resource 1 and utilize resource 3 as much as possible. Of course, that will depend on how changing those resources affects failure rates.

Table 12.3 shows the failure meantime and standard deviation, repair time, one-time material costs if a repair is needed, as well as total labor costs for that job. Labor costs presented here are those if a failure should occur and are different from task labor costs.

12.2 General Approach

This section will discuss important aspects of our general approach in solving this problem. These concern equipment age, repair standard deviation, and how each equipment's availability will be modeled based on varying failure characteristics.

Two pieces of data were left out of this problem formulation that many might think are important – Equipment Age and Repair Standard Deviation. Let's address each of those issues separately.

12.2.1 Equipment Age

Equipment age definitely influences particular task frequency. However, for this problem, we wanted to have an "apples to apples" comparison of the other variables, so we set all ages to zero.

What age really does is change when we start our first task, rather than anything else. For example, if our ultimate task frequency is once a year, we wouldn't likely wait a year before starting our program if the part was not brand new. Rather, depending on the exact age, we might initiate our first task after only three months and then continue at a once-a-year frequency. To simplify the optimization, we left that nuance out of the problem definition. Of course, in real life, that's important, and it has to be included, but it was an unnecessary complication for our example.

Another important consideration regarding age has to do with how failure characteristics may change over time. For example, we might assume that the failure mean time is 15 years when a part is new but decreases to 10 years over some time horizon. The natural way to incorporate the effect of age and the necessity of increased task frequency as a part of age is through changing failure mean time frequency. We didn't do that here in this optimization, but it's easily doable.

Now, let's talk a few words about repair standard deviation.

12.2.2 Repair Standard Deviation

We also didn't include the effect of repair time standard deviation in our calculations. Strictly speaking, we would want to include that to be accurate, but that's going to have a fairly insignificant effect on our final answer for these examples. For that reason, we excluded it.

Now that we have addressed the two pieces of data that we have intentionally decided to exclude from our problem formulation let's consider an important issue. That is, how we build the relationship between task timing and equipment reliability.

12.2.3 Relationship Between Task Timing/Availability

To avoid problems with numerical accuracy that occur naturally with Monte Carlo mathematical modeling, we separated the optimizer from directly interacting with a Monte Carlo algorithm. That was necessary, as noted in previous chapters since GRG2 requires precise and unalterable derivatives when determining a search direction. Monte Carlo's methodology can't provide that because it is a statistical methodology, and there's natural variation from one run to another.

Actually, separating the problem in this fashion is a more efficient way of solving the problem, anyway, since it removes the calculation overhead that exists when using Monte Carlo. Hence, the problem solution runs faster with this approach. In addition, there is no loss of accuracy. Therefore, it's a "win-win."

We achieved this goal by using Excel to mimic a Monte Carlo simulation through the use of Data Tables. We then used the failure mean time and standard deviation from that exercise to determine an approximate availability as monthly task timing was varied. The resulting data set allowed us to regress an exact equation between availability and task timing, which was then used directly in the optimization.

12.2.4 Availability Constraints

One of the main constraints for many optimization problems in this book has been minimal process unit availability. While it makes theoretical sense to include it for these problems since if one reduces task timing low enough availability might be impacted, it's not practically required given the data set used for this problem. After evaluating the results from all the optimization runs in this chapter, what one learns is that due to the relatively high repair costs for a failure, the optimization determines task frequencies that result in exceedingly high equipment availability – especially when compared to the much cheaper task labor costs. Nonetheless, while we didn't include availability constraints here, there's no reason why it might not make sense for other problem formulations.

Now, let's detail the optimization definition for each of our three different optimization problems – Single Equipment, Intra-Unit, and Inter-Unit Optimization.

12.3 Single Equipment Task Optimization

In this particular example, we would like to optimize task frequency for equipment C – the Isom recycle gas compressor.

12.3.1 Independent Variables

The independent variable for this problem is:

- The task frequency for equipment C.

12.3.2 Dependent Variables

There is one dependent variable equation used to create the relationship between task timing and equipment availability for equipment C, given its failure characteristics. There are also dependent variable calculations associated with determining the cost associated with each resource pool.

12.3.3 Constraints

Since we are only evaluating one equipment task, constraints won't be included in this example. However, we will incorporate constraints for the remaining examples in this chapter.

12.3.4 Objective Function

The goal of this optimization is to determine the maximum profit.

Labor costs will be balanced against the possibility of revenue loss due to lost production. Variables that influence those are unit margin and labor costs.

12.3.5 Results

Eight model runs were ultimately conducted with results shown in Table 12.4. The rows in the table are broken into five major segments.

In the first segment, task frequency and time to complete the task are shown. In addition, the number of failures recorded over a 20-year time frame, as well as the total mean repair time. It should be noted that task frequency is the optimizer output.

In the second segment, total failure repair costs (material, labor, and production loss) are calculated based on the number of failures over 20 years, as noted from the previous segment. The second segment also shows the total task costs and then sums these two different cost values together to produce a final total. This second value is what the optimizer is trying to minimize to produce maximum profit.

The third segment shows the material and labor values for a repair if the equipment should fail. In addition, the unit margin is noted.

The last two segments are related to resources. The first shows the percentage of each resource required for the task, while the second shows the optimizer-calculated total costs for that resource.

Table 12.4 Equipment C Optimization Results

	Base	Case 1	Case 2	Case 3	Case 4	Case 5	Case 6	Case 7
Task Frequency (months)	9.42	7.06	9.83	9.58	11.05	9.16	9.94	10.67
Task Duration (days)	2	1	2	2	2	2	2	2
Failures (20 years)	0.429	0.378	0.439	0.433	0.469	0.423	0.442	0.460
Repair Time	4	4	4	3	4	4	4	4
Failure Cost	$176,089	$154,902	$180,002	$170,069	$126,714	$186,327	$181,104	$188,400
Total Task Costs	$89,683	$59,860	$95,745	$88,154	$76,425	$92,238	$85,000	$79,200
Total Costs	$265,772	$214,762	$275,747	$258,223	$203,139	$278,565	$266,104	$267,600
Material Cost	$340,000	$340,000	$340,000	$340,000	$200,000	$340,000	$340,000	$340,000
Labor Cost	$40,000	$40,000	$40,000	$30,000	$40,000	$40,000	$40,000	$40,000
Margin	$7,500	$7,500	$7,500	$7,500	$7,500	$15,000	$7,500	$7,500
Resource 1 %	5%	5%	10%	5%	5%	5%	5%	5%
Resource 2 %	35%	35%	40%	35%	35%	35%	35%	35%
Resource 3 %	60%	60%	50%	60%	60%	60%	60%	60%
Resource 1 Cost	$10,191	$6,802	$19,540	$10,017	$8,685	$10,482	$9,659	$9,000
Resource 2 Cost	$42,803	$28,570	$46,895	$42,073	$36,476	$44,023	$40,568	$37,800
Resource 3 Cost	$36,688	$24,488	$29,310	$36,063	$31,265	$37,734	$34,773	$32,400

The first column is the base case, while the seven successive columns are cases where a certain variable is modified, as highlighted, to understand its sensitivity to the base case results. The particulars for those cases are described right after the table is explained.

The seven cases were as follows:

- Case 1 – Reduce task time from two days to one
- Case 2 – Change resource allocation percentages
- Case 3 – Shorten task repair time from four days to three
- Case 4 – Reduce failure material cost from $340,000 to $200,000
- Case 5 – Increase Isom margin from $7500 to $15,000 per day
- Case 6 – Impose a total task cost constraint of $85,000
- Case 7 – Add Resource 1 constraint of $9000

Each of the individual case results are discussed below.

12.3.5.1 Case 1 – Reduction in Task Frequency Time

In this case, task time is reduced from two days to one. In essence, reducing the time to complete the task reduces its cost to the organization. In this particular case, the optimizer responds by increasing task frequency, which has the effect of reducing the number of failures over 20 years. However, even with an increased task frequency, the total cost to the organization (failures plus task costs) is less than the base case.

12.3.5.2 Case 2 – Increase Resource Allocation of More Expensive Resources

In this case, we increase the resource allocation percentages for both resources 1 and 2 while reducing resource 3. This has the net effect of making this task more expensive for the organization since those resources are more costly on a relative basis. The optimizer responds by slightly delaying task frequency. Unfortunately, the number of failures goes up, but that's due to a more expensive resource pool. In this case, the optimizer tries to do the best it can when confronted with more expensive resources.

12.3.5.3 Case 3 – Shorten Task Repair Time

Here, we reduce task repair time from four days to three. It's important to note that while there is a repair labor savings, the material cost required for the repair remains the same. Therefore, the benefit is not as significant as one might imagine. The optimizer stretches out the task frequency very slightly, which has the effect of reducing task costs. There's actually a slight increase in failures since they aren't as costly anymore.

12.3.5.4 Case 4 – Reduce Failure Material Cost

In this case, we reduce failure material cost from $340,000 to $200,000. This has the effect of significantly reducing the pain of a failure. So, the optimizer responds by increasing task timing. Failures go up, but the total cost is still substantially less than the base case.

12.3.5.5 Case 5 – Increase Isom Margin

In this case, we will increase Isom's margin from $7,500 to $15,000 per day. Now, equipment failures are more painful. The optimizer responds by performing the task more frequently, which reduces the number of failures and is exactly what one would have expected.

12.3.5.6 Case 6 – Impose Total Resource Cost Constraint

In this case, we impose a total task constraint of $85,000 – which is less than the value from the base case. The optimizer rides this constraint and adjusts by reducing task frequency, thus saving money. There's a slight increase in failures, but there's nothing the optimizer can do about it.

12.3.5.7 Case 7 – Impose Resource 1 Constraint

In this case, we add a Resource 1 constraint of $9000 to the previous $85,000 total task cost constraint. Even though they appear different, in many ways, cases 6 and 7 are the same. That's because with only one piece of equipment to work with, the optimizer doesn't have another independent handle to manipulate. In this case, since the Resource 1 constraint is more demanding, the optimizer rides the Resource 1 constraint.

12.4 Intra-Unit Equipment Task Optimization

Intra-Unit Task Optimization optimizes task frequencies performed on different equipment within a given process unit. This might be representative of a limited resource available to a process unit, whereby the objective is to determine the best allocation for that scarce resource. The tasks could be the same task performed on multiple equipment. On the other hand, it could also be more than one task type performed on one or more equipment.

In this example, we will optimize the task frequencies for tasks associated with the three equipment in the Hydrocracker.

12.4.1 Independent Variables

The independent variables for this problem are:

- The task frequency for equipment F–H

12.4.2 Dependent Variables

There are dependent variable equations for each piece of equipment that detail the relationship between task timing and equipment availability, given their failure characteristics.

In addition, there are also calculations associated with determining the cost associated with each resource for each piece of equipment. Last, there are simple equations that sum those resource totals together.

12.4.3 Constraints

In this example, we introduce different cost constraints for total task cost, as well as individual constraints for each resource pool.

12.4.4 Objective Function

The goal of this optimization is to determine the maximum profit, subject to cost constraints.

12.4.5 Results

Eight model runs were conducted with results shown in Table 12.5.

The first segment shows the task frequency for the three equipment in the Hydrocracker, which are the optimizer outputs.

Table 12.5 Hydrocracker Unit Optimization Results

	Base	Case 1	Case 2	Case 3	Case 4	Case 5	Case 6	Case 7
Task Frequency – F	7.08	7.08	6.37	7.88	8.13	8.61	9.25	7.30
Task Frequency – G	4.96	4.96	4.10	5.55	5.73	6.09	6.57	5.34
Task Frequency – H	10.37	12.97	9.75	11.44	11.95	12.60	12.29	10.89
Failures (20 years) – F	0.355	0.355	0.343	0.368	0.372	0.381	0.392	0.358
Failures (20 years) – G	0.870	0.870	0.837	0.893	0.900	0.914	0.934	0.885
Failures (20 years) – H	0.133	0.154	0.129	0.141	0.145	0.151	0.148	0.137
Total Task Cost	$266,614	$245,872	$300,757	$240,000	$231,434	$218,517	$211,253	$285,067
Total Cost	$1,085,808	$1,000,425	$1,377,262	$1,089,201	$1,092,056	$1,098,400	$1,103,710	$1,118,964
Material Cost – H	$1,100,000	$500,000	$1,100,000	$1,100,000	$1,100,000	$1,100,000	$1,100,000	$1,100,000
Margin	$50,000	$50,000	$100,000	$50,000	$50,000	$50,000	$50,000	$50,000
Resource 1 Cost	$86,402	$78,994	$97,107	$77,813	$75,000	$70,833	$68,751	$124,145
Resource 2 Cost	$103,682	$94,793	$116,528	$93,376	$90,000	$85,000	$82,501	$103,745
Resource 3 Cost	$76,530	$72,085	$87,121	$68,810	$66,434	$62,683	$60,000	$57,177
Resource Allocation Change	No	No	No	No	No	No	No	Yes

The second segment is the number of failures over a 20-year time frame for that same equipment.

The third segment shows the task costs and the total cost, which includes total failure costs as well as task costs. The total cost is what the optimizer is trying to minimize to produce maximum profit.

The fourth section details material costs for equipment H only – if it should have a failure.

The fifth section shows the resource costs for all three resource pools. The last line is whether the resource allocation for the three types of labor has changed for any of the three equipment.

The first column is the base case, while the seven successive columns are cases where a certain variable is modified, as highlighted, to understand its sensitivity to the base case results. The particulars for those cases are described right after the table is explained.

The seven cases were as follows:

- Case 1 – Reduce equipment H failure material cost from $1,100,000 to $500,000
- Case 2 – Increase Hydrocracker margin from $50,000 to $100,000 per day
- Case 3 – Impose a total task cost constraint of $240,000
- Case 4 – Add Resource 1 constraint of $75,000
- Case 5 – Add Resource 2 constraint of $85,000
- Case 6 – Add Resource 3 constraint of $60,000
- Case 7 – Change resource allocation percentages (shown in the Table 12.6)

Each of the individual case results are discussed below.

12.4.5.1 Case 1 – Reduce Failure Material Cost

In this case, we reduce equipment H failure material cost from $110,000 to $500,000.

This has the effect of significantly reducing the pain of a failure – but only for equipment H. Similar to the Isom example, the optimizer responds by reducing task frequency for equipment H to minimize total task cost. Failures go up, but the total cost is still substantially less than the base case.

What's interesting, and maybe not intuitive, is that the optimization results for the two other equipment remained the same. That's because there were no constraints that affected all three pieces of equipment simultaneously that would have driven the optimizer to a different solution. In effect, the three pieces of equipment were independent of one another. However, as we'll see later, that can change once the optimizer starts to ride various constraints.

12.4.5.2 Case 2 – Increase Hydrocracker Margin

In this case, we increase Hydrocracker's margin from $50,000 to $100,000 per day.

Now, equipment failures are more painful for any of the three equipment, if they should fail. The optimizer responds by performing tasks more frequently for all three equipment, which increases total task costs. It determines the best combination of task frequencies that drives the total cost to a minimum, given this new margin.

12.4.5.3 Case 3 – Impose Total Resource Cost Constraint

In this case, we impose a total task constraint of $240,000. The optimizer rides this constraint and adjusts by reducing task frequency for all three equipment in the most optimal combination. Failures increase for all three equipment, but the optimizer is doing the best job it can give the constraint.

12.4.5.4 Case 4 – Add Resource 1 Cost Constraint

In this case, we add a Resource 1 constraint of $75,000 to the previous total task cost constraint. Here, the Resource 1 constraint is more demanding, so the optimizer rides this constraint, but it makes changes to all three equipment task frequencies to do so.

Table 12.6 Hydrocracker Unit New Task Resource Allocations

Original Values	Resource 1 (%)	Resource 2 (%)	Resource 3 (%)
Equipment F	15	30	55
Equipment G	15	30	55
Equipment H	20	40	40
New Allocations			
Equipment F	20	30	50
Equipment G	25	35	40
Equipment H	30	40	30

12.4.5.5 Case 5 – Add Resource 2 Cost Constraint

In this case, we add another resource constraint, this time a Resource 2 cost constraint of $85,000, while the two previous constraints remain in place. In this case, the Resource 2 constraint is the most demanding of the three constraints, and the optimizer responds by riding that constraint.

12.4.5.6 Case 6 – Add Resource 3 Cost Constraint

In this case, we add a cost constraint of $60,000 to resource 3 while keeping the three other constraints in place. Once again, this new constraint is the most limiting of the four constraints.

It's important to point out that while the addition of the three resource constraints shifted the solution to the new constraint just added, there's nothing inherent that forces that to happen. It's simply due to the values chosen.

Also, the optimizer only hit one constraint during these examples. This is somewhat of a fluke because the optimizer could have easily ridden multiple constraints at once in order to drive the cost to a minimum. In these cases, that wasn't necessary to find the minimum cost.

12.4.5.7 Case 7 – Increase Resource Allocation of More Expensive Resources

In this case, resource allocation percentages for Resources 1/2/3 have been changed for all three equipment. Table 12.6 shows the resource percentages for the prior runs versus the new allocations considered for this case.

After examining the table, what one notes is that allocations rise for resource 1 and 2, while decreasing for resource 3. Since resource 1 is more expensive than resource 2 and resource 2 is more expensive than resource 3, the allocation changes have the net effect of making a task more expensive to execute.

Thus, the optimizer responds by reducing task frequency. That has the unfortunate effect of increasing the number of failures, but that's the best the optimizer can do when confronted with more expensive resources.

12.5 Inter-Unit Equipment Task Optimization

This is the most general task optimization problem.

Inter-Unit Equipment Task Optimization optimizes the task frequencies performed on different pieces of equipment across multiple process units. This might be representative of a limited resource at a plant site, whereby the objective is to determine the best allocation of that scarce resource among several process units. Similar to the previous problem, the tasks could be the same task performed on multiple equipment. On the other hand, it could also be more than one task type performed on one or more equipment.

12.5.1 Independent Variables

The independent variables for this problem are:

- Task frequency for equipment A–H

12.5.2 Dependent Variables

There are dependent variable equations for each equipment that detail the relationship between task timing and equipment availability given the equipment's failure characteristics.

In addition, there are also calculations associated with determining the cost for each resource for each equipment. Last, there are simple equations that sum those resource totals together.

12.5.3 Constraints

In this example, we will conduct model runs that constrain total task cost as well as individual resource pools.

12.5.4 Objective Function

The goal of this optimization is to determine the maximum profit, subject to constraints.

12.5.5 Results

Ten model runs were conducted consisting of a base case and nine cases with various variables altered, as highighted, relative to the base case. The results presented in Table 12.7 are the model runs that were unconstrained. On the other hand, Table 12.8 shows the results where the optimizer runs up against various constraints.

Table 12.7 is broken into six different segments.

The first segment shows the task frequency for all equipment, which are the optimizer outputs.

Next is the number of failures over a 20-year time frame for the same equipment.

The third segment details material costs associated with a failure for equipment H, while the fourth section shows the optimizer-calculated resource costs for all three resource pools.

The fifth segment shows the task costs and the total cost, which includes total failure costs as well as task costs. The total cost is what the optimizer is trying to minimize to produce maximum profit.

The last segment is the margins for both the Reformer and Hydrocracker.

Next, we will discuss the results for the first three cases, which are unconstrained. They are as follows:

- Case 1 – Reduce equipment H failure material cost from $1,100,000 to $500,000
- Case 2 – Increase Reformer margin from $35,000 to $70,000 per day
- Case 3 – Increase Hydrocracker margin from $50,000 to $100,000 per day

Each of the individual case results are discussed below.

Table 12.7 All Equipment Optimization Results – Unconstrained Cases

	Base	Case 1	Case 2	Case 3
Task Frequency – A	4.99	4.99	4.99	4.99
Task Frequency – B	5.84	5.84	5.84	5.84
Task Frequency – C	9.42	9.42	9.42	9.42
Task Frequency – D	12.17	12.17	11.76	12.17
Task Frequency – E	4.21	4.21	3.80	4.21
Task Frequency – F	7.08	7.08	7.08	6.37
Task Frequency – G	4.96	4.96	4.96	4.10
Task Frequency – H	10.37	12.97	10.37	9.75
Failures (20 years) – A	0.515	0.515	0.515	0.515
Failures (20 years) – B	0.436	0.436	0.436	0.436
Failures (20 years) – C	0.429	0.429	0.429	0.429
Failures (20 years) – D	0.121	0.121	0.118	0.121
Failures (20 years) – E	0.342	0.342	0.332	0.342
Failures (20 years) – F	0.355	0.355	0.355	0.343
Failures (20 years) – G	0.870	0.870	0.870	0.837
Failures (20 years) – H	0.133	0.154	0.133	0.129
Material Cost – H	$1,100,000	$500,000	$1,100,000	$1,100,000
Resource 1 Cost	$186,196	$178,788	$189,063	$196,901
Resource 2 Cost	$303,935	$295,045	$307,621	$316,781
Resource 3 Cost	$242,180	$237,735	$244,392	$252,772
Total Resource Cost	$732,311	$711,569	$741,076	$766,453
Total Cost	$2,500,464	$2,415,080	$2,560,190	$2,791,918
Margin – Reformer	$35,000	$35,000	$70,000	$35,000
Margin – Hydrocracker	$50,000	$50,000	$50,000	$100,000

12.5.5.1 Case 1 – Reduce Failure Material Cost

In this case, equipment H failure material costs were reduced from $110,000 to $500,000.

Similar to the discussion for the Hydrocracker, this has the effect of significantly reducing the pain of a failure for equipment H. Once again, the optimizer responds by reducing task frequency, but only for equipment H, to minimize that particular task cost. Failures go up for that equipment, but the total cost is still substantially less than the base case. The results for all other equipment remain the same.

12.5.5.2 Case 2 – Increase Reformer Margin

In this case, we increase the Reformer margin from $35,000 to $70,000 per day.

Now, equipment failures are more painful for equipment within the Reformer, but there's no change for the equipment in the other process units. The optimizer responds by performing tasks more frequently for equipment in the Reformer, which increases total task costs for that equipment. It leaves all other equipment unchanged.

12.5.5.3 Case 3 – Increase Hydrocracker Margin

In this case, we increase the Hydrocracker's margin from $50,000 to $100,000 per day. The same thing happened this time for equipment in the Hydrocracker, as just discussed for the Reformer. That is, the optimizer decides to perform tasks more frequently for all Hydrocracker equipment, which increases total task costs. All other equipment results remain the same.

Now, let's turn our attention to the cases run where the optimizer rides different constraint limits. The first four segments of Table 12.8 are exactly the same as in the previous table. However, segment five shows the actual constraint limits. Those highlighted are variables where the optimizer hits constraint limits. The cases are as follows:

- Case 4 – Impose a total task cost constraint of $700,000
- Case 5 – Add Resource 1 constraint of $170,000
- Case 6 – Add Resource 2 constraint of $285,000
- Case 7 – Add Resource 3 constraint of $200,000
- Case 8 – Change equipment H material cost per repair while maintaining total task cost constraint
- Case 9 – Change resource allocation percentages while maintaining total task cost constraint

Each of the individual case results are discussed below.

12.5.5.4 Case 4 – Impose Total Resource Cost Constraint

In this case, a total task constraint of $240,000 is imposed. The optimizer rides this constraint by reducing task frequency for all equipment in the most optimal combination, thus saving the most amount of money while minimizing the effect of equipment failures. Failures increase for all equipment, but the optimizer does the best job it can give the constraint.

12.5.5.5 Case 5 – Add Resource 1 Cost Constraint

In this case, a Resource 1 constraint of $170,000 is added to the previous total task cost constraint. Here, the Resource 1 constraint is more demanding, so the optimizer rides this constraint but makes changes to all equipment to minimize total costs.

12.5.5.6 Case 6 – Add Resource 2 Cost Constraint

In this case, a Resource 2 constraint is added to the total task constraint. (The Resource 1 constraint has been removed.) In this case, similar to the previous case, the Resource 2 constraint is the most demanding constraint.

12.5.5.7 Case 7 – Add Resource 3 Cost Constraint

In this case, a Resource 3 cost constraint of $200,000 is added while keeping the total cost and Resource 2 constraints in place. Once again, this new constraint is the most limiting of the three constraints.

12.5.5.8 Case 8 – Reduce Failure Material Cost while Riding Total Task Constraint Limit

In this case, we revisit case 1, where we reduced equipment H material cost to $500,000. However, this time, we impose a total task cost constraint. The optimizer responds by increasing task frequency for all equipment, slightly increasing failure frequency, while riding the total cost constraint. However, the increase in failure cost is less than the savings in task costs.

12.5.5.9 Case 9 – Increase Resource Allocation of More Expensive Resources

In the last case, the resource allocation percentages for Resources 1/2/3 have been changed for all equipment, as shown in Table 12.9.

Table 12.8 All Equipment Optimization Results – Constrained Cases

	Base	Case 4	Case 5	Case 6	Case 7	Case 8	Case 9
Task Frequency – A	4.99	5.32	5.55	5.44	6.72	5.11	5.78
Task Frequency – B	5.84	6.12	6.28	6.31	6.98	5.94	6.36
Task Frequency – C	9.42	9.76	9.73	9.98	11.21	9.55	10.28
Task Frequency – D	12.17	12.40	12.56	12.49	13.30	12.25	12.42
Task Frequency – E	4.21	4.40	4.74	4.50	4.80	4.28	4.17
Task Frequency – F	7.08	7.37	7.76	7.45	8.39	7.18	7.56
Task Frequency – G	4.96	5.17	5.46	5.23	5.92	5.04	5.19
Task Frequency – H	10.37	10.78	11.43	10.97	11.54	13.11	10.43
Failures (20 years) – A	0.515	0.535	0.548	0.542	0.611	0.522	0.561
Failures (20 years) – B	0.436	0.446	0.451	0.452	0.475	0.440	0.454
Failures (20 years) – C	0.429	0.438	0.437	0.443	0.473	0.432	0.450
Failures (20 years) – D	0.121	0.122	0.124	0.123	0.130	0.121	0.123
Failures (20 years) – E	0.342	0.347	0.355	0.349	0.357	0.344	0.341
Failures (20 years) – F	0.355	0.360	0.366	0.361	0.377	0.356	0.363
Failures (20 years) – G	0.870	0.878	0.889	0.880	0.908	0.873	0.879
Failures (20 years) – H	0.133	0.136	0.141	0.138	0.142	0.155	0.134
Material Cost – H	$1,100,000	$1,100,000	$1,100,000	$1,100,000	$1,100,000	$500,000	$1,100,000
Resource 1 Cost	$186,196	$178,032	$170,000	$174,839	$156,491	$175,923	$187,503
Resource 2 Cost	$303,935	$290,590	$279,564	$285,000	$253,626	$290,276	$299,307
Resource 3 Cost	$242,180	$231,378	$223,094	$227,055	$200,000	$233,801	$213,191

	Base	Case 4	Case 5	Case 6	Case 7	Case 8	Case 9
Total Resource Cost	$732,311	$700,000		$686,894	$610,117	$700,000	$700,001
Total Cost	$2,500,464	$2,501,964	$2,506,243	$2,503,523	$2,526,660	$2,415,275	$2,520,955
Resource 1 Cost Limit		$200,000	$170,000	$200,000	$200,000	$200,000	$200,000
Resource 2 Cost		$325,000	$285,000	$285,000	$285,000	$325,000	$325,000
Resource 3 Cost		$300,000	$300,000	$300,000	$200,000	$300,000	$300,000
Total Resource Cost Limit		$700,000	$700,000	$700,000	$700,000	$700,000	$700,000
Resource Allocation Change	No	No	No	No	No	No	Yes

Table 12.9 All Equipment New Task Resource Allocations

Original Values	Resource 1 (%)	Resource 2 (%)	Resource 3 (%)
Equipment A	10	30	60
Equipment B	10	40	50
Equipment C	5	35	60
Equipment D	10	30	60
Equipment E	20	40	40
Equipment F	15	30	55
Equipment G	15	30	55
Equipment H	20	40	40
New Allocations			
Equipment A	15	35	50
Equipment B	10	50	40
Equipment C	15	30	55
Equipment D	5	40	55
Equipment E	15	30	55
Equipment F	20	25	55
Equipment G	10	40	50
Equipment H	15	35	50

There's no apparent pattern with these changes. In some cases, resource allocations for more expensive resources increase for various equipment while decreasing for others.

For the most part, the optimizer responds by decreasing task frequency, which appears to be in response to the $700,000 total task limit. In the end, the total cost is slightly more than the base case, but that difference is actually less than the total cost task reduction. Once again, the optimizer responds by doing the best job it can give the limitations.

12.6 Conclusion

Maintenance task frequency timings are usually set once and rarely, if ever, changed after that initial setting. However, it makes sense to reevaluate task timings periodically since plants and equipment age, unit configuration or equipment within the unit could change, and condition monitoring technologies may evolve, which potentially allows a reduction in task frequencies to achieve a given level of risk reduction.

What the examples in this chapter demonstrated was that task/resource optimization determines the best balance between task costs and possible failure costs. It also does an excellent job of allocating scarce resources among tasks or determining what resources are actually required to maintain equipment at an acceptable risk level.

Task Optimization can be performed at the unit level, across several units – especially where there are people or other resource constraints, or even at the individual equipment level.

13

Tankage Determination/Optimization

Earlier in the book, an entire chapter was devoted to determining the optimal process unit availability for each process unit within a plant. In that example problem, there were intermediate storage tanks located in the usual places for the typical oil refinery. For example, there were intermediate feed tanks for the Isomerization unit and Reformer, after processing in the Naphtha Splitter. In addition, there was a storage tank for atmospheric gas oil (AGO) after the crude unit, in essence, a feed tank for the FCC.

As expected, storage tanks were assumed to be of a fixed size. That's normal since a tank is designed and constructed once. It is not variable. Of course, a tank can be repurposed in certain situations. However, most of the time, we are normally stuck with what we have.

Nonetheless, understanding how tankage affects overall production is important, and shouldn't be taken for granted. Let's explore that topic further.

13.1 Why Tankage Size Matters

From an operations perspective, tankage is desirable since it is a buffer. It smooths out process issues when a unit runs into problems. It is fundamentally a buffer against unreliability.

For example, consider the situation when unit A feeds unit B directly. If both units had 100% availability, then the combined system together would also have 100% availability. However, in this arrangement, even if unit B retains 100% availability, overall system availability declines as unit A's availability declines. For example, if unit A's availability decreases to 90%, then the overall system availability will be 90%.

Let's take another example where neither unit has perfect availability. Let's assume that unit A has an availability of 94.3%, unit B's availability is 92.0%, and the nominal feed flow rate for each unit is 10,000 barrels per day (BPD).

In a series unit configuration system availability can never be greater than the final unit's availability. In the previous example, the system could get to 100%, but that was only because unit B had an availability of 100%. However, in this new example, unit B's availability has now decreased to 92%. Therefore, in this example, system availability can never exceed 92%. However, it will be less because system A does not have 100% available. For example, if there is no tankage between units A and B, then system availability is simply the product of the two availabilities multiplied together:

$$94.3\% \times 92\% = 86.8\%$$

Plant Optimization in the Process Industries: Incorporating Equipment/Assets in the Decision-Making Process, First Edition. Marty Moran.

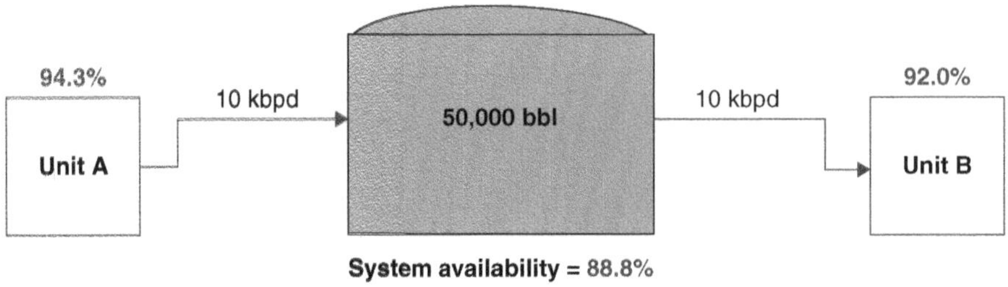

Figure 13.1 50,000 bbl Tank

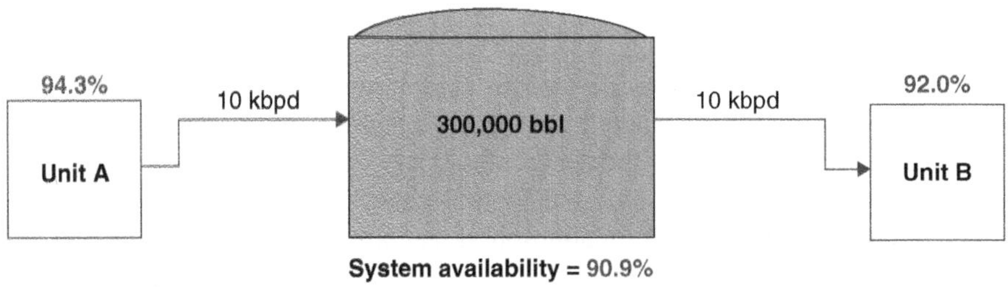

Figure 13.2 300,000 bbl Tank

Since 86.8% availability isn't very impressive, so if we want to increase the overall system availability, and it isn't practical to improve the individual unit availabilities, then another option is to build a storage tank between the units. Of course, this requires a significant capital cost.

If we install a 50,000 bbl tank between the units, or 5 days of storage capacity, as shown in Figure 13.1,[8] then unit B can continue operating in situations when unit A goes down. In this arrangement, overall system availability could increase to 88.8% with this new tankage, which is a significant improvement over the previous arrangement.

If the tankage size increased to 300,000 barrels (30 days of storage), then system availability would correspondingly increase to 90.9%, as shown in Figure 13.2.[9]

It's fairly obvious that as tankage size increases, system availability also increases. The limit to increasing system availability is always the availability of the last unit (in this case, unit B at 92%), though we would need infinite tankage to achieve that goal. In this case, unit A would fundamentally be entirely isolated from unit B. The downside is that large tanks cost a considerable amount of money. For that reason, there truly is an optimum where one can achieve the greatest production for the lowest tank capital cost.

13.2 Example Problem Overview

While we have broken many of the previous optimization problems into single equipment, intra-unit, and inter-unit optimization cases, we will take a different approach to tankage determination/optimization. Rather, we will show more general cases to illustrate the basic tradeoffs.

Our example problem will be similar to those just discussed, where there is an upstream process unit, tank, and downstream process unit in series. It's important to note that once one chooses upstream availability, tankage size, and downstream availability, the quantity of level excursions either below a specified minimum or above a desired maximum is fixed. Since a plant can only tolerate so many of these events, just plotting the Monte Carlo results can be enough to make informed decisions on what size tankage your facility needs. Since this is so important, direct optimization isn't always needed. Therefore, our example problems in this chapter will be a series of improvised Monte Carlo runs performed using Excel's Data Tables. We say "improvised" because fundamentally, we are using statistical analysis via Excel's Data Tables to "mimic" what a Monte Carlo type of analysis would have determined.

There is one place where optimization has a role and will be discussed in the second set of cases. It occurs typically when there is a significant mismatch between the upstream and downstream process unit availabilities. In those instances when we want to improve system availability, there is a choice as to whether to implement just an improvement project, just tankage, or some combination of the two. This is a classic optimization problem.

13.2.1 Two Cases

The examples in this chapter will consist of two different cases with multiple runs in each case.

In the first set, upstream and downstream process units will have the same availability with tankage size being adjusted to show how often tankage level exceeds 10%, 30%, 70%, and 90% for each tankage size, as well if we encounter either a tankage empty or overflow event. In subsequent model runs, we will vary the availability, performing the same exercise, though the availability will remain the same for both the upstream and downstream units. Ultimately, we will examine availability cases ranging from 95% to 99% in 0.5% increments.

In the second set of cases, we will maintain an upstream availability of 99% for all cases but vary downstream availability. Once again, we will adjust tankage size to show how often we encounter level events. Downstream availability will be varied from 95% to 99% in 1% increments.

The second set of cases demonstrates the basic tradeoff that exists between an improvement project that improves the downstream unit's availability versus a project to increase tankage size. This becomes especially important as downstream availability decreases.

13.2.2 Tankage Problem Details

The example problem was built in Excel and consisted of an incoming/outgoing flow from the upstream/downstream process units. Tankage level was calculated on a daily basis over a five-year time frame based on the difference between incoming and outgoing flows, of course taking into consideration assumed tankage size. Level excursions of <0%, <10%, <30%, >70%, >90%, and >100% were recorded. A Monte Carlo simulation of 1000 iterations was performed using Excel's Data Tables.

Process unit failures were assumed to follow a lognormal distribution with the mean being 14 hours and a few hour variation around the mean. It's important to note just how drastically failure characteristics affect how many level events are experienced. When using this technology for your plant, you will want to ensure that these values are extremely accurate, so you can choose the correct tankage size. For the examples in this chapter, what's more important are not the absolute values, but rather the visualization of the relative tradeoffs.

13.3 Same Availability for both Upstream and Downstream Process Units

In this case, upstream and downstream process units had the same availability. Tankage size was then varied to create the trend lines in Figure 13.3. Values on the Y-axis are the number of days in the five-year simulation time period where there were level excursions below 30%. (We could have also chosen events above 70% also, since they are essentially equal to one another.)

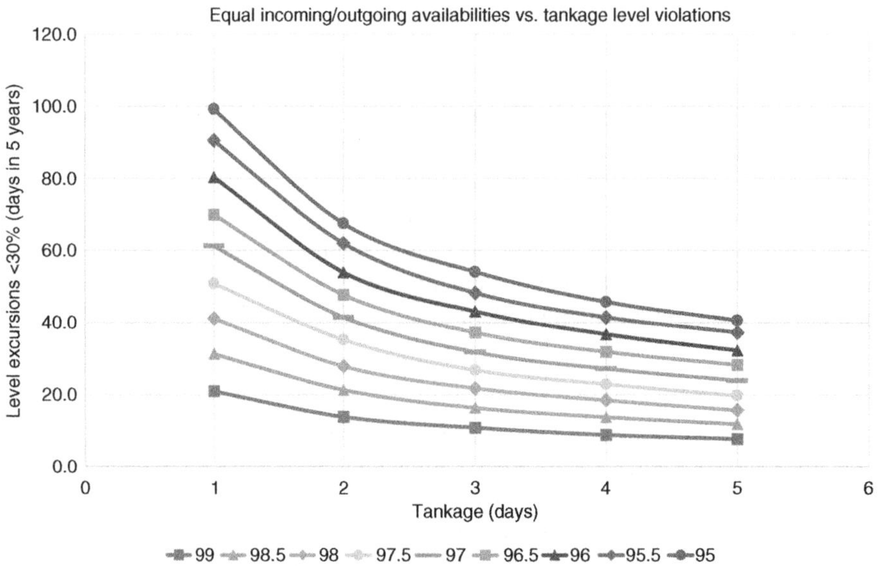

Figure 13.3 Tankage Level Violations – Equal In/Out Availabilities

While the values of <30% level excursion were plotted in the graphs, results were also tabulated for excursions below 10% and tank empty events (not shown). In general, those events disappeared altogether when <30% level events dropped below 10 and increased correspondingly as <30% events increased. For a more thorough analysis, one would want to plot all the relevant tankage levels.

Obviously, the <10% and tank empty are more dramatic events, and no plant can realistically tolerate either, so they would normally predominate our concerns for our analysis process. However, the number of <30% events can serve as a useful proxy for those other more extreme events, since they tend to correlate well with one another.

As would be expected, there are fewer level excursions at the higher availabilities. In addition, the range of excursions is narrower than the lower availabilities cases. For the 99% case, there is only a range of 10 events over a four-day tankage volume range, whereas for 95% availability, the range is 60 events.

A plant would likely target a maximum number of excursions and then determine what tankage size would be required. For example, if the plant agreed that they would tolerate no more than 20 excursions, which would be a horizontal line on the graph above, it's clear that only a single day of storage would be required for the 99% availability case. On the other hand, two days would be required for 98.5% availability; three days for 98%, and five days for 97.5%. All availabilities below that would require an improvement project on either the upstream or downstream unit, in addition to possible tankage changes.

13.4 Downstream Availability Variable with Constant Upstream Availability

In this case, the availability of the upstream process unit was held constant at 99%, while the downstream unit's availability was altered. Tankage size was then varied, as shown on the X-axis in Figure 13.4.

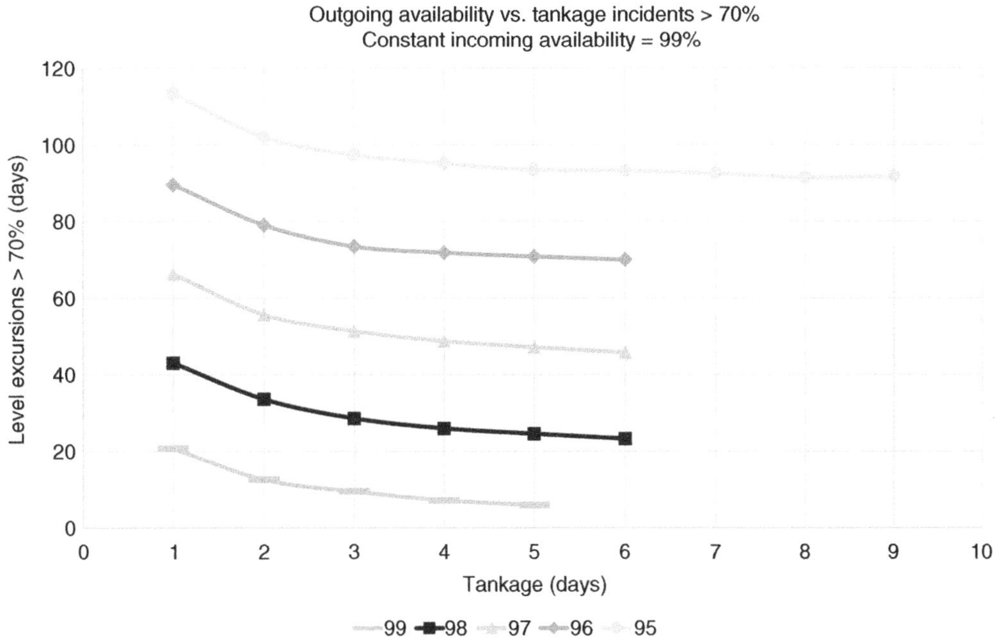

Figure 13.4 Tankage Level Violations – Variable Outgoing Availabilities

Values on the Y-axis are the number of days in the five-year simulation time period where there were level excursions above 70%.

In this case, monitoring higher levels is far more important than lowering tank levels because system failures are much more likely to be in the downstream unit since the upstream process unit has such a high availability. Thus, the tank backing up is much more likely than draining.

As would be expected, as downstream availability increases, the number of level excursions decreases. What's interesting to note is how dramatic the difference is between the 99% and 95% downstream cases.

Also, there is relatively little variation in level excursions, even over a fairly wide tankage range for an upstream/downstream availability combination. That's because the effects of increasing tankage size have smaller and smaller incremental effects as tankage size increases. Of course, level excursions could be eliminated altogether if tankage was increased to extraordinary sizes. That is neither practical nor cost-effective.

However, in the normal ranges, there are limits. For example, for the 95% availability runs, excursion values change from a low in the lower 60s to a high in the mid-80s for reasonable tankage sizes. That's a fairly small range. Another thing to note in the graph is that there's no overlap in the excursion range between 95% and 99% availabilities.

While we have been able to make decisions about tankage size so far just by looking at Monte Carlo results, for the lower availabilities, there is one optimization possibility. Consider the 95% availability case. Does it make more financial sense to increase tankage size, incurring its cost, when it can only add so much incremental value, or does it make more sense to implement an improvement project to increase unit availability to 97% or something similar? As always, this is an economic decision.

On the other hand, when looking at the 97% case, maybe it truly is more economical to find a tankage alternative to avoid a costly improvement project to increase unit availability much beyond its current 97% value. Once again, it's purely an economic decision based on the difference between a tankage increase or the cost of an improvement project.

13.5 Conclusion

Intermediate storage tankage provides both operational flexibility and the ability to "buffer" the unreliability of two process units operating in series. Tankage between two process units will shield the downstream from the unreliability of the upstream unit. For example, there may be times if unit A comes down that unit B will still be able to continue processing since there is still feed material in the intermediate tank. If the tank had not been there, then unit B would have shut down. However, tankage is an added expense that needs to be cost-justified.

In the case of similar upstream and downstream availabilities, examples showed that tankage requirements to maintain a given level of tankage level excursions increase as both process unit's availability decreases.

In the case of an upstream process unit with high availability, the number of level excursions increases quite dramatically as the downstream availability is reduced. In fact, in these cases, there are optimization possibilities between increased tankage and/or a process unit improvement project – especially as downstream availability decreases.

14

Improving Availability

At this point in the book, we've been introduced to the most important asset decisions many process industry companies face – when to take a turnaround and what equipment to restore, how to optimize maintenance task resources, what improvement projects can be cost justified, the best mix of spare parts, the optimal choice of availability targets for each process unit, and how to properly allocate storage capacity, among others.

So, what if your company is facing a situation where one or more of your process units cannot achieve the availability targets chosen for them? If so, how do you proceed?

This chapter will discuss your options.

While we have referred to it largely as Monte Carlo simulation technology up to this point in time, it is also commonly known as Reliability, Availability, and Maintainability (RAM) modeling. Each of the three words is important.

Reliability refers to the ability of a single piece of equipment, a collection of equipment, or even an individual subcomponent within one piece of equipment to be able to perform its intended function. That's a fancy way of saying that it works like you expect it should.

On the other hand, **Availability** is the percentage of time that a collection of assets is available to make product.

Last, **Maintainability** is the time that it takes to restore a single piece of equipment, a collection of equipment, or an individual subcomponent within one piece of equipment back to achieve its intended function or work as it should be.

Of these three, Availability is the net result of a system of equipment with given Reliability and Maintainability characteristics. In other words, Reliability and Maintainability are inputs, while Availability is an output. Other inputs include the system configuration along with the quantity of tankage.

We will learn that there will be times when one can improve the Reliability and/or Maintainability of equipment. When we do that, Availability will also improve.

However, in other situations, it might take a configuration change (not always possible) or an increase in storage capacity (once again, often not possible) to improve availability.

So, what are our options if our system cannot meet its availability targets?

Plant Optimization in the Process Industries: Incorporating Equipment/Assets in the Decision-Making Process, First Edition. Marty Moran.

14.1 Options to Improve Availability

There are six fundamental ways to potentially improve availability. Four have previously been mentioned:

1) Increase Equipment Reliability
2) Decrease Maintainability Time
3) Increase Spare Parts Availability
4) Increase Storage Capacity

The other two are related to configuration – adding redundancy or altering the configuration.

Here are a few words about each option before we discuss how to evaluate each one and which one might be the best scenario for a given situation.

14.1.1 Increase Equipment Reliability

Most people believe that when availability doesn't meet its target, it's because equipment reliability is the issue. In other words, the equipment is failing too often. Sometimes that is true, sometimes not. It all depends on the situation.

In fact, when equipment actually does fail, it may not be the equipment that is the issue, but rather how the equipment is being operated. To the outside person, it may seem like the same thing, but in reality, it's due to equipment maloperation.

These are two vastly different issues that must be addressed in two different manners. For example, maybe there is nothing wrong with the equipment, but the operational philosophy is incorrect. If operated correctly, it would have performed its intended function.

The good news is that equipment operates better today than it ever has. That's important to put into a historical perspective.

14.1.1.1 How Equipment Reliability Has Improved over Time – Implications for Asset Management

I always remember my father telling me a story about how poorly cars performed in his youth back in the 1930s/1940s. He said that engineers eventually solved many of the design problems he saw when he was growing up. However, it took time. As the industry progressed, problems that were once unsolvable were able to be addressed as new materials and techniques became available. He mentioned that I was spoiled since cars were much more reliable in my high school years during the 1970s than in his day.

But I also saw automobile performance improve considerably during my life. When I was in high school, cars needed a yearly tune-up, at a minimum, to replace spark plugs, the condenser, and adjust the timing. Now, a car can go 100,000 miles before it needs a tune-up.

Rust was a constant problem and often the limiting factor in the lifetime of many cars of that era – especially in the northern climates where salt was used to melt ice and snow. Today, nobody remembers those problems.

In fact, when looking at industry overall, equipment reliability for virtually all types of equipment has improved over time. That is as true in the process industries as it was for automobiles – especially as we gained more experience operating that equipment.

14.1.1.2 Equipment Reliability Issues that Are really People Issues

When one reflects on the actions from many Root Cause Analyses (RCA) – especially those in mature plants – one realizes that they are often people issues and not fundamental equipment issues.

For example, one common action from an RCAs is the failure to follow a procedure. Other times, it might be that a procedure was incorrect to start with.

Inexperience is another area. The reality is that a process plant is an extraordinarily complex operating environment, and it often takes years of time to become proficient. Sometimes, pure training can bridge the gap.

However, my experience is that the real bottleneck is gaining the innate hands-on experience that allows a person to "connect the dots" to really understand plant operation in their "gut." That usually takes calendar time but is essential if we want to achieve the production and availability goals that management desires.

Not too long ago, I was working with a board operator on a process unit where he was relatively new, and he had made a change in tower reflux to counteract a rising overhead impurity. It was absolutely the right action, but he overdid it.

Unfortunately, he introduced an even worse upset into the tower, which affected a number of downstream units. Fundamentally, it was due to a lack of process understanding specific to that distillation tower.

In his case, the incident occurred because he was a new board operator. However, there are other cases, where the same thing might have occurred, but for different reasons.

During one portion of my career, I was an Advanced Process Control and Optimization engineer. This technology adjusted the major operating variables in a unit to maintain compositions. It would also run the plant up against its most important operating constraints to maximize profitability. That's certainly the right business objective, but it comes at a certain cost.

That is, when advanced technology is in the background making many of the important moves, the operator slowly becomes rusty since they are not always tweaking the knobs as they did in previous eras.

Would we want to go back to that earlier time period? Absolutely not, since our units run better than ever. However, we must recognize our dependency and guard against this loss of experience by building even better simulation and training environments than those used in the past. Given digital capabilities nowadays, that's an attainable goal.

What does this mean for equipment availability? While there's a tendency to blame the equipment, the reality nowadays is that people are often our weak link.

I always remember a presentation that I attended where gentlemen mentioned that in the airline industry, equipment is rarely an issue when accidents occur nowadays. Rather, it's "pilot error." The same is becoming truer in the process industries as our equipment has become more reliable. Therefore, we must learn to manage people even better than we did in the past.

What that means as we build out our asset management program is that we will need to include factors in our models to properly account for procedure issues, lack of experience or training, routine maintenance that wasn't done properly, or a lack of operational understanding.

14.1.2 Decrease Maintainability

Another way to increase Availability is to increase Maintainability. For many, that implies the time to repair the equipment. That's certainly one important aspect of maintainability. However, decreasing the time to restore equipment and returning it to operation is more than decreasing wrench time. In fact, it consists of several different facets, as noted below:

- Cool the plant down
- Isolate the equipment

- Repair the equipment
- Return it to the service
- Bring it back up to the temperature

Anything that can be done to reduce any one of these aspects of a failure will be a step in the right direction. Sometimes, this can be prepositioning critical equipment, making more spares available onsite, etc.

14.1.3 Increase Spare Parts Availability

If equipment fails, it's important to have adequate spare parts on hand to deal with the failure. With increasing failures, one has to buffer those failures with more spare parts – either directly in stock from a warehouse, having access to spares through another distributor, or in process through increased reorders.

Sometimes, this will involve retaining more parts on hand for a part that we were already stocking. On the other hand, it may sometimes require stocking a part that we ordinarily didn't in the past. All these options cost money, but depending on the situation, they can be an important backstop against failures. As always, this is an optimization problem, as discussed in the chapter on Spare Parts Optimization.

14.1.4 Add Storage or Buffering Capacity

The previous chapter on Storage Tankage Determination/Optimization provided a thorough introduction to how tankage improves the system availability of a group of process units operating in series. Summarizing what we learned:

- When two units are in series, system availability can never be greater than the final unit's availability.
  - Therefore, if unit B has an availability of 92%, even if unit A has an availability of 100%, then the maximum system availability of the two units in series would be 92%.
  - If there is a requirement that system availability be greater than 92%, it would be necessary, at the very least, to implement an improvement project for unit B.
- If two process units are interconnected without tankage between the units, system availability is simply the product of the two availabilities multiplied together.
  - Assuming that unit A has an availability of 94.3% and unit B is 92.0%, then system availability would be $94.3\% \times 92\% = 86.8\%$
  - If 86.8% is not adequate, we could attempt to improve either unit A or unit B's availability. However, if after any improvements, we still have not achieved the desired result, then one might consider installing storage tankage between the units.
- The only way to determine the proper amount of storage is through a Monte Carlo simulation. Assuming nominal feedrate of 10 KBPD for both Unit A and Unit B:
  - If we install a 50,000 bbl tank, system availability will increase to 88.8%.[9]
  - If the tank size is increased to 300,000 bbl, system availability will increase to 90.9%.[10]
  - Continued increases in tank size would asymptotically approach 92% availability – unit B's availability.
  - The downside of large tanks is that they cost a lot of money.
  - There is truly an optimum that balances increased production against lower tank capital cost.

14.2 How Reliability and Process Configuration Effects Availability Results

It's important to understand how some standard configurations, such as equipment connected in series or in parallel, affect reliability results. After you understand the concepts, you'll see they are similar to the same principles in electrical circuits.

14.2.1 Series Configuration

As shown in the previous section, if you have two different process units connected in series, then system availability is simply the product of the two individual availabilities, as shown in Figure 14.1.

Since availability is simply the product of individual blocks multiplied by one another in a pure series arrangement, the overall system availability will inevitably decline as more blocks are added. For example, if there are five blocks in a series, as shown in Figure 14.2, with each block at 95% availability, then system availability will be 77.4%.

14.2.2 Parallel Configuration

A parallel configuration will always be fundamentally superior to a series configuration. In a series configuration, adding more blocks will always reduce system availability, unless the newly added block has 100% reliability. However, the opposite occurs in parallel configuration, where adding more blocks increases system availability. For example, if unit A and unit B were connected in parallel instead of series, then system availability rises from 86.8% to over 99.5%, as shown in Figure 14.3.

14.2.3 Change Configurations from Series to Parallel

The more components in a series, the lower the total system availability. There's no way around this simple mathematical fact.

So, wherever stuck with a series arrangement, one should ask themselves if there is any possibility of converting that series arrangement to a parallel arrangement. If so, it should be strongly considered.

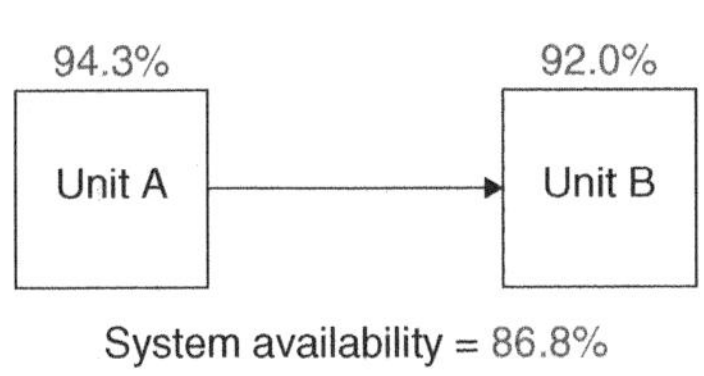

Figure 14.1 Example Series Availability – Two Units

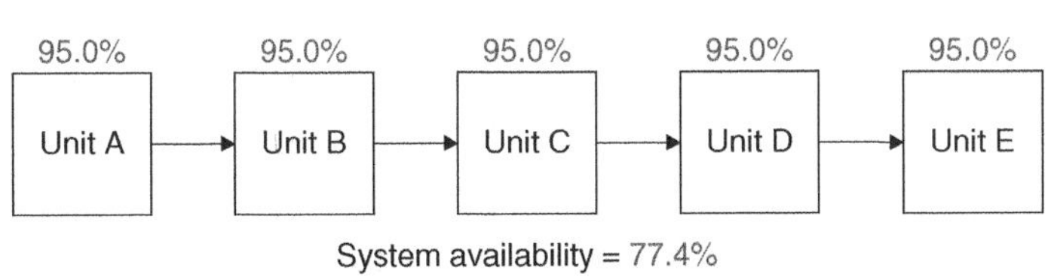

Figure 14.2 Example Series Availability – Five Units

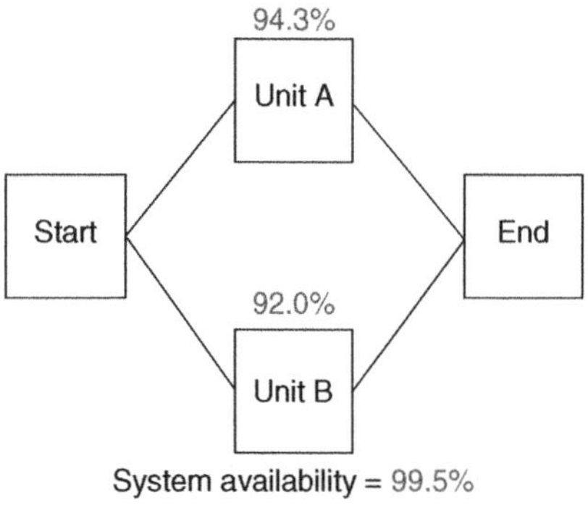

Figure 14.3 Example Parallel Availability

Sometimes, what appears to be a parallel arrangement from a process standpoint is actually a series arrangement from a reliability standpoint. For example, I have seen heat exchanger circuits where the exchanger trains were physically in parallel, but from a reliability standpoint, it was actually a lengthy series arrangement. That's because if any exchanger in either parallel circuit failed, both circuits would be shut down. If that's the case, a plant should strongly consider creating interconnections between the exchanger circuits so that both don't have to be shut down when there is a problem in one of the trains. In essence, the plant should have changed it to be a true parallel arrangement.

I also saw this once in a parallel compressor arrangement. In that case, there was logic that would bring down the entire set of parallel compressors if any of the compressors had an issue. So, while they were physically in parallel, from a reliability standpoint, it was actually a series arrangement. In that case, the logic should have been changed so that the other compressors could have continued to function if one of the others had failed.

14.2.4 Redundancy Configuration

Another way to increase reliability is through redundancy. This is almost always practiced for liquid pumps, but the general ideas there can be extended to all equipment.

Let's take the obvious case. If one assumes that a given pump fails once a year for 14 days, then the availability of that pump is roughly 96.2%. However, if there's a spare pump with identical failure characteristics, then the availability of the two pumps together rises to nearly 100%.

For practical purposes, redundancy is just a parallel arrangement, with one of the paths on standby, waiting to go into service, if needed.

14.3 Which Option Is the Best?

So, at this point, which of the six is the best option for a given situation?

Some things are obvious.

Decreasing maintenance time will almost always be a positive thing, and it should be done irrespective of any desire just to improve availability. Directionally, it's always the correct thing to do.

Configuration often cannot be altered. In fact, the majority of the time, the process arrangement cannot be changed. However, one should always be on the lookout for potential parallel arrangements that are, in reality, a series configuration from a reliability standpoint – such as that pointed out in previous examples. Wherever it is possible to create a parallel path, it should be strongly considered.

Increasing storage tankage is only possible between two process units and is ordinarily not practical within a single process unit – except in unique circumstances. So, it only has limited potential.

The remaining options – increasing equipment reliability, holding more spare parts, and certain maintenance options – are part of an overall optimization where the individual portions cannot be easily isolated from one another. In other words, there is no "one size fits all" answer on how to proceed with any of these three options. They must be determined for each case individually. In fact, in those situations, it's best if they are part of one, overall combined optimization. Ideas similar to these are explored in a future chapter.

14.4 Conclusion

There are six ways to possibly increase availability. Two are related to configuration – moving from a series configuration to a parallel one and adding redundancy. The other four are listed below:

1) Increase Equipment Reliability
2) Decrease Maintainability Time
3) Increase Spare Parts Availability
4) Increase Storage Capacity

Increasing equipment reliability can be one option and will be discussed in more detail in a future chapter.

Anything that can reduce any aspect of total "feed out" to "feed in" time is another possibility and should always be on the table for discussion.

There are times when holding more spares can be helpful, but one must run the optimization to determine whether it is the right answer.

Last, if none of these options have been helpful, another idea might be to increase the intermediate storage capacity between a set of process units.

Deciding the best option is not trivial and often requires a more comprehensive analytical problem. This will be the subject of a future chapter on combined optimization.

15

Equipment Reliability Optimization

As noted in the previous chapter, when contemplating options to increase availability, one should always consider whether spending additional money on more reliable equipment is the right economic decision – or if there are other, better options. In this sense, this is the "Reliability" analog to the "Maintainability" discussion earlier. It's the "R" in "RAM," whereas "Maintainability" was the "M." In both cases, increases in either variable tended to increase the "A" – availability.

In considering improvements to equipment reliability, we are approaching the problem from the opposite direction. That is, we are considering options where equipment fails less often. However, more reliable equipment usually costs more. Thus, the real question is, where is the point of diminishing returns?

From a modeling perspective, many of the same ideas that apply to this optimization type were already covered in the Opportunistic Maintenance section, just from the opposite direction. Thus, let's review some of that material, updated for the reliability options, before getting to the example problems. Plus, since we are evaluating reliability during the design phase, we will refer to these optimization problems as 'Design Optimization' from this point forward.

As in the Opportunistic Maintenance chapter, the example problems here will be broken down into three main categories:

1) Single Equipment Reliability Optimization
2) Intra-Unit Design Optimization
3) Inter-Unit Design Optimization

The first category is simple. As previously mentioned, this problem rarely needs a full-blown optimization but does a good job of demonstrating the basic tradeoffs, which only get more complicated as more equipment gets added. Intra-Unit Design Optimization is a fancy way of saying that we are evaluating equipment design options that affect numerous pieces of equipment within a single process unit. On the other hand, Inter-Unit Design Optimization considers numerous pieces of equipment having various design options across different process units.

As noted in earlier chapters, it's important to remember that all figures used in the scenarios/examples in the chapter are relative. Thus, the absolute values of any one figure are not all that important. Rather, it's the concepts behind the numbers that we are trying to convey in this discussion.

Plant Optimization in the Process Industries: Incorporating Equipment/Assets in the Decision-Making Process, First Edition. Marty Moran.

15.1 General Approach

This section will discuss two important aspects of our general approach to solving this problem. The first will concern how to best incorporate the various design options, while the second concerns how each equipment's availability will be modeled based on varying failure characteristics.

Let's start first with the discussion of design options.

15.1.1 Design Options Modeling

From a modeling perspective, we will need to transform the discrete options under evaluation for equipment reliability improvements into a form easily understood by the optimizer. For practical purposes, we'll need to regress those options into one equation with design cost as a function of failure time.

We'll also need to assume that all options are ordered into a list, starting with the smallest failure time, and assumed largest cost, progressing to options of increasing failure time with smaller costs. Options not in line with that assumption should be reevaluated. For example, we normally wouldn't want to consider less reliable options, which are more expensive. It's just not a practical possibility, and nobody would choose that route anyway.

Of course, costs could potentially vary significantly from one option to another and end up being quite nonlinear as we progress through our options, as shown in the graph below. That's OK since we will be using a nonlinear optimizer, which easily handles nonlinear equations.

For example, Figure 15.1 shows a discrete list of four different equipment reliability options with varying failure times and costs. You'll see later that these options are those for equipment D in our examples.

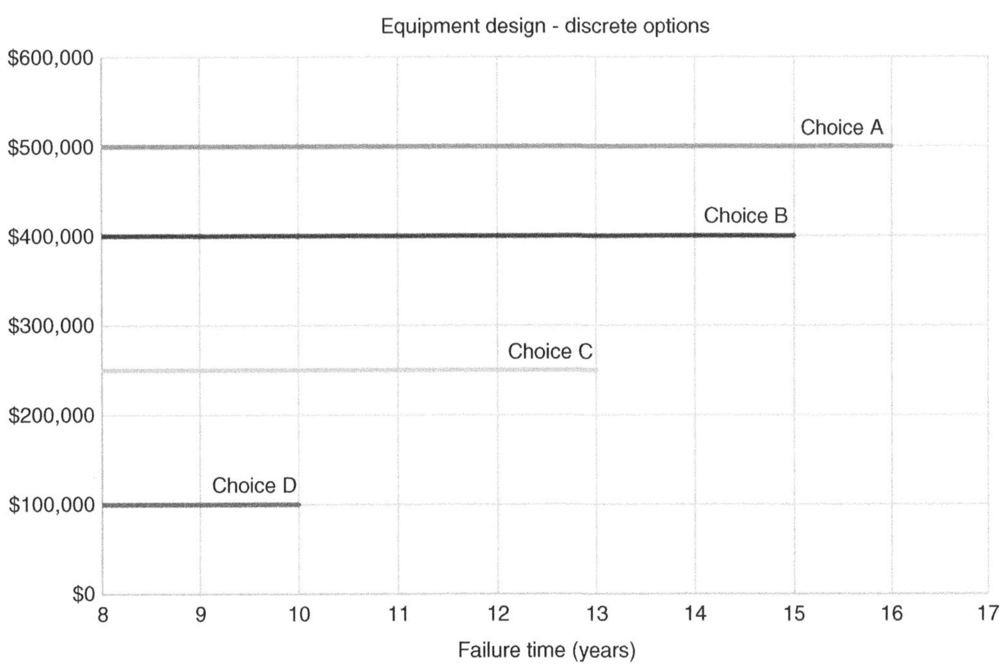

Figure 15.1 Design Options – Discrete Choices

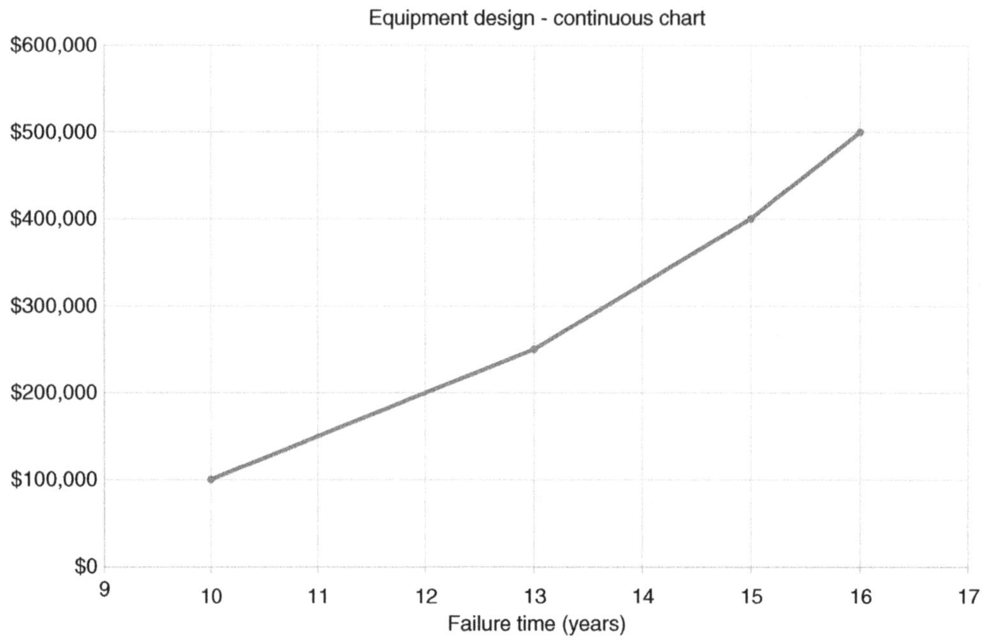

Figure 15.2 Design Options – Continuous Choices

The graph demonstrates that the more expensive options have less frequent failure times (longer numerical mean failure time value). This gives the optimizer options to consider where it can evaluate a more expensive choice that has higher equipment availabilities.

The equipment would cost $500,000 for choice A but, on average, would fail every 16 years. On the other hand, the cost would only be $100,000 for choice D. However, its failure rate would be more frequent – every 10 years. Choices B and C are in between these.

If the optimizer you are using doesn't handle mixed integer problems well, you'll need to convert the discrete options listed above into its continuous analog, as shown in Figure 15.2. The results won't be quite as accurate, but it would normally still yield acceptable results. That will be the approach taken for the scenarios/examples in this chapter.

15.1.2 Availability Based on Differing Failure Characteristics

Another important aspect of our solution approach is how we handle estimating equipment availability based on varying failure characteristics. In general, that's done using Monte Carlo modeling. However, the downside of the Monte Carlo methodology is that it is a statistical approach and produces inherently "noisy" results with some randomness from run to run. In general, that's a positive since it gives one an idea of the variation around the mean.

However, when working with GRG2, our principle optimization technology used for the examples in this book, that ends up being a negative, since GRG2 requires absolute precision when it takes its partial derivatives for determining search directions. If there is variation in those derivative results, GRG2 gets seriously confused. For that reason, we'll need to take a different approach.

To avoid that problem, we used Excel's Data Table features to simulate Monte Carlo within Excel with the goal of generating an availability for each year for the design options being considered. It's important to note that not only are mean failure times important, their failure time standard deviation is just as important. More about that nuance shortly.

Using those results, we then generated a regressed equation that had one, and only one, result for a given failure time. So, in essence, we simulated Monte Carlo through Excel Data Tables to generate initial results and then smoothed it out by regressing the final results.

15.2 Example Problem Data

To illustrate the design option tradeoffs, let's create a fictitious example that will serve as the data for our examples in this chapter.

The examples will consider equipment A–G, which are located in the Naphtha Splitter, Isom unit, and Reformer. These units are all intertwined with one another since the Naphtha Splitter separates out feed to the two downstream conversion units. Therefore, one might want to think of this example as a complete project to revamp this section of the plant.

Table 15.1 shows how much each design option costs for each piece of equipment (shown in columns) as a function of the failure time (shown in rows), measured in years. Some equipment, like equipment A, will have three different options to choose from, whereas equipment B will only have two. Equipment C and D both have four, and so on.

For equipment A's three options, the cheapest is $250,000, and its mean failure time is 10 years. On the other hand, the most expensive option is $600,000 and is estimated to have a failure time of 15 years. There's also an intermediate option that costs $350,000, predicted to last 13 years. The values for all the other equipment in the table can be read in a similar manner.

Here's the same information except from the table now shown in a graphical form in Figure 15.3 using continuous functions.

Failure time standard deviations are also important, as will be demonstrated shortly. The values, measured in years, for each equipment are listed in Table 15.2.

Table 15.1 Example Problem Data

	Naphtha Splitter	Isom	Isom	Reformer	Reformer	Reformer	Reformer
Failure Time	**A ($)**	**B ($)**	**C ($)**	**D ($)**	**E ($)**	**F ($)**	**G ($)**
7.5						400,000	
10	250,000	175,000		100,000		500,000	225,000
13	350,000	500,000	150,000	250,000	75,000	800,000	
15	600,000		200,000	400,000	150,000		350,000
16			300,000	500,000	200,000		
17.5							
18			500,000				

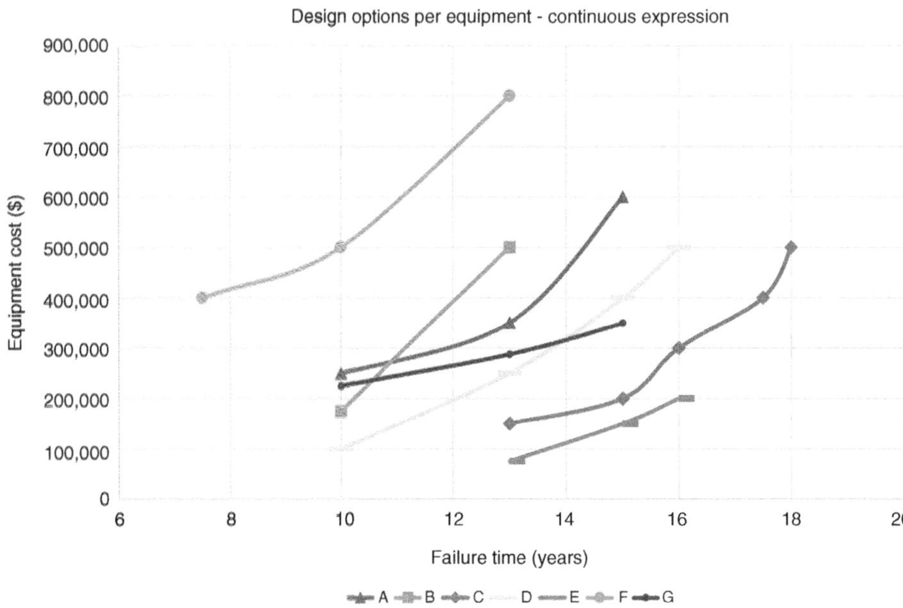

Figure 15.3 Design Options per Equipment – Continuous

Table 15.2 Equipment Failure Standard Deviation

Equipment	Failure Std. Dev
A	2.5
B	1.5
C	3
D	2
E	2
F	2
G	1.5

15.3 First Impressions of Example Data – Impact on Problem Solution

Before we actually examine the results from our different optimization runs, it's instructive to first think through the problem to understand the results that we might anticipate.

To state the obvious, as failure time increases, equipment lasts longer and the option become more expensive. However, it's not really reliable equipment we are striving for, but rather how increased reliability can potentially impact availability for a given particular process unit. That's important because the equipment doesn't exist on its own but rather as part of a comprehensive whole. Thus, sometimes increasing reliability for one piece of equipment can increase availability within the processing unit enough to make that expenditure worth our while. On the other hand, there are other times when we are already past a point of diminishing returns, and it doesn't make sense to spend the money.

Nonetheless, the optimizer will want to shorten the failure time as much as possible to reduce total costs. However, per the previous discussion, there's the possibility that it can only go so far because it might eventually negatively affect availability. On the other hand, there is also a point of higher availability values where the additional availability isn't worth that extra investment. Of course, that depends on the unit's margin, though it may not be the dominant effect for a particular case.

The optimization runs in this chapter are based on attempting to maximize profit over a 10-year investment horizon. The choice of 10 years is important as one examines specific data used in the examples.

For example, we observe that equipment F has options at 7.5, 10, and 13 years. One has to wonder about F's first option, since if the mean failure value time is only 7.5 years, then how likely is to make it to 10 years? Not very.

What's also instructive about equipment F is that the other design options are quite expensive – considerably more expensive than when compared to other equipment. Equipment F is also part of the Reformer unit, where there are three other equipment, so it will be interesting to see how the optimizer trades off one or more of those other equipment options in order to satisfy minimum Reformer availability constraints at a reasonable cost.

Moving on to other equipment besides F, there are four others whose first option is at 10 years – equipment A, B, D, and G. One has to wonder how often the optimizer will choose the 10-year option or will it settle for more expensive options with better availability given that 10 years may be on the ragged side of equipment availability? We'll see.

On the other hand, there are two equipment that have options farther out, which are relatively cheap – equipment C and E. Having those options may make it more likely that the optimizer can reduce the cost for the other equipment, since equipment C and E will have better equipment availability.

15.4 Effect of Failure Standard Deviation

As we examine the options, we must also consider the pernicious effect of failure time standard deviation. To illustrate its effects, Table 15.3 was generated from a Monte Carlo simulation using Excel's Data Tables, which shows a representative set of values for the likely availability (expressed as a fraction here) as a function of failure standard deviation, expressed in years, in the 7.5–13-year range.

Table 15.3 Standard Deviation vs. Failure Times

Std Dev	7.5	8.75	10	11	12	13
0.5	0.7504	0.8832	0.9973	1.0000	1.0000	1.0000
1	0.7394	0.8694	0.9404	0.9952	0.9978	0.9999
1.5	0.7322	0.8806	0.9417	0.9767	0.9953	0.999
2	0.7496	0.8614	0.9164	0.9583	0.9983	0.9983

As noted earlier, since these values are generated from a statistical analysis, they can be inherently noisy. That's why you see some places where a particular value will be higher or lower than expected. However, the trends in the data are clear.

For example, take a look at the 10-year column. See how the values rapidly decline from 0.9973 to 0.9164 as the failure standard deviation increases from 0.5 years to the 2-years. What we will see is that this will have a huge effect on how the optimization iterates to its solution. In fact, this will be the subject of one set of optimization runs later in the chapter. Another thing to note is how quickly the 0.5 standard deviation row plateaus at unity, yet that only starts occurring at 12 years when the standard deviation is 2 years.

It's also noteworthy that the values for both 7.5 and 8.75 years are so poor. At the other end of the spectrum, all the values for year 13 are very good. What this discussion illustrates is that design choices for 10 years are right on the ragged edge. That's where the failure standard deviation difference starts to become quite significant.

The same ideas just presented here for equipment F's standard deviation also apply to the other equipment. As just illustrated, the effects vary dramatically depending on what the starting and ending years are for the design options.

The bottom line is that it isn't intuitively obvious what the optimal design choices should be for a given set of equipment where there is a range of design choices/costs, how availability is constrained, failure standard deviation, and unit margin.

So, now that we have a basic background of what factors are likely to affect the results let's proceed to discuss each of our three different examples: Single Equipment, Intra-Unit, and Inter-Unit Desgin Optimization.

15.5 Single Equipment Design Optimization

In this example, we present a problem where we choose the best design option for a single piece of equipment within one process unit, subject to constraints. Equipment A is in the Naphtha Splitter, and its design options are shown in Table 15.4.

What's significant about the data in this table is that there is only a $100,000 difference between years 10 and 13 – a 4-year time difference. Given the discussion in the previous section, it will be interesting to see if the optimizer ever chooses year 10, given the small incremental cost difference between these options.

On the other hand, there's a much bigger cost gap between years 13 and 15 – $250,000 over only a 2-year time frame. So, it may be hard for the optimizer to justify moving much beyond the 13-year time choice.

Table 15.4 Equipment A Failure Data

Failure Time	A ($)
10	250,000
13	350,000
15	600,000

15.5.1 Independent Variables

The independent variable for this problem is:

- Failure time for equipment A

15.5.2 Dependent Variables

The major dependent variable would be the equation that converts equipment A's three different design options into a single internal equation used by the optimizer. In essence, cost will be a function of failure time. This smooth equation (not discrete) allows the prediction of design cost as a function of failure time between the values of 10 and 15 years.

15.5.3 Constraints

The most important constraint for this problem will be minimum process unit availability, which forces the optimizer to evaluate options beyond the cheapest design cost by including consideration of the production impact of that choice.

In this particular problem, we will not impose design cost constraints. It's only one piece of equipment, so the answer would be obvious. However, that constraint will be added for more complex problems later in this chapter.

Since this is only one piece of equipment, the plot of failure time versus availability is actually straightforward. In fact, this curve was generated from a simple Monte Carlo analysis, performed in Excel, using Data Tables. After varying failure times, the resulting data was regressed to remove the noise inherent within Monte Carlo analyses. This is shown in Figure 15.4.

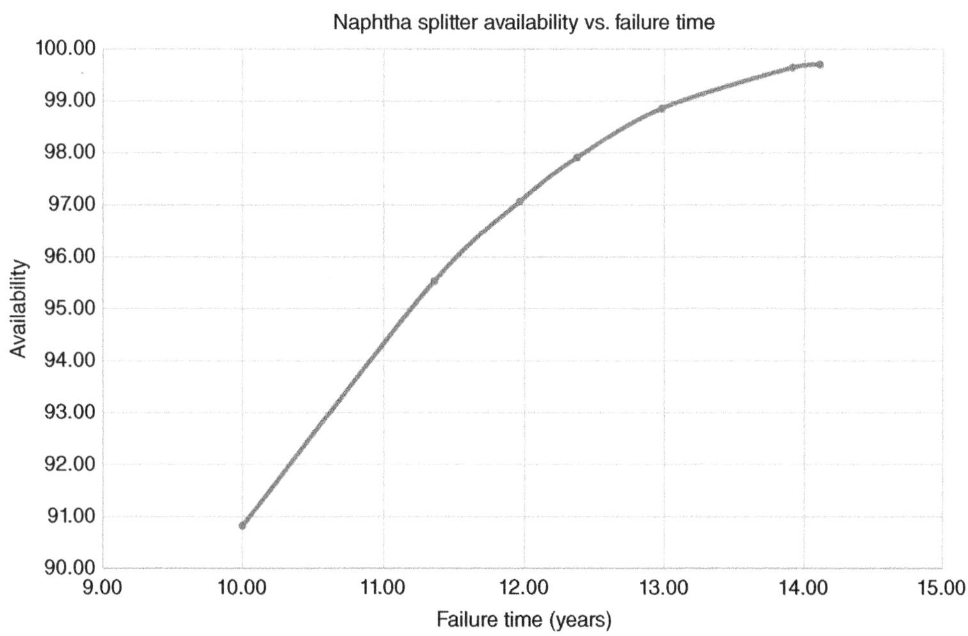

Figure 15.4 Naphtha Splitter Availability vs. Failure Time

In our case, there are only three options – 10 years at 91% availability, which is almost never acceptable, 13 years at 99% availability, and 15 years, which almost never makes sense for reasons that will be discussed below. So, the answer to this problem would almost always be 13 years. However, it's still instructive to go through the analysis to see the effect of various input variables.

15.5.4 Objective Function

The goal of this optimization is to determine maximum profit. The optimizer will balance equipment design options and choose the best option subject to availability constraints.

15.5.5 Results

This problem was simulated in 10 different model runs using the GRG2 option within Excel's Solver. Results are shown in Figure 15.5. In the first four runs, the margin was set at $7500, and the minimum constraint varied. Since the optimizer ended up riding the minimum constraint in each of the four runs, it just recreated the graph in Figure 15.4. In other words, the margin doesn't matter if we are riding the availability constraint.

However, the margin does matter if we are not hitting the constraint. In the remaining model runs, the minimum availability constraint was set at 98.5%, and the margin varied, which produced the following graph. The optimizer did ride that constraint for the two lowest margin data points but produced a true "optimum" for all the other margin values.

What this analysis shows is that margin does play a role in the selection of the best option. Here, we see that at roughly $15,000, the optimizer chooses the 13-year option. However, as the margin increases, the optimizer wants to move toward a more reliable equipment option, though it

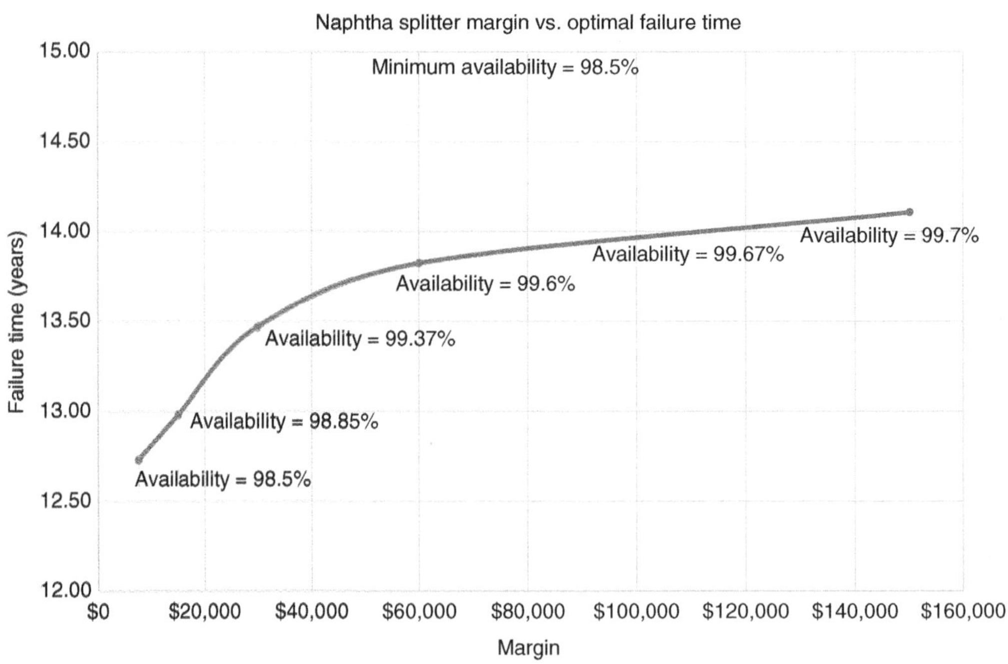

Figure 15.5 Naphtha Splitter Margin vs. Optimal Failure Time

progressively reaches a point of diminishing returns since the smaller incremental production gains from smaller availability increases start to level off.

In other words, reliability starts to plateau at a certain point, which realistically is at about 14 years for this example. That's not due to anything within the example, but rather because we choose to evaluate options at 10 years. Fundamentally, what we are witnessing is the effect of what happens as one moves farther and farther away from 10 years. In our case, we hit a point of diminishing returns as we approach 14 years.

It's important to note that whereas the optimization problem was modeled with a continuous function, the user is forced to make a binary choice between the two choices – 13 or 15 years. Nonetheless, the continuous graph clearly shows the trade-off as we progress through different margin choices.

Now, let's move to the cases where there are multiple equipment within one process unit.

15.6 Intra-Unit Design Optimization

Intra-Unit Design Optimization evaluates different equipment design options for different equipment within a given process unit. In this section, we will evaluate two different cases: two pieces of equipment for the Isom unit and the four pieces of equipment within the Reformer.

Let's start with the Isom unit.

15.6.1 Isom Unit

The equipment choices in the Isom are shown in Table 15.5.

These are two really different choices.

For equipment B, there's a big difference between the options – both from availability and cost consideration. However, equipment C has a number of choices within a rather narrow range.

Another important consideration for this problem is we will assume that this equipment is in series, which means that the Isom system availability is the product of their individual availabilities. That has the effect of lowering overall system availability.

15.6.1.1 Independent Variables

The independent variables for this problem are:

- Failure times for equipment B/C

Table 15.5 Isom Equipment Failure Data

Failure Time	B ($)	C ($)
10	175,000	
13	500,000	150,000
15		200,000
16		300,000
17.5		
18		500,000

15.6.1.2 Dependent Variables

The major dependent variables are the equations that convert equipment B/C's different design options into two equations, one for each piece of equipment, and used internally by the optimizer. Cost will be a function of failure time for both equations.

We will also include a total budget constraint, which is the sum of the two equipment costs.

15.6.1.3 Constraints

There are two constraints defined for this problem – a minimal Isom process unit availability, which will force the optimizer not to simply choose the cheapest design cost all the time. In addition, a total design cost constraint will be included.

15.6.1.4 Objective Function

The goal of this optimization is to determine the maximum profit, subject to meeting the constraints. Generally, the optimizer will balance increased costs for more reliable equipment against production increases that might result from those equipment choices.

15.6.1.5 Results

There were 13 different model runs for this problem, which were concentrated into three different scenarios – constrained availability, constrained total design cost, and the effect of margin in unconstrained cases.

15.6.1.5.1 Constrained Availability

For the Constrained Availability scenario, one should read Figure 15.6 by choosing an availability value and examining the two design costs for that vertical. There are two curves corresponding to

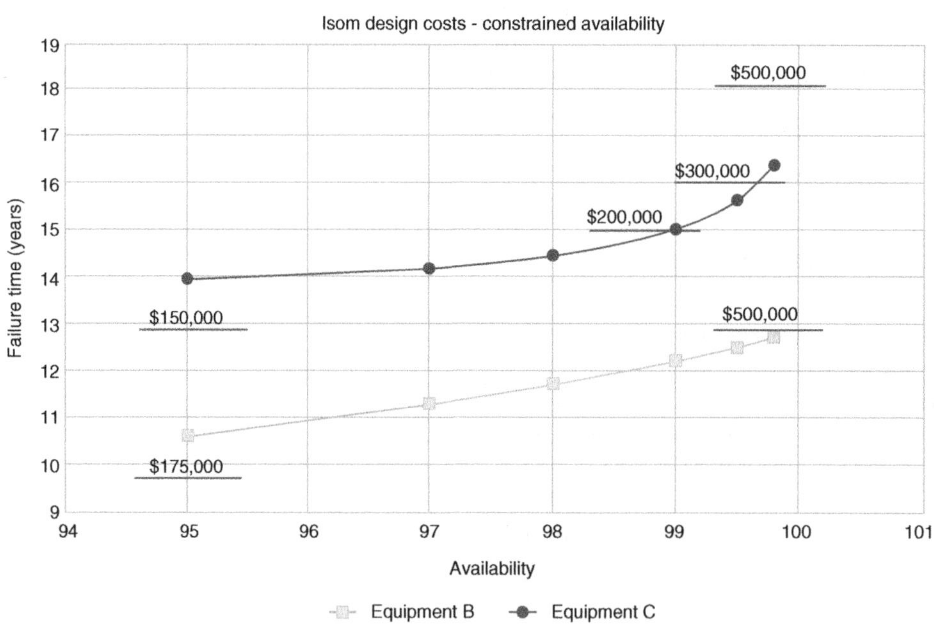

Figure 15.6 Isom Design Costs – Constrained Availability

the two different equipment. The short horizontal lines show the discrete design choices for that equipment.

In this scenario, we note that the two cheapest design options for both equipment together result in an availability below the 95% constraint limit, so we can effectively discount that combination.

Since equipment B only has two options, we note that above 97.5% availability, the optimizer is already choosing the largest design cost value of $500,000. So, if we desire an ultimate availability of 97.5% or greater all the way up to almost 100%, then we'll need to go with the $500,000 option.

On the other hand, the final choice for equipment C is a little bit more complicated. By choosing the $200,000 option, we are already at 99% availability. If we want to spend another $100,000, then we can achieve all the way up to 99.7%. If we truly desire 100% availability, then we could even choose the 18-year failure time option. How each organization makes the final decision of these discrete choices would be a judgment call based on many different factors.

Even though making the actual discrete choices for each piece of equipment is somewhat more complicated for this example than when compared with the Naphtha Splitter example, it's still instructive to see the choices presented as smooth graphs since it's easier to visualize the tradeoffs.

15.6.1.5.2 Total Design Cost Constraints

In the Total Design Cost scenario, the cost was constrained at three different values: $500,000, $600,000, and $700,000 with results shown in Figure 15.7.

To make it easier to understand the results, in addition to the equipment trends, the graph adds a third curve – for the related availability at a given cost constraint. Thus, one can see what the optimizer chooses for each equipment design cost, as well as the associated availability for that cost constraint. In all three cases, the optimizer was able to stay above the minimum availability

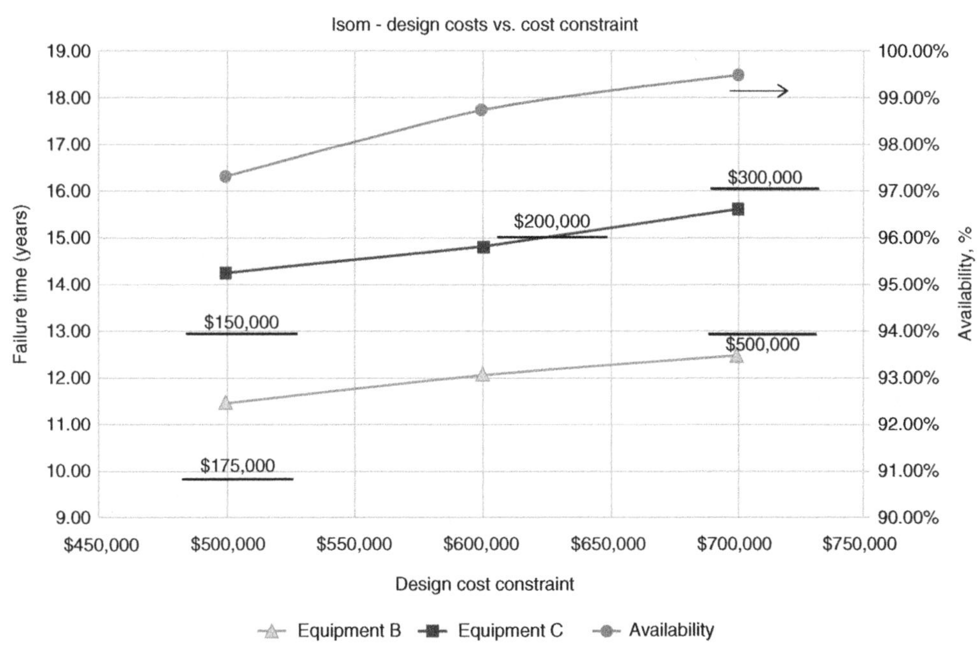

Figure 15.7 Isom Design Costs – Cost Constraint

Table 15.6 Relation Between Optimizer and Discrete Choices – Cost Constraint

Cost Constraint	$500,000	$600,000	$700,000
Equipment B – Optimizer	11.41	12.06	12.46
Equipment B – Discrete Choice	10	13	13
Equipment C – Optimizer	14.26	14.81	15.57
Equipment C – Discrete Choice	15	15	16

constraint of 97%. But what this example illustrates is if the cost is too restricted, the optimizer is forced to choose poorer availability options.

As we evaluate options for our final binary choices from the curves presented above, it's instructive to see where equipment flips from one value to another, as shown in Table 15.6.

What we observe is that equipment B flips from 10 to 13 years as we approach the $600,000 limit. On the other hand, equipment C flips as we get nearer to the $700,000 value.

Once again, while the optimizer works with continuous functions, the tradeoffs are clearly shown as we move through the different total cost constraints.

15.6.1.5.3 The Effect of Margin

The last scenario examines how changing the margin for an unconstrained case affects the results. In all cases, the calculated availability was above the lower limit of 97%, as shown in Figure 15.8.

What one notices is that the effects of margin plateau fairly quickly for equipment B – not changing much after $40,000. However, the effect on margin or equipment C is more gradual.

The difference between these two equipment curves is how individual reliability affects process unit availability. After a certain point, there's simply not much more production that can be squeezed from the unit. In cases where it can be, it tends to favor equipment design options that are cheaper.

The same general pattern is observed here for the margin scenario as compared to the Total Design Cost scenario already discussed. That is, the optimizer flips equipment B from 10 to 13 years between a margin of $15,000 and $30,000. On the other hand, it requires a higher margin for equipment C to flip, which occurs somewhere between $40,000 and $60,000. These are shown in Table 15.7.

15.6.2 Reformer Unit

The following shown in Table 15.8 are the design options for the four pieces of equipment in the Reformer.

As we review our choices, we note that equipment F has one option for 7.5 years. Given we are evaluating the system from a 10-year perspective, a value of 7.5 years probably isn't going to be very realistic. Other than that observation, it's hard to draw any obvious conclusions. What the likely optimum value should be isn't quite obvious, since we have many choices.

Similar to the Isom example, we will assume that all four pieces of equipment are in series, which means that Reformer system availability is the product of their individual availabilities.

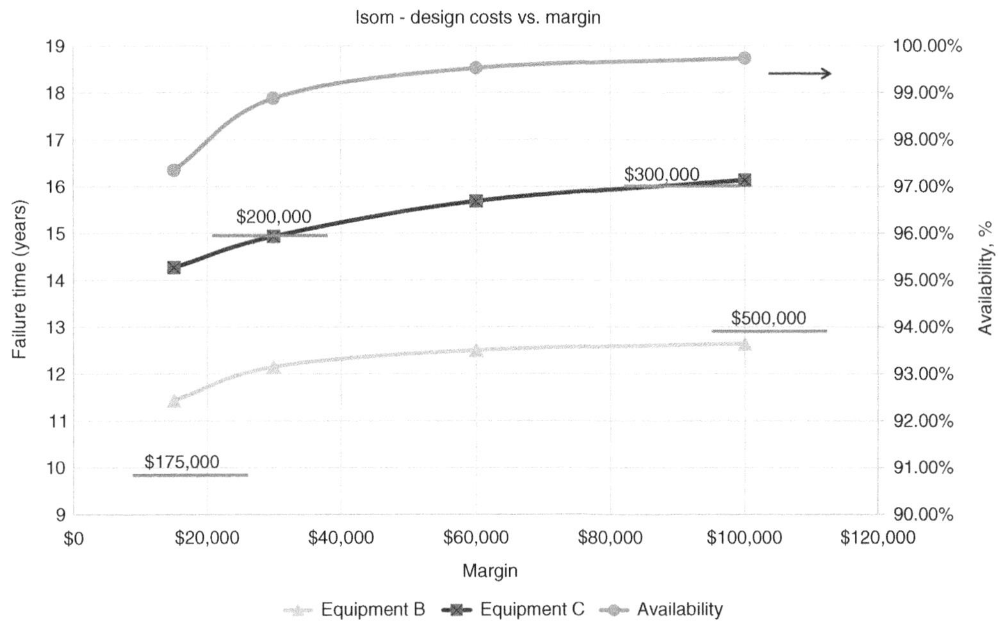

Figure 15.8 Isom Design Costs – Margin

Table 15.7 Relation Between Optimizer and Discrete Choices – Margin

Margin	$15,000	$30,000	$60,000	$100,000
Equipment B – Optimizer	11.44	12.15	12.50	12.64
Equipment B – Discrete Choice	10	13	13	13
Equipment C – Optimizer	14.27	14.93	15.68	16.14
Equipment C – Discrete Choice	15	15	16	16

Table 15.8 Reformer Equipment Failure Data

Failure Time	D ($)	E ($)	F ($)	G ($)
7.5			400,000	
10	100,000		500,000	225,000
13	250,000	75,000	800,000	
15	400,000	150,000		350,000
16	500,000	200,000		

This is even more restricting than in the Isom case, because as more equipment gets added, system availability ends up being lower due to the multiplication of four values, all less than unity. Conversely, it implies that we will need even more reliable equipment choices to meet the same availability target levels as in the Isom example.

15.6.2.1 Independent Variables

The independent variables for this problem are:

- Failure times for equipment D–G

15.6.2.2 Dependent Variables

The major dependent variables are the equations that convert equipment D–G's different design options into four internal equations, one for each piece of equipment, which will be used by the optimizer. In essence, cost will be a function of failure time.

An equation will also be added for a budget constraint, which is a sum of the four equipment costs.

15.6.2.3 Constraints

There are two constraints defined for this problem.

One will be a minimum availability constraint that will force the optimizer to consider Reformer production and not just choose the cheapest design option for each piece of equipment.

The other constraint is a maximum total design cost constraint.

15.6.2.4 Objective Function

The goal of this optimization is to determine the maximum profit, subject to the two constraints.

15.6.2.5 Results

There were 12 cases run for 4 different scenarios. These scenarios examined margin variations, different design cost limits, one situation where only equipment D and E were considered, as well as some model runs with different equipment F failure characteristics.

It should be noted that equipment G is not shown on the graph for the first three scenarios, primarily because the optimizer always pegged it against its upper design option limit of 15 years in those model runs. In other words, equipment G didn't influence the final optimization decision in those three scenarios.

After reflecting on that result, it's likely because of the failure distribution standard deviation for equipment F per the discussion earlier in this chapter. Even though equipment F has the standard deviation issue, the optimizer was forced to peg equipment G against the 15-year value since there are only two equipment G choices. In essence, the optimizer used that choice to give it the room necessary to meet the overall availability constraint. Plus, choosing the 15-year option for Equipment G was a relativley cheap choice.

In fact, later in this section, we'll challenge the equipment F standard deviation failure assumption. Ultimately, we'll see that it has a large effect on the results.

15.6.2.5.1 Varying Margin

In the first four cases, the margin was varied while always enforcing a minimum availability constraint of 97%. Results are shown in Figure 15.9. At margins greater than $40,000, there was little variability primarily because the predicted availability was already high – 99.1% or higher. So, there was little room to increase production. On the other hand, as margins decreased below $40,000, predicted availability decreased quite rapidly. In fact, at a margin of $15,000, the availability had already dropped to 97.4% – not far from the constraint limit.

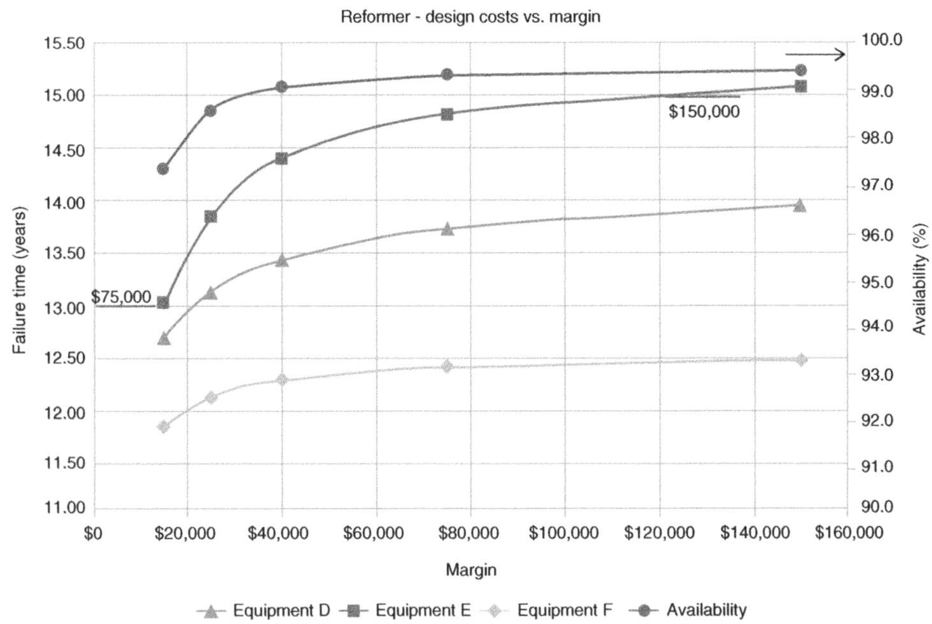

Figure 15.9 Reformer Design Costs – Margin

Nonetheless, over the entire margin of $15,000–$150,000, the optimizer always chose the same design options for equipment D and F.

It's only equipment E, where the optimizer flipped. Between the values of $25,000 and $40,000, the optimizer transitions from 13 to 15 years, which is a design difference of $75,000. Only in that time period was there is enough availability change to justify the extra investment.

15.6.2.5.2 Varying Total Design Cost Constraint

There were three cases run where the optimizer rode the total design cost constraint. For this scenario, the problem definition was set up to enforce a minimum availability of 95%. Results are shown in Figure 15.10.

What one observes is that availability is quite sensitive to the total design cost constraint when the constraint limit drops below $1,300,000. Interestingly, it then rapidly heads toward the minimum availability constraint over a small additional drop of just $50,000.

The optimizer chooses the same design option values for equipment D and E in all three cost constraint limit runs. It's only equipment F where the optimizer flips and chooses a lower design value of $400,000, rather than $500,000, which occurs just below $1,300,000. As noted earlier, equipment G shows no sensitivity and is not shown on the graph. It is pegged at 15 years for all three cases.

If we compare this scenario with the previous one, what's interesting is that the optimizer flips on two different equipment. For this scenario, the optimizer flips on equipment F. On the other hand, in the previous scenario, it flipped on equipment E. Thus, the optimizer responds to the appropriate aspects of each problem and chooses the correct design options in each case.

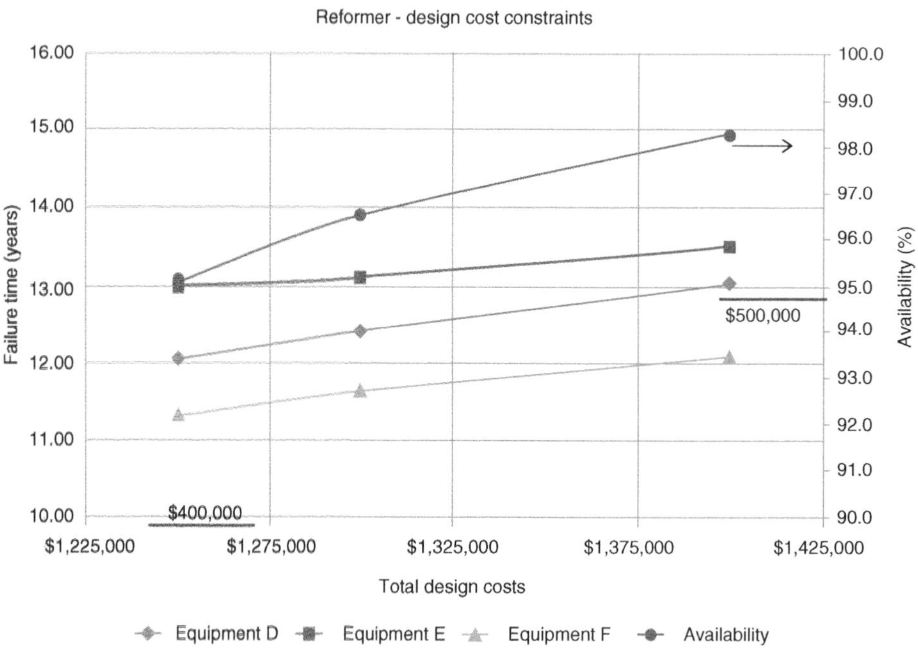

Figure 15.10 Reformer Design Costs – Cost Constraints

15.6.2.5.3 Equipment D/E Only

One of the downsides of including four equipment at once is we are assuming they are all in series, which implies that the availabilities must multiplied by one another to produce the overall Reformer availability. Thus, system availability can decrease quickly – even if only one of those four equipment availability values starts to drop off.

For that reason, we decided to include only equipment D and E in this scenario and then rerun the total design cost constraint to determine if only including two pieces of equipment significantly affects the final solution. Once again, availability is constrained at a lower limit of 95%. Results are shown in Figure 15.11.

There were only two cases run in this arrangement. However, what's interesting is that even in the narrow total design cost range of $225,000–$300,000, the optimizer quickly flips from 13 to 10 years for equipment D. (Equipment E is not shown in the graph since it was pegged at 13 years and did not change.). What's also noteworthy is that in the same range availability rapidly drops off from over 98% at the $300,000 total design cost constraint limit to just above 95% at $225,000.

15.6.2.5.4 Changing Equipment F's Failure Characteristics

Up to this point in time, we have not varied any equipment failure characteristics to see how that influences final optimization decisions. That's the purpose of this scenario.

We will reevaluate the margin variation scenario, after changing equipment F's standard deviation from 2 to 0.5 years per the discussion earlier in this chapter. All four pieces of equipment will be modeled in this example.

What we learn is that, surprisingly, optimizer choices can be quite sensitive to even small changes in standard deviation values.

The results, as shown in Figure 15.12, are dramatically different in this margin scenario than the previous one. In both situations, there is little difference concerning equipment D with the

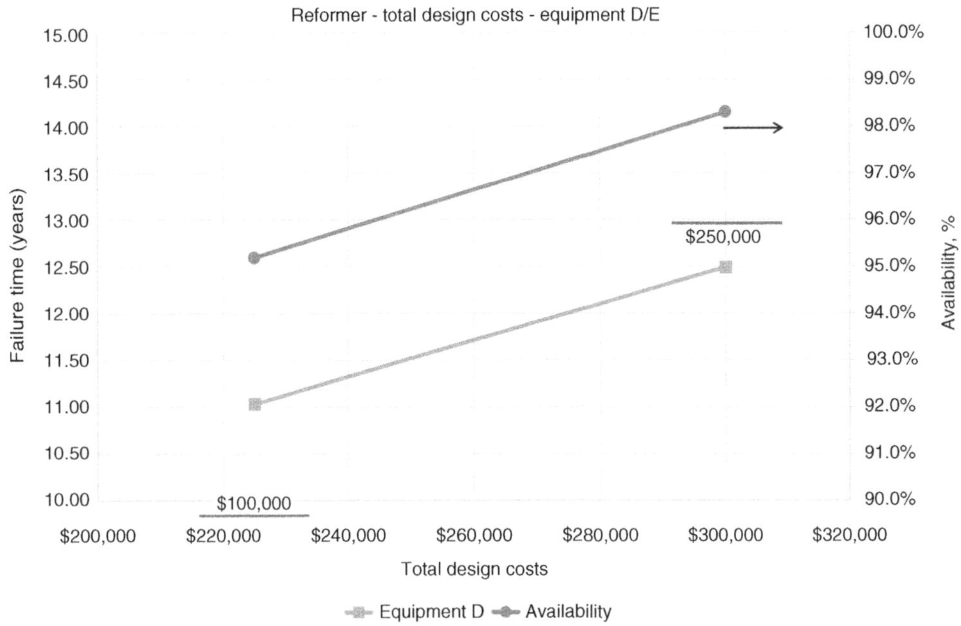

Figure 15.11 Reformer Equipment D/E Total Design Costs

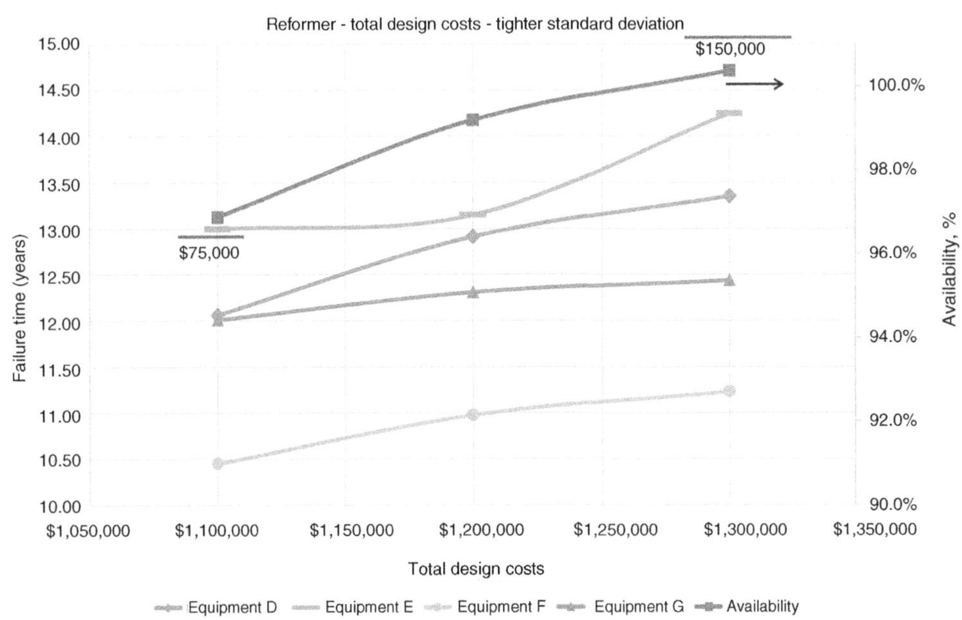

Figure 15.12 Reformer Total Design Costs – Tighter Standard Deviation

optimizer choosing the 13-year option in all cases. However, the situation for all other equipment is quite different.

Whereas in the previous scenario, the optimizer always pegged equipment G to 15 years, it now considers equipment G. In retrospect, that's because the optimizer had to compensate for equipment F's poor failure characteristics through the use of another handle. In those cases, it used equipment G. Here, it now has more flexibility.

Not only is the optimizer now considering equipment G, but it has also flipped and is now choosing the 10-year option (it's a close call since it is roughly halfway in between the 10- and 15-year values).

In the previous example, equipment E did not flip at all. However, with the new failure characteristics, the optimizer chooses the 15-year option for equipment E, as cost constraints become less restrictive. This requires an increased investment of $75,000, but the production gain must be enough to justify that expenditure.

On the other hand, there is enough of a difference for equipment F that the optimizer now prefers the 10-year option in all cases. It's not as dramatic as it might sound because the optimizer is working with continuous functions. But it causes it to flip, nonetheless.

The bottom line is that this scenario really shows that what the optimizer ultimately decides is a function of ALL the variables included in the problem. In this case, a relatively small change in failure distribution characteristics for one equipment had a fairly significant effect on the final results. That won't always be the case, but for this example, it was enough to tip the problem into an entirely different solution regime.

15.7 Inter-Unit Design Optimization

Inter-Unit Design Optimization evaluates a series of equipment design options for one or more equipment across multiple process units. In our case, we will now examine all seven pieces of equipment across the three different process units – Naphtha Splitter, Isom, and Reformer. This is the most general equipment design optimization problem.

15.7.1 Independent Variables

The independent variables are:

- Failure times for equipment A–G

15.7.2 Dependent Variables

The major dependent variables would be equations that convert equipment A–G's different design options into seven internal equations, one for each piece of equipment. These are the same equations used in the earlier examples.

A total budget constraint will also be included, which is a sum of all the equipment costs.

15.7.3 Constraints

Similar to the previous scenarios, there will be minimum process unit availability constraints for all three process units at the same time – Reformer, Isom, and Naphtha Splitter. Whereas we might think that this gives the problem a lot of flexibility, as will be shown, it's less than one might think.

The most important constraint for this scenario is the total cost constraint since the optimizer can now trade off cost possibilities among several different process units.

15.7.4 Objective Function

The goal of this optimization is to determine the maximum profit, subject to meeting all availability constraints, as well as the total budget constraint.

15.7.5 Results

There were eight case runs for this scenario, split into two different scenarios – Total Design Cost Constraint and Margin.

15.7.5.1 Total Design Cost

In the Total Design Cost Constraint scenario, a minimum availability constraint of 97% was imposed for all three process units. The results show that the optimizer always rides the 97% availability constraint for both the Naphtha Splitter and Isom unit. In fact, in both scenarios for all 8 cases runs, the optimizer always chooses the same value for the single piece of equipment in the Naphtha Splitter and the two in the Isom. The only variation in results is in the Reformer.

As shown in Figure 15.13, the most significant changes occur below the total cost constraint of $2,100,000. In fact, at values above $1,980,000, the optimizer flips equipment E choosing 15 years instead of 13 years.

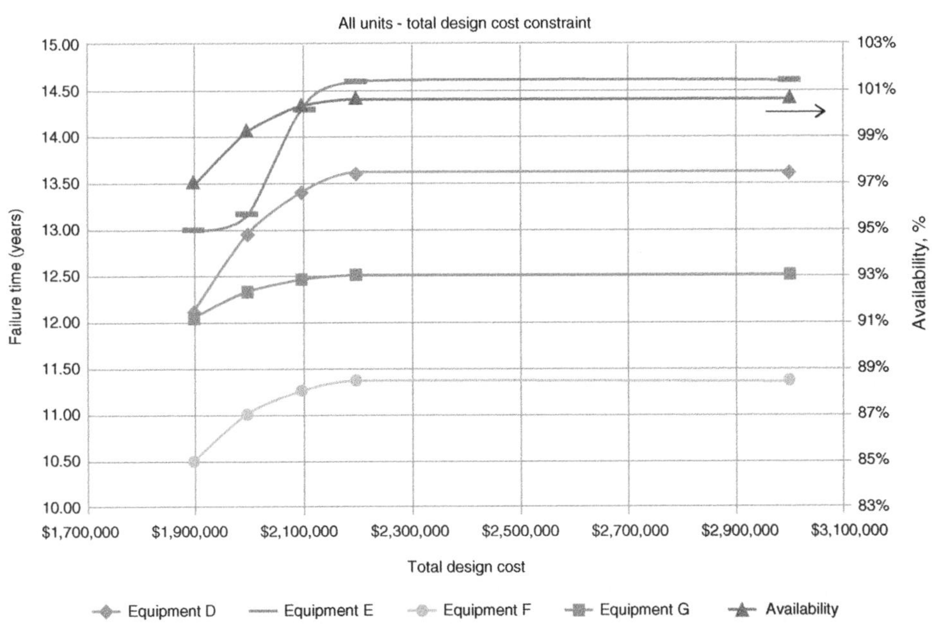

Figure 15.13 All Units – Total Design Cost Constraint

15.7.5.2 Margin

There were three case runs where the margin was varied linearly. The first case was $10,000 for the Naphtha Splitter; $10,000 for the Isom unit; and $50,000 for the Reformer. Those values were then doubled for the next case and incremented linearly one more time for the last case. In all those cases, the optimizer rode the Naphtha Splitter and Isom unit low availability limits. In addition, the Reformer final availability values were considerably higher, but little changed from one run to another. In all cases, the optimizer chose the same design options for the four Reformer equipment. There's no graph shown for this case because the results don't warrant it. The overall results are surprisingly unremarkable, in fact.

On the surface, it's somewhat unusual that the Inter-Unit Equipment Design Option cases showed such little variability. In fact, they were decidedly less interesting than the individual process units – especially in the Reformer. That's likely because, in retrospect, availability constraints were more important for the final decision than total design costs.

As a friend once commented about sports, "That's why they play the games." No matter how many people analyze the game beforehand, it's what ultimately happens on the field that matters most.

Optimization is similar. It's not always easy to anticipate the results until you actually perform the model runs. Once you do, it's possible to then dig into the "why" behind the results.

15.8 Scenario Final Thoughts

The scenarios in this chapter allowed us to draw a number of interesting conclusions:

- Minimum availability constraints are crucial in the formation of the problem. In fact, they probably dominate problem results more than any other single factor.
- Failure standard deviations can have a significant effect on the final solution since seemingly small changes can dramatically affect the future reliability of equipment, depending on the actual equipment data, such as in the case of equipment F during the Reformer model runs. Thus, getting accurate data, or estimates via our equipment experts, is quite important. Or we need to be cognizant of those situations when they might have an effect and ensure that we make the best estimates for those specific cases. It doesn't occur in every situation.
- One needs to understand whether equipment should be in "series" or "parallel" from a reliability standpoint because this can have a significant effect on the final solution.
- While it might not have been intuitively obvious before working through the example problems, the optimal solutions for the Inter and Intra-unit examples were not all that much different from one another.
  - The only factor that makes the Inter example more complicated is that the total cost constraint extends across all equipment in the Inter case, as opposed to the Intra case. However, even there, the final solution for the individual units was largely the same as for the Intra-case.
  - On the other hand, the availability constraints don't change much, if at all, since there are no cross effects between units.

15.9 Conclusion

The decision of whether to spend more money on equipment reliability is an optimization problem that balances the incremental cost of the equipment under consideration against minimum availability constraints.

In order to make the proper determination, one needs to construct a table with different cost options showing how mean failure times vary with the associated cost for that particular failure time.

In order to ultimately make the best decision, one should also consider other alternatives to increasing availability, such as increased storage tankage, changes in equipment configuration, if possible. All reasonable options should be on the table. These can be combined into one overall optimization, as desired. Those ideas will be considered in a future chapter.

16

Plant Optimization Within the Design Process

Up to this point in time, this book has focused on how to use Asset Optimization for existing plants. This makes sense because most process plants have already been built, and thus, will be where the majority of applications are focused. However, we shouldn't forget we can apply these same principles during the Design Phase.

Of course, Design is a matter of degree. For most companies, the Design Phase is mostly incremental Improvement projects, or possibly a Debottlenecking project. There will also be times when a company will build an entirely new process unit, though it doesn't occur nearly as frequently as either Improvement projects or Debottlenecking projects, which are much more common.

Even fewer companies will actually construct a brand-new grassroots facility, consisting of numerous process units. For that reason, we won't cover that case in this book. It represents such a small percentage of total applications, so we aren't leaving out much by omitting it. Plus, the methodology for a new grassroots facility is considerably more complicated than what will be presented in this chapter and is beyond the scope of this book.

The good news is that Asset Optimization can be employed for all the different types of projects: Improvement, Debottlenecking, and Process Unit Projects, but how it gets done may be slightly different from one type of design project to another.

Before we tackle each of those types of projects individually, let's touch on some important points that are common to all Design cases. More specifically, how Asset Modeling integrates with traditional process simulation modeling, as well as budget incentives for capital projects.

16.1 Combining Process Simulation with Monte Carlo Simulation

Even though it is not widely appreciated, it will be shown that Monte Carlo simulation modeling plays a role in the design process on a scale consistent with traditional process simulation modeling. In fact, this type of modeling is the natural "equipment" counterpart and the compliment to the material, energy, stream composition, and thermodynamic modeling found in process simulation modeling.

Monte Carlo modeling seeks to estimate the availability of a given plant, process unit, equipment, or other system given the equipment layout and failure/repair characteristics. When used properly, it allows a client the ability to determine an optimal solution that balances the initial capital costs against the long-term operating and maintenance costs.

Plant Optimization in the Process Industries: Incorporating Equipment/Assets in the Decision-Making Process, First Edition. Marty Moran.

On the other hand, traditional process simulation modeling packages assume that the system is 100% available to make a product since they do not consider individual equipment availability. Of course, as we all know, this is an unrealistic assumption.

That's where asset modeling comes into the picture. It takes the theoretical equipment shown as a symbol on a PFD within a process simulator and transforms it into equipment with real-life characteristics. It's where the rubber hits the road.

Since equipment cannot always be 100% available, it gives engineers options as to how much money they might want to spend to achieve a given incremental availability. In other words, if a given piece of equipment can achieve 99% availability and currently costs $50,000, how much more are we willing to spend to achieve 99.5% availability? Or is it even necessary or desirable? Could we live with 98% availability at a $30,000 cost?

It also gives engineers the ability to choose the desired level of redundancy designed into the process. For example, should we design a single train with a high individual equipment availability at a higher cost, or should we have two trains with lower individual train availabilities? As always, that's an economic decision.

16.2 Balancing the Short/Long Term within the Design Process

Designing "reliability" into a system seems intuitively obvious. Yet, as unbelievable as it may sound, it often takes a back seat to the main project goal of meeting capital budget requirements. That's unfortunate because total operating/maintenance costs are almost always several times the total initial capital costs when evaluated over the history of the process unit/plant being designed. Unfortunately, the traditional desire to minimize capital expenses during the design phase virtually guarantees higher lifecycle costs.

The reason for this suboptimal performance is misaligned incentives. The reality is that most internal project teams or engineering and consulting projects are compensated on budget adherence. Their budget is immediate and visible, while long-term operating expenses are more uncertain, and significantly farther in the future. In addition, those costs are virtually guaranteed to occur under a different manager. Furthermore, tools to accurately estimate availability only recently became available. Thus, the true long-term costs were not well known. Last, the thought leadership necessary on how to handle this challenge has been lacking.

The good news is that determining the optimal level of up-front capital expense to minimize lifecycle costs is an optimization problem that Monte Carlo simulation combined with Optimization is well positioned to solve.

16.3 Improvement Project

The difference between Improvement and Debottlenecking projects is a matter of degree. For the purposes of this discussion, let's assume that an Improvement project is normally driven by a bad actor type analysis, or similar process. The project would originate from those normally in charge of the daily operation of the unit – Operations, Maintenance, and Reliability. In general, Process Engineering personnel or upper management might be cognitive of the project, and be supportive, but they would not be driving the project.

By design standards, Improvement projects normally result in relatively small changes to the physical design of the plant. In other words, the unit physically largely remains the same after the project has been implemented. There are some changes, but in the overall scheme of things, it's not significant from a process engineering viewpoint.

That has important implications for our analysis process. First, it's not likely that the unit material balance will change much, if at all. In addition, it's likely that any prior asset optimization analysis can be reused again with minor modifications.

Improvement projects would have the following individual project steps:

- Perform Optimal Process Unit Availability Analysis Again?
- Kickoff Meeting
- Detailed Overview Meeting
- Proposed Changes and Alternatives
- Update RBD
- Consultation Process
- Optimizer Objective Function
- Model Update/Runs
- Model Review and Updates

These will be discussed in more detail below.

16.3.1 Perform Optimal Process Unit Availability Analysis Again?

The analysis process for determining optimal process unit availability will highlight projects with the greatest cost/benefit impact on the company. It's entirely possible that this particular Improvement project was driven by that analysis. If that's the case, then we already know the availability target we need to maintain for this unit and can concentrate solely on design details within the unit.

However, on the other hand, if this Improvement project was chosen for other reasons, it might be helpful to rerun the Optimal Process Unit Availability at the beginning of this project to determine updated targets.

The Asset optimization team should discuss this issue among themselves first and then bring this topic up again during the Kickoff meeting. If everybody agrees it's worthwhile to revisit determining what the unit availability target should be, then the project schedule could be amended appropriately.

16.3.2 Kickoff Meeting

The process begins with a kickoff meeting. The Facilitator, Monte Carlo modeler, as well as those with different expertise within the unit – relevant Operations, Maintenance, and Reliability staff should attend. Process Engineering isn't required, though they may decide to attend if it seems pertinent.

The purpose of the meeting is to provide a project overview of what will happen and the reason for each step. Deliverables and timelines will be discussed, with an emphasis on who has responsibility for each major phase of the project.

The Facilitator or Technical Lead will describe the role of the Asset Optimization team in the project. The Asset team will ask for any technical documents they might need for future preparation.

16.3.3 Detailed Overview Meeting

The purpose of the Detailed Overview meeting is to review technical details beyond the scope of those discussed during the kickoff meeting, which might assist the Asset team in being better prepared for future project steps. For example, they might need additional material to better understand what changes are being considered. At a minimum, this meeting should include the technical details behind the Proposed Changes and Alternatives, as discussed in the next section.

16.3.4 Proposed Changes and Alternatives

This task is shown as a separate task but may be discussed during the Detailed Overview meeting.

Every project will have a base set of proposed changes. However, in order to perform true unit optimization, we will need to evaluate a set of alternatives. These can be as simple as different types of the same piece of equipment, perhaps with differing reliabilities and costs – either initial purchase price or long-term maintenance/operational costs, such as presented in previous chapters. For each set of alternatives, there will likely need to be intermediate calculations that allow the optimizer to see how changes in reliability affect profit (since both initial purchase price will change, as well as potentially long-term operational/maintenance costs, in addition to overall production volumes.)

Of course, some design changes might be more significant. Those might be handled within one optimization run if they do not fundamentally affect the RBD. However, if the RBD must be changed, then they likely will need to be done in an iterative fashion since there is no easy way to adjust the RBD within a particular optimization run.

16.3.5 Update RBD

The Asset Team will update the current RBD if it currently exists. On the other hand, if this is a new project, it will need to create the RBD. In addition, the spreadsheet required for the Consultation process should be prepared. The goal of this activity is not perfection but rather to be well enough prepared to conduct the Consultation meeting.

16.3.6 Consultation Process

The Consultation Process will play a similar and significant role in an Improvement project – similar to what it did during previous analyses described earlier in the book.

If this unit has already participated in a previous analysis, then we would likely simply be updating information from a prior Consultation Process. In that case, the new consultation meeting would be relatively brief since most of the important information already exists.

On the other hand, if this is the first time that this unit has been involved in the asset optimization program, we can use this Improvement project as the impetus for creating basic information for this unit. In that case, the Consultation Process will be more time consuming.

It's important to point out that we need to do the analysis not only for the base set of proposed changes but also for all possible alternatives. This will be important so that the modeler can properly set up the model and any internal calculations, in addition to any information required for the optimizer.

16.3.7 Optimizer Objective Function

The optimization objective function will be set up to maximize profit over a given time horizon subject to meeting all constraints. One of the more important constraints will be various availability constraints.

For example, one availability constraint would be to define a max/min mean availability value. However, since the tails of a probability distribution can vary widely, it's likely that one would want to define additional constraints for the absolute maximum and minimum from the probability distribution function.

The reason this is important is that by honoring the minimum constraint, we will guarantee, under all circumstances, we never dip below this value. That's probably the most important constraint outside the minimum mean availability value. However, we will also want to know how the upper tail behaves since by skewing that to the upside, we may be spending money unnecessarily, given that costs increase exponentially as we approach 100% availability.

16.3.8 Model Update/Runs

The Monte Carlo simulation modeler will take input from the Consultation Process and convert that information into mathematical distributions so the model can be run. This will include running all "What If" cases.

16.3.9 Model Review and Updates

After all initial model runs have been completed, a meeting is held to discuss the results. Based on the discussion, the group may decide whether additional model runs are needed. If so, those changes are agreed upon, and the modeler runs those models. This can be done during the meeting if the changes are easy, or they can be done offline, and another meeting can be convened to discuss the results.

This process is repeated until a final agreement is reached.

16.4 Debottlenecking Project

As opposed to an Improvement project, which normally arises out of a bad actor type analysis and by those normally in charge of the daily operation of the unit – Operations, Maintenance, and Reliability, a Debottlenecking project is usually driven more by Process Engineering or senior management. Debottlenecking projects normally address a market need that might not necessarily be apparent to those with daily responsibilities in the unit.

In this case, as opposed to an Improvement project, there may be more significant equipment changes. In addition, the material balance is more likely to change.

With regard to previous Consulting exercise results, it doesn't mean that we have to start all over. Rather, it's more likely that there will be more changes than what normally occurs during an Improvement project.

Debottlenecking projects have the same following individual project steps as Improvement projects, as shown below:

- Perform Optimal Process Unit Availability Analysis Again?
- Kickoff Meeting

- Detailed Overview Meeting
- Specific Changes and Alternatives
- Update RBD
- Consultation Process
- Optimizer Objective Function
- Model Update/Runs
- Model Review and Updates

These will be discussed in more detail below, with differences noted between the project types.

16.4.1 Perform Optimal Process Unit Availability Analysis Again?

Given that the plant material balance may change due to the project, it's reasonable to expect we may need to rerun the optimal process unit target availability exercise. In fact, as we evaluate different alternatives as described in this section, it may need to be reevaluated several times as each scenario is considered.

The asset team should discuss this first among themselves, then revisit the topic during the Kickoff meeting. If everybody agrees, then the schedule should be properly amended to redetermine the proper process unit availability target.

16.4.2 Kickoff Meeting

The kickoff meeting for a Debottlenecking project will be similar in scope to that of an Improvement project. The only difference is that process engineers will likely play a much more significant role in the meeting and for the Asset Optimization project in general since they will likely be driving the Debottlenecking project.

During the meeting, the Process Engineering department will discuss an overview of the proposed changes. Detailed discussion could occur in a separate meeting.

16.4.3 Detailed Overview Meeting

The Detailed Overview Meeting is even more important for a Debottlenecking project since one will want to understand both the reason for the changes and equipment impacted, as well as how the material/energy balance will change due to the project. The purpose of this meeting is to review technical details so that the Asset team can prepare better for future project steps. While shown as a separate step below, this meeting should discuss what possible alternatives are being considered so that the optimizer can be designed properly.

16.4.4 Specific Changes and Alternatives

This task is shown as a separate task. If time permits, it could be discussed during the Detailed Overview meeting. If not, it should occur soon afterward.

The idea behind this meeting is similar to that discussed for Improvement projects. The only difference is that the range of possible changes may be more significant. In addition, one would need to evaluate if the optimal process unit availability target exercise needs to be reevaluated. This would especially be the case if one of the alternative cases materially affects the units surrounding it.

16.4.5 Update RBD

This step is the same as for an Improvement project. Based on the results from all other meetings up to this point in time, the Asset Optimization team would update the RBD, as well as any alternatives. The only difference is that there may be more changes than for most improvement projects.

16.4.6 Consultation Process

The Consultation process will play a significant role, as in an Improvement project, though it may require more work.

If this unit has already participated in a previous analysis, then the role of most equipment will likely be similar to that in the current operation. In this case, the default failure/repair values would be the discussion starting point.

One thing that we'd want to be aware of as we discuss previous failures/repairs is to evaluate whether there are any changes based on the new design. The majority of the time, the values would remain the same, but there are certain circumstances where they would change based on a new operational regime. For example, there may have been feed quality or feed rate changes that might influence equipment performance.

Then, we would examine any new equipment and assign failure/repair information based on typical values that the group would suggest. However, not only do we need to do the analysis for the base set of proposed changes, but we also need to perform the Consultation Process for all possible alternatives. This is important so the modeler can properly prepare the model, any constraints, required internal equations, and the optimization objective function.

Of course, if this is the first time this unit has been involved in the Asset Optimization program, we can use this Debottlenecking project as the impetus for creating all the required information for the unit.

16.4.7 Optimizer Objective Function

Similar to an improvement project, the optimization objective function will be to maximize profit over a given time horizon, subject to meeting all constraints. At a minimum, constraints would be process unit availability, constructed in a similar way as for Improvement projects. However, it is possible that we need to consider additional constraints that are project-specific.

16.4.8 Model Update/Runs

The Monte Carlo simulation modeler will take all input from the Consultation Process and convert those assumptions into mathematical distributions so the model can be run. This will include running all "What If" cases.

16.4.9 Model Review and Updates

After all initial model runs have been completed, a meeting is held to discuss the results. Based on that discussion, the group may decide that additional model runs are needed. If so, those changes are agreed upon, and the modeler performs those model runs. This can be done during the meeting if the changes are easy, or they can be done offline if more time-consuming, and another meeting convened to discuss the results.

This process is repeated until a final agreement is reached.

16.5 Changes to Plant-Level Model for Grassroots Process Design

Designing, engineering, and constructing a new process unit is one of the most complicated and demanding Asset Optimization projects. Not only would modeling need to be performed at the individual process unit level, but it's also likely we would need to revisit our plant-level analyses. More than likely, we would need to repeat those tasks iteratively.

At the plant level model, if none of the current units in the plant has ever participated in an Asset Optimization project, then each of the process units would need to be abstracted as a simple block with one overall failure time and repair time. "Failure time" would be the average unit uptime, while repair time would be the time to bring the unit back up to normal production rates.

On the other hand, if any of the units had already been part of a past Asset Optimization project, we could then use its submodel, which would contain all detailed equipment data.

Of course, the new unit to be constructed would initially be modeled as a simple block in the plant model with an estimated failure and repair time. These would approximate its expected overall availability and time to bring it back on stream after it had been shut down.

16.6 Grassroots Process Unit Design

Of the three projects: Improvement, Debottlenecking, and Grassroots process unit – the grassroots design is the most complicated. Not only would it require a complete Consulting exercise to determine initial estimates of failure and repair time for each major piece of equipment in its flowsheet, it would also require changes to the plant-level model as detailed in the previous section. In fact, both of these steps would need to be repeated in an iterative fashion. Thus, our project methodology would need to reflect that likelihood in our project steps.

The steps for a Grassroots Process Unit Design project are somewhat different from the other project types, as shown below:

- Kickoff Meeting
- Detailed Overview Meeting
- Review Technical Documentation
- Create Initial RBD
- Perform Initial Plant Level Model Run
- Initial Consultation Process
- Initial Model Update/Runs
- Proposed Changes and Alternatives
- Optimizer Objective Function
- Model Update/Detailed Model Runs
- Model Review and Updates

These will all be discussed in more detail below.

16.6.1 Kickoff Meeting

The kickoff meeting for a new process unit design project would be quite different than for either an Improvement or Debottlenecking project. In this meeting, the process licensor and his team will describe the overall project's detailed deliverables, timelines, and who is responsible for each step. Details would need to wait until the Detailed Overview Meeting.

16.6.2 Detailed Overview Meeting

In addition to an overview of the new unit design, in this meeting, the Asset Optimization team would be introduced to all project technical documentation they may need to access for the project. In addition, the Asset Optimization team would be briefed on the initial estimate of the design availability for the unit, along with a breakdown of major asset item costs. These would become necessary later on during the modeling and optimization phases.

16.6.3 Review Technical Documentation

The Asset Optimization project team would scrutinize the new process unit documentation to discern details necessary for the Asset Optimization project.

16.6.4 Create Initial RBD

Based on the previous technical documentation review, the Asset Optimization team will create an initial RBD for the new unit, as well as the initial spreadsheet to be used for future Consultation meetings.

16.6.5 Perform Initial Plant Level Model Run

The starting point for the Asset Optimization project would be the plant-level model. As discussed previously, these would be simple blocks, if no prior analysis had been done on that unit, with assumptions for uptime and restart time.

Or it would be the submodel corresponding to that unit if there had been an earlier analysis. However, before directly using a submodel, one should cross-check to see how well uptime and restart times are in predicting current unit performance (in a single block) against the submodel results. If those results are not consistent with one another, then it's probably better to use a single block with estimates for average unit availability and normal restart time in the plant model.

At this point, we would need to use the design or estimated availability that we learned during the Detailed Overview Meeting. If no availability target was provided, we could just assume a value. While we need a decent starting point, the final target availability for the unit will almost certainly be different from this value as we move through the optimization.

The model would now be run with those assumptions to generate a base case.

16.6.6 Initial Consultation Process

We will need to complete an entire Consulting exercise, reviewing all proposed equipment in the new unit to generate initial estimates of failure/repair time and production loss, if the block should fail. In this sense, it is no different than what was described during the initial discussion of the Consulting approach.

That could be scheduled separately in the same fashion described in earlier chapters. However, it's also possible that we might be able to piggyback onto other meetings where the right personnel were already attending a separate review, such as a HazOp. During a new process unit project, there are many reviews, so there might be an opportunity to attend one of those other meetings to generate the data for our analyses.

16.6.7 Initial Model Update/Runs

Based on the input from the Consultation Process, the Monte Carlo simulation modeler takes input from those meetings and converts those initial estimates into mathematical distributions. The modeler would then run the model once to determine how well the calculated availability matches the proposed design availability. This is only a cursory check, more a sanity check, than anything else. More detailed model runs, typical for an asset optimization project, follow in later steps.

16.6.8 Proposed Changes and Alternatives

Now that we have a base plant and unit model, we would want to conduct a meeting with the design team to understand what various alternatives are available. The idea is similar to what was discussed during Improvement or Debottlenecking projects. That is, what are the possible range of changes that we could make? We would want to start with items the design team considered to be critical paths, high-cost equipment, or any equipment that had major question marks associated with them. Failure and repair estimates would be made for these items as a part of this meeting.

16.6.9 Optimizer Objective Function

The optimization objective function is similar to what was discussed previously in the Debottlenecking section. It would be set up to maximize profit over a given time horizon subject to meeting various unit constraints.

Some of these will be process unit availability constraints, as presented for the other project types. We would be including max/min mean availability, as well as an absolute maximum and minimum from the probability distribution function.

However, we may desire to include other more equipment-related constraints.

16.6.10 Model Update/Detailed Model Runs

The Monte Carlo simulation modeler takes the input from the Proposed Changes and Alternatives meeting and converts all possibilities into mathematical distributions.

16.6.11 Model Review and Updates

The model would be rerun, and the group presented with the results. If they were not satisfactory, the group could review the Pareto chart to determine what equipment would be a good candidate to change and/or improve. We would also like to review how these results fit into the plant-level model. While further updates may only be required for the new grassroots unit model, there's always the possibility that we need to rerun the plant model again. This iterative approach may require several cycles before we converge to an answer.

This process would continue until the project met the cost and material balance requirements.

16.7 Design Considerations

As you proceed with your design, it's important to keep in mind that increasing reliability doesn't always mean purchasing individual equipment with better reliability. While that is probably the principal route to higher plant or unit availability, it is by no means the only path, as discussed in a previous chapter.

In fact, there are certain practices that are inherently part of a process plant that are there solely to protect against "unreliability." However, it's likely we never thought of them in that manner. The two things that immediately spring to mind are increased storage capacity and retaining more spare parts. However, we shouldn't forget about the impact of how process configuration could affect availability within a design, as articulated previously.

For completeness, let's review those other ideas.

16.7.1 Storage Capacity

While operational people might politely disagree, intermediate storage tanks are a hedge against an upstream unit's unavailability. As shown mathematically in a previous chapter, storage tankage between units can help increase the system availability of several units combined together over the case when one unit runs down straight to another.

In fact, it might be a more economical solution than trying to increase the upstream unit's availability. From a purely economic standpoint, you are trading off the cost of tankage and additional tank maintenance versus trying to improve the availability of various equipment in the upstream unit.

When making the final decision, one should be open to Operations viewpoint about extra operational flexibility. If it's a close call, consider that Operations might like to have the extra tankage from that vantage point; it gives them flexibility that they might not have now.

16.7.2 Spare Parts

Spare parts are inherently a buffer against unreliability. Think about it this way. If equipment were always available, you would never need spare parts. However, that's clearly unrealistic, so we stock up. As equipment becomes less reliable, we stock up even more. Plus, we retain more if we don't trust the vendor to deliver the part in a reasonable time frame, or if there are prohibitively long lead times.

Ironically, there may be times when storing more spares is a more economical solution than trying to increase equipment availability within a unit. However, it often takes a more detailed analysis before a final call can be made, as described in Chapter 11.

16.7.3 Process Configuration Matters

Normally, we think of process configuration as the path the process fluid traverses as it moves through the unit. However, from a system reliability standpoint, process configuration is different than moving through equipment on a process flow diagram – even if there is overlap.

As explained in detail in the Appendix on Monte Carlo simulation, a system is available when all the blocks (units or equipment) are available such that you can trace a path from the beginning of the reliability block diagram through to the ending block. From theory, and pure common sense for that matter, we know that if units (or equipment) are connected in series with no buffering capacity, then if any of the units (or equipment) fail, the entire system fails. In that case, system availability is simply the product of the individual units (or equipment).

However, providing a parallel path can increase system availability significantly. It could be pure redundancy, such as a spare pump, or it could be an entire train.

Thus, one aspect of the design process should be to evaluate where and when to add either a parallel path or pure redundancy.

16.7.4 Improve Maintainability

While probably less important in a design scenario, there are times when one can improve system availability if it is possible to reduce the time from when the requipment is shut down the time that it is put back into service. Anything that can reduce the time for any of the following activities will move us directionally toward improved profitability:

- Cool the plant down
- Isolate the equipment
- Repair the equipment
- Return it to the service
- Bring it back up to the temperature

16.8 Conclusion

"Reliability in Design" is an important concept and is part of Optimization via better asset decisions. In fact, good design decisions can have a more positive impact than virtually anything else discussed in this book. That's because one will live with design decisions throughout the equipment's life cycle. As the old saying goes – "Decisions have Consequences." In the design world, nothing could be truer.

This applies to all three different types of design projects discussed in this chapter – Improvement, Debottlenecking, or Grassroots units. Each of these projects proceeds thoroughly roughly through the same process:

- Kickoff Meeting
- Update RBD
- Perform Optimal Process Unit Availability Analysis
- Optimizer Objective Function
- Proposed Changes and Alternatives
- Consultation Process
- Model Update
- Model Review and Updates

During the design phase, it is important to be able to determine how different process configuration and equipment layouts affect availability and asset utilization. This is not easy, even for experienced engineers, given highly integrated process unit and equipment configurations and intermediate storage tankage. This is clearly an optimization problem.

Improving unit or plant availability is more than just improving individual equipment availability. In fact, increased storage tankage or increased holding of spare parts can be an effective way to increase availability, but it is situation-dependent. In addition, parallel processing paths or more redundancy can also be a cost-effective way of increasing profitability.

Table 16.1 summarizes important aspects for the three different types of projects.

If we examine the rows in the table, we see that as one proceeds from an Improvement project to a Grassroots design, there is much more process engineering involvement. In addition, the material balance changes much more significantly. Improvement projects are largely local unit projects. However, projects become more complicated as we move to the right. For the Debottlenecking and Grassroot Design project, one has to be much more cognizant of how the project under consideration affects surrounding units.

Table 16.1 Summary of Project Type Differences

	Improvement Project	Debottlenecking Project	Grassroots Design
Material Balance Change	Not much, if any	More significant	Major change
Process Engineering Involvement	Depends on project	Yes	Necessary
Iterate between plant view and unit view	Not much	Maybe some	Necessary
Consultation Meeting Results Reusable?	Yes	Decent starting point	Need new analysis
Redo Optimal Process Unit Availability Analysis	Normally not needed	Probably a good idea	Necessary
Amount of asset optimization work	Largely update work	Medium project	Large project

Obviously, the amount of Asset Optimization work increases significantly as one moves from left to right. While Improvement projects can reuse a significant portion of their Consultation effort, we have to allocate more time to that effort as we move toward a Grassroots project.

Another way to think about this table is to analyze it from a project-type basis, as shown below.

16.8.1 Improvement Project

In general, an Improvement project is a local unit project that doesn't even impact the unit material balance much, if at all. Hence, it won't change the overall plant product flows, so normally, one doesn't need to be concerned about how this affects the remainder of the plant. Local plant personnel, such as appropriate Operations, Maintenance, and Reliability are usually sufficient to be able to execute the Asset Optimization aspect of the project, supplemented with Process Engineering, if necessary.

From an Asset Optimization perspective, these are not normally large projects. Rather, it's usually a matter of updating a previous project unless it doesn't exist. If it is a new effort, an Improvement project is a good excuse to piggyback an Asset Optimization project for this unit to help build out the Asset Optimization infrastructure for that process unit.

16.8.2 Debottlenecking Project

A Debottlenecking project is more complicated than an Improvement project because the unit and plant material balances usually change more as a result of the project. For that reason, one has to pay attention to how a Debottlenecking project impacts the units surrounding it. Since it will likely affect more equipment in the unit, process engineers may play a more important role.

While the Asset Optimization project efforts for a Debottlenecking project can be significant, it's likely that the majority of any previous Consultation work can be reused. If a previous project has not been executed on this unit(s), then it's a good idea to use this opportunity to achieve that goal.

16.8.3 Grassroots Project

A Grassroots design is the most complicated project since it involves new process units. By definition, the material balance changes significantly. Of course, process engineers are heavily involved.

From an Asset Optimization perspective, these are the largest projects. One has to start from scratch since the previous material doesn't exist. As one works on a given process unit for a Grassroots design, one has to be cognizant of how this affects other units surrounding it, as well as how the plant material balance changes. Often, this requires an iterative approach, since when one does some work at the unit level, there may be a requirment to go back to the plant level to see how that previous work impacted the plant as a whole. In this sense, this type of Asset Optimization project is quite different from either of the other two project types.

17

Combined Optimization

Up until this point in time, we have discussed the following optimization problems:

- Turnaround timing/duration, as well as what equipment should be serviced in a given turnaround
- Optimal Improvement Projects (optimal process unit availability)
- Maintenance Plan Options Optimization
- Spares Optimization
- Task Optimization
- Storage Tankage Determination/Optimization
- Asset Design Optimization

Those discussions in previous chapters presented these as separate, independent problems. However, that was done only in the spirit of making those ideas easy to understand. It was simply easier to convey the concepts if split into individual problems. In addition, whenever learning something new, it's easier to learn those concepts in smaller chunks and only combining them when you have had some time to absorb the basics.

The same thing will happen as the adoption of these ideas occurs in our industry. Borrowing the analogy of crawl, walk, run, realistically, our industry has been crawling when considering whether we have been including the important asset information in plant optimization problems. That's because, for the most part, optimizations such as those detailed above, are not being done in any significant way.

If there is any attempt at all, it's that they are being executed as individual Monte Carlo simulation primarily evaluating availability. That's not the same thing as optimization because those problems are not being done with a profit-based optimizer sitting above them, nor with constraints included in the problem definition.

How often have you encountered a world without constraints? No budget constraints, no labor or material constraints? That's not the world I live in. Constraints are omnipresent in the real world. They must be part of the problem definition.

So, once the industry has been crawling for a while, we will eventually progress toward walking, then running, and last sprinting. What will that look like when we get to that stage?

The answer is probably obvious. We will begin to break down the barriers around each of these individual optimizations and combine them into one or more overall optimization(s). Up to this point in time, the boundaries around the optimizations have been intentionally artificial. But that's not the way that the real world actually works.

Plant Optimization in the Process Industries: Incorporating Equipment/Assets in the Decision-Making Process,
First Edition. Marty Moran.

We can now remove some of those artificial distinctions and group the optimizations into more logical sets. The two listed below spring to mind and have the potential to add real value:

- An optimization that would extend across the entire plant and combine the following optimization problems together:
  - Optimal Improvement Project (Optimal Process Unit Availability)
  - Crude Mix
    - In other words, this question challenges the organization to consider whether they would change the improvement project mix (and resulting process unit availability targets) in light of a change in crude charge types. This will allow us to compare and contrast our results here with those earlier in the book to see how they might be different.
- The second combined optimization problem is similar to the previous one in that we will also be determining the optimal crude mix while considering what equipment we would be restoring in the crude unit turnaround. Similar to the previous problem, we'll now be able to look back and see if the decisions we made previously when not considering a feed mix change would be the same. Maybe. Maybe not.

We could have only chosen examples that were only "asset-centric," such as combined Spares and Tasks optimization (as your teacher told you in school, that example is left for the reader). However, I would like to challenge our industry to start combining the process and asset viewpoints together. The two example problems in this chapter were specifically chosen to show how optimization can bridge the gap between these worldviews.

It's important to note that while these are fairly logical groupings, there's nothing that forces our hand toward just these two problems. They can actually be grouped in any logical fashion to solve the issue at hand. The groupings above are just some fairly obvious choices.

At this point, let's work through each of those examples.

17.1 Combination of Improvement Projects and Crude Feed Mix Optimization

The first example will be a refinery-wide optimization problem. It will choose the best improvement projects (i.e., availability) for each of the major process units, while also selecting the optimal crude mix of crudes A/B.

Including crude A/B in the optimization problem is important if the organization is considering this change because each crude has a different feed cost and different yield structure, but also affects equipment performance in significantly different ways, primarily due to its differing sulfur content.

For example, if we only ran crude A, we may have come to a certain set of conclusions about the most favorable improvement projects and tankage choices. On the other hand, if we only ran crude B, we might have come to an entirely different set of decisions.

What's interesting about this problem is that the optimizer will be adjusting independent variables that have no direct relationship to the Monte Carlo function. While we are including only one process condition change in this manner, there's no reason why we couldn't include more.

Thus, the objective of this optimization problem is to choose the best combination of improvement projects along with the best crude mix to optimize profitability. As in the previous optimizations involving improvement projects, choosing the best improvement projects will also determine the optimal availability targets for each process unit.

17.1.1 Process Unit Availability/Improvement Project Data

The historical availabilities for each of our refinery process units is shown in Table 17.1.

The cost to improve a given unit's availability from its current value to some future availability target is shown in Figure 17.1.

Let's assume the following market prices as shown in Table 17.2.

Table 17.1 Process Unit Historical Availabilities

Process Unit	Availability (%)
Crude	97
Vacuum	96.5
Coker	95
Sat Gas Plant	98
Naphtha Splitter	98
Isomerization	98
Reformer	97.5
Diesel Hydrotreater	97
FCC	96.5
Alkylation	97
Hydrocracker	97

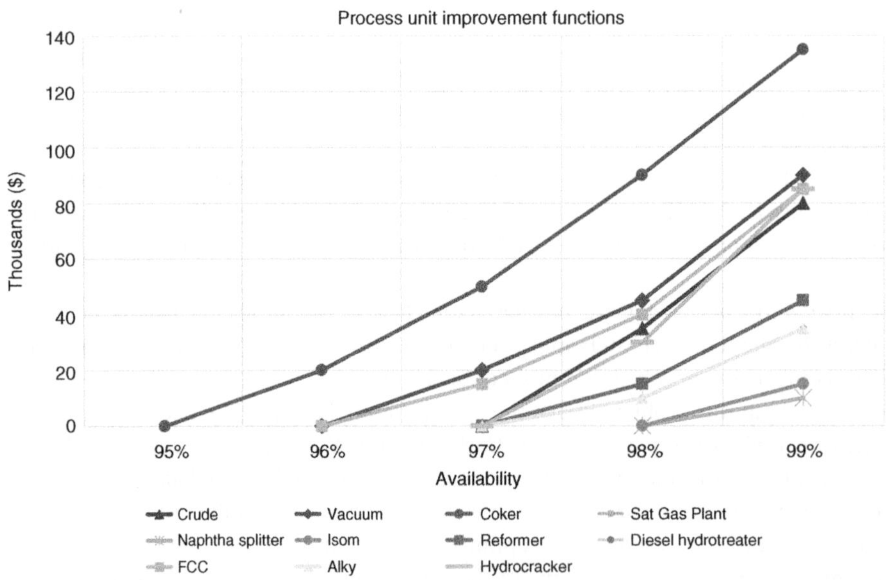

Figure 17.1 Process Unit Total Project Cost vs. Availability

Table 17.2 Assumed Market Prices

Stream	Price ($)
Crude	40/bbl
Gasoline	2.00/gal
Kerosine	2.00/gal
Diesel	2.00/gal

17.1.2 Crude A/B Characteristics

The data for the two crudes is the same as that presented in an earlier chapter.

Crude A contains more sulfur than crude B and is considered to be a "sour" crude. On the other hand, crude B is a "sweet" crude with a significantly lower sulfur content.

The major difference between crude A and B is that crude B is a lighter crude, as indicated by its API number. Thus, crude B will produce somewhat more lighter products when initially distilled. The good news is that our example plant is a complex refinery with considerable "conversion" capacity to convert heavier streams into more valuable light products. In other words, the fact that crude A is heavier will have less of an impact than it could in other refineries, given our operational flexibility. In fact, after processing each crude through all of our 11 process units, the final product distribution for each crude is shown in Table 17.3 based on a feed rate of 100,000 barrels per day (BPD).

What this clearly indicates is that from a pure material balance standpoint, it's more desirable to process crude B since it produces more valuable products. However, price also needs to be included since crudes are priced based on several factors, with the distribution of lighter products being only one of the more important factors. Another is sulfur content.

In that sense, another downside is that crude A has significantly more sulfur than crude B. Crude A is what we refer to as a "sour" crude with a sulfur content of 2% – which is significant. On the other hand, crude B is considered a "sweet" crude since it only has a sulfur content of 0.25%.

Most people in the refining industry are aware that higher sulfur levels can play havoc with refinery operations. So, as sulfur content increases, they become concerned it might negatively and significantly affect equipment life. In this instance, the planning department has reached out to our team, since they have become aware of our program, to ask if they could evaluate what might be the upper limit for running crude A before they make any final decisions.

The planning department has informed us that their strategy would be to alternate the crude run roughly every week due to crude tankage limitations. They expect the percentage of each crude would remain roughly the same each year over a multi-year time frame.

As in our previous example, we will confine the equipment being affected by the different crudes to just furnace tubes. In reality, more equipment other than just furnace tubes would likely be affected.

Table 17.3 Crude A/B Product Material Balance

	Gas	Naphtha	Kerosine	Diesel	AGO	Crude Resid	LVGO HVGO	Vacuum Resid
Crude A	1,300	18,000	12,900	11,900	13,400	42,300	18,697	23,519
Crude B	1,600	25,000	11,000	11,933	12,400	38,000	19,000	19,000

17.1.3 Equipment Impact

While there are likely many different pieces of equipment that could potentially be impacted by the difference in crude types, let's demonstrate the concept by simulating furnace tubes as our example.

If there is only one crude in our problem definition, we would have assigned one, and only one, failure rate to furnace tubes during the Consulting exercise. However, now that we will be using two crudes, we need to reevaluate how changing the percentage of different crudes being run can potentially affect the furnace tube failure rate. To achieve that goal, we will need a dynamic calculation across a range of feed compositions. That's where both our equipment experts will be a significant help.

First, our equipment experts will need to estimate how many failures would occur if only crude A were running. They've told us that they expect the furnace tubes to last 20 years based on solely running crude A with a standard deviation of 5 years, so there's plenty of variability when the tubes could fail. (BTW, these figures are totally made up for illustration purposes to make the math easy for the problem and are not intended to be linked to any realistic equipment estimates observed in refinery operation.)

Now, they will also need to do the same for crude B. For this case, they have decided that crude B would decrease the failure rate by 50% over and above crude A. Those two bookends define the different ends of the spectrum.

Now, our experts would need to estimate how the failure rate would change as the percentage of different crude changes. In other words, they would need to estimate the failure rate across the entire range from 100% crude A and 0% crude B, all the way up to the opposite end – 0% crude A/100% crude B.

While a linear extrapolation would seem to be the obvious answer, and might be one possibility, the equipment experts conclude that there are nonlinear characteristics, as shown graphically in Figure 17.2.

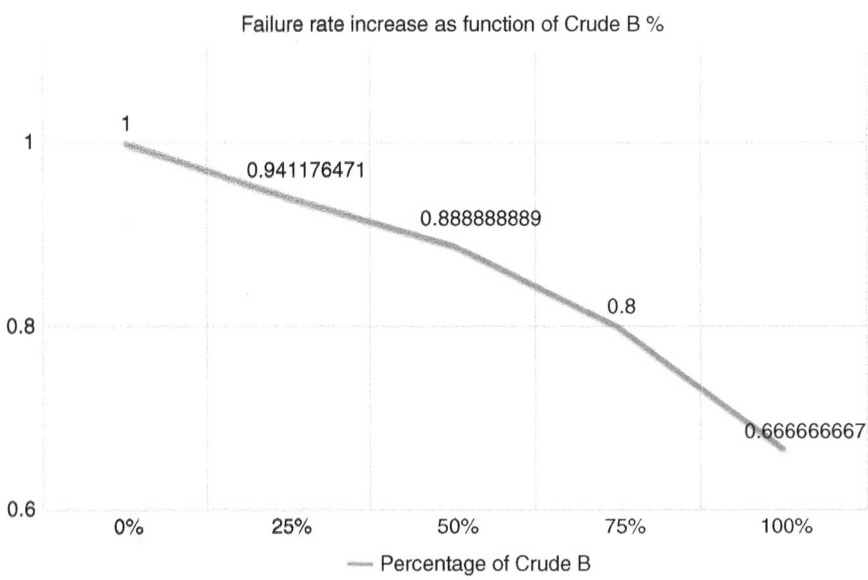

Figure 17.2 Failure Rate Increase vs. Crude B%

17.1.4 Optimizer Definition

This problem will be performed across the main refinery process units. Fundamentally, the optimizer will choose each process unit's target availability and crude A/B split. The availability choices will then allow us to calculate the cost of the improvement projects necessary to achieve those availabilities. The crude mix will also allow the determination of the furnace tube failure rate. After the Monte Carlo engine runs, we will have determined system availability, which will allow us to calculate the product distributions and hence revenue. Profit is calculated as the difference between the product revenue and the sum of the total crude feed and total project improvement costs.

17.1.5 Independent Variables

The independent variables for this problem are:

- Each process unit's availability target except the crude unit
- Percentage of crude A being run
- Once the percentage of crude A has been determined, we know the percentage of crude B since they must sum to 100%. Therefore, crude A and B are not independent of one another.

The reason that the crude unit availability target is not directly manipulated by the optimizer is that it is not truly independent. That's because as soon as we choose a crude A/B mix percentage, it becomes possible to calculate the crude unit availability.

17.1.6 Dependent Variables

There are seven main dependent variables for this problem:

1) Each unit's improvement function
2) Individual gasoline component totals
3) Total product volume increase
4) Total product revenue
5) Total improvement project expenditure
6) Furnace tube failure rate
7) Total crude feed cost

Each time that the optimizer chooses a process unit availability target, it will use the improvement project graphs to determine how much it must spend to achieve that new availability target. There is one equation for each different process unit.

The refinery material balance in the Appendix allows us to calculate gasoline component production for each new availability target. The final gasoline component streams are then added together and multiplied by the wholesale gasoline price to determine the total revenue increase for the new availability targets.

The total values for each of the individual gasoline component streams are also included as constraints since the planning department can estimate the maximum/minimum values that it needs to blend gasoline successfully later.

Another calculation is the total improvement project expenditure. This will be used in two different places. First, it will be one of the major dependent variable constraints. In addition, it will be an important aspect of the objective function calculation.

Then, after the Monte Carlo program has run, production values are calculated based on overall system availability. These are then multiplied by their appropriate wholesale prices to determine the total revenue increase for the new optimizer choices.

To account for how different crude mixes affect our furnace equipment, we need to include dependent variable equations for the furnace tube failure rate determined based on the crude mix being run. In other words, different crude choices can dramatically affect equipment performance, which shows up as differing system availability. Equipment failure rates would be based on estimates provided by equipment experts.

There will also be a dependent variable calculation for total crude cost since the price of each crude varies in the marketplace.

17.1.7 Constraints

There will be two constraints defined for the problem.

The first will be the minimum gasoline pool quantity since we are contractually obligated to deliver certain gasoline volumes.

The second will be a minimum crude unit availability, which will be calculated to reflect the poorer long-term availability when running higher quantities of crude A. This is due to its higher sulfur values and related higher failure rates, as shown in the graph earlier in the chapter.

17.1.8 Objective Function

The objective of this problem is to maximize profit, which is the difference between incremental production revenue minus the combination of incremental improvement project spend and increased crude feed costs.

17.1.9 Results from Previous Problems

Before we actually detail the results from the combined optimization, it's instructive to step back and review the individual optimizations from previous chapters so we are better able to predict how those might change for the combined case.

In the individual project improvement optimization case, the major finding was that the optimizer favored crude unit projects over all other projects. It spent money on those projects first, and if it came time to reduce spending, to cut them last. Second, it would also spend money on FCC projects (a major gasoline producer) and less on the Coker and Hydrocracker. So, in other words, it focused on feed rate through the crude unit and then on the major gasoline conversion units in a prioritized fashion.

On the other hand, in the case of the crude mix optimization, we found that the optimizer was extremely sensitive to crude B's price relative to crude A. As crude B's price increased on a relative basis, and over a fairly narrow price range, it would significantly decrease the percentage of crude B. In addition, there were limits to how much of crude A could be run since it would start to run up against minimum availability limits.

17.1.10 Results

So, now let's turn our attention to the combined case results. Figure 17.3 shows the relationship between changes in crude B's feed price at a constant crude A feed price of $40/bbl. What one notices is

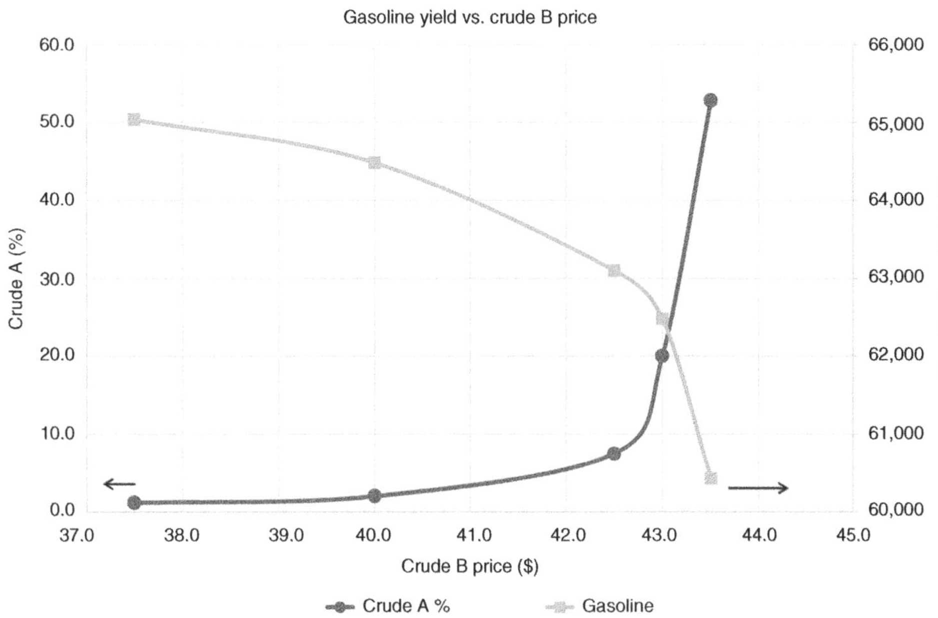

Figure 17.3 Gasoline Yield vs. Crude B Price

that there is a very steep change in the optimizer crude B choice as one approaches a price of $45 for crude B. In essence, it flips above $43.

Plus, since there is a yield difference between the two crudes, the optimal gasoline quantity decreases dramatically exactly in sync with the crude price increase and rapidly approaches the gasoline pool constraint limit of 60,000 barrels per day above a crude B price of $43. That makes sense since crude B produces considerably more naphtha material, so backing it out reduces gasoline yield. (Naphtha is principally gasoline material.)

Figure 17.4 shows the pure financials for this optimization. At a crude B price of $43, the optimizer chooses to spend less money on improvement projects. From the previous graph, we know that gasoline yields are lower at higher crude B prices. Combining those two thoughts together, the optimizer can make more money by reducing crude B feedstock costs, even though it reduces gasoline yield with the bonus that it can reduce improvement project investments at the same time.

Figure 17.5 shows how the optimizer decides to spend the improvement project money as the amount of crude B increases. Above $43 crude B feedstock price, the optimizer dramatically decreases its improvement project spend – especially with regard to the crude unit. This really shows us that crude feedstock price is the driving determinant of this problem. It doesn't care as much about gasoline production, so the optimizer backs out investment since it saves so much money on total crude feed acquisition costs.

Figure 17.6 shows how crude unit availability changes as a result of the project improvement investments.

What this combined optimization example showed is how significantly different optimization results can be when we start combining all the various variables into the optimizer definition.

In this particular case, the optimizer only chose to change crude unit project investments since crude acquisition costs were the major determinant driving profit. In the previous example, when the crude mix wasn't changing, the optimizer selected a set of investments that favored conversion

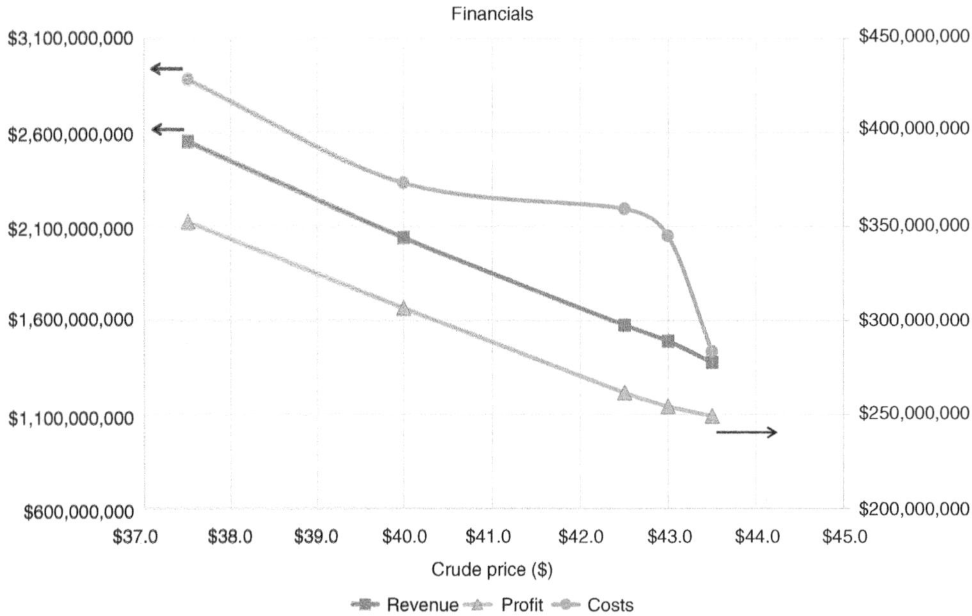

Figure 17.4 Financial Results vs. Crude B Price

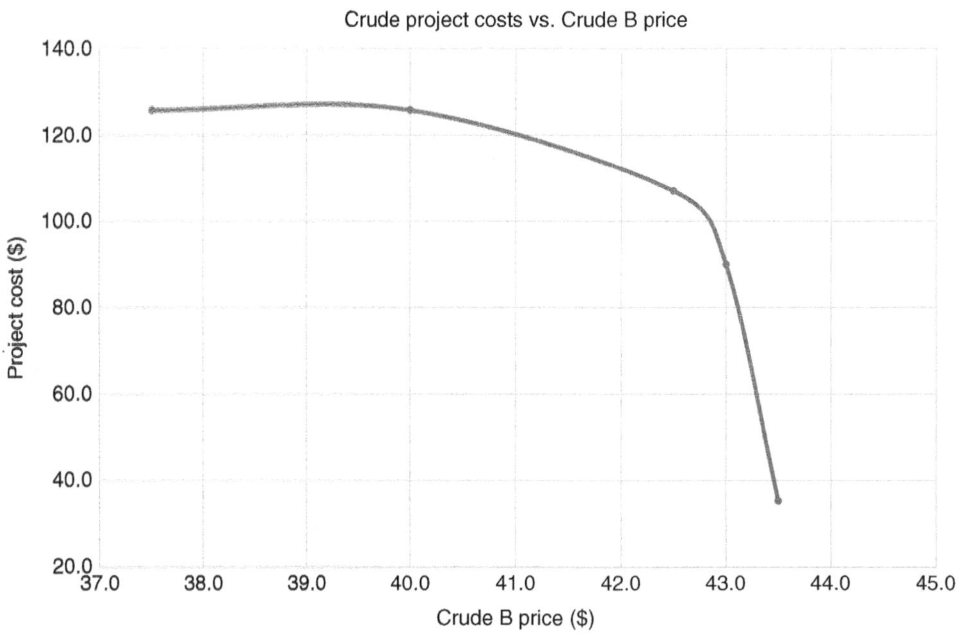

Figure 17.5 Project Costs vs. Crude B Price

units downstream of the crude unit. In this case, those were minor contributors to the profit function and were overlooked by the optimizer. What this also shows, and is well known to those familiar with the refining industry, is that crude selection is one of the biggest contributors to profitability and significantly greater than small changes in incremental conversion yields. However, it's difficult to truly anticipate all these results before one runs the optimizer. You have to run the numbers.

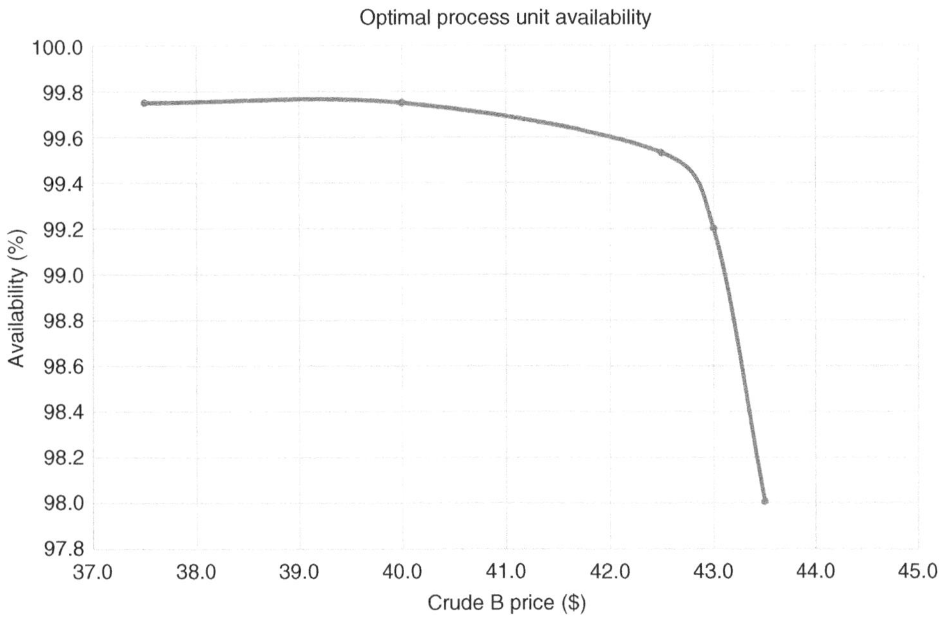

Figure 17.6 Optimal Process Unit Availability vs. Crude B Price

17.2 Combining Turnaround and Future Feed Composition

In this example, let's assume we are four years into a crude unit run, and we have tentatively scheduled a turnaround one year from today. The duration will be 21 days, and the equipment we will consider in our problem is not on the critical path. There will be another turnaround five years from the time this next turnaround is completed.

We are unsure whether to include 12 equipment, labeled equipment A–L, with failure, repair, and production loss characteristics as described in the next section. Our job is to perform an optimization to determine which equipment we recommend be included, and which we do not.

Up to this point in time, this problem is exactly similar to the one we solved back in the Turnaround Considerations chapter. However, we are now going to introduce a nuance where the plant is considering running two crudes in the future (crude A/B from the previous section), as opposed to just crude A as it does right now.

Therefore, the optimizer will need to determine both the optimal crude split and the proper equipment to restore. Note that what equipment is to be restored will potentially be different depending on the crudes selected.

17.2.1 Basic Turnaround Problem Data

The data for this problem will be exactly the same equipment list that was presented for the problems associated with the Turnaround chapter earlier in the book. Let's review that data once again and assume that the failure rates presented in the table are those when running crude A. All values listed in Table 17.4 are in years.

Reviewing the data, we see that the second column, **Current Age**, lists how long the equipment has been running since last fully repaired, with the third column, **Mean Time Failure**, being the mean time to failure estimated by our equipment experts. Column 4, **Estimated Failure Date**,

Table 17.4 Equipment to be Considered for Turnaround

Equipment	Current Age	Mean Time Failure	Estimated Failure Date	Std Dev Failure	Earliest Failure	Latest Failure
A	3.5	10	6.5	1	5.5	7.5
B	3	17	14	3	11	17
C	3.5	14	10.5	1.5	9	12
D	5.5	14	8.5	1.5	7	10
E	0.5	10	9.5	1.5	8	11
F	3	8	5	1.6	3.4	6.6
G	3.5	12	8.5	1.5	7	10
H	3	13	10	2	8	12
I	5	15	10	2	8	12
J	7	16	9	2	7	11
K	4	12	8	1.5	6.5	9.5
L	7.5	18	10.5	3	7.5	13.5

represents the approximate time when we expect the equipment to fail based on the mean failure time using some simple math.

The next column, **Std Dev Failure**, is the best estimate from our equipment experts of the time when 68% of the likely failures would occur concentrated around the mean. In other words, 32% of the failures will be less than the mean, while the other 32% will be greater. From our perspective, we are only worried about those that are less than the mean since those would occur before the end of the next turnaround.

So, if we want to know the range where the equipment would fail, it would be one standard deviation around the **Estimated Failure Date**. Thus, the last two columns represent the earliest and latest time when each of the equipment would fail, assuming those failures occurred within one standard deviation around the mean.

So, equipment A would fail somewhere between 5.5 and 7.5 years from now. But we are still one year from the turnaround. Thus, one year down the road, those same figures would be 4.5 and 6.5 years, assuming that it doesn't fail between now and then. Similar figures for equipment F would be 2.4 and 5.6 years when the turnaround is scheduled to start.

17.2.2 Rearranged Turnaround Data – Crude A

We are going to skip some intermediate steps from the turnaround data explanation in the Turnaround Considerations chapter and jump to the final table. If one needs to refresh one's mind, return to the chapter, "Turnaround Considerations" and read through the complete introductory description.

The following table is prioritized based on the column **"Earliest Failure (at TA)"** with the smallest values shown first extending to the largest values. The column **"Earliest Failure (at TA)"** is similar to the column **"Earliest Failure."** The only difference is that all values in the **"Earliest Failure (at TA)"**

Table 17.5 Crude A Rearranged Information

Equipment	Std Dev Failure	Earliest Failure 3 Std Dev	Earliest Failure 2 Std Dev	Earliest Failure (at TA)	FOFI Time (Days)	Repair Time (Days)	Repair Costs
K	1.5	2.5	4	5.5	1.5	0.5	High
J	2	2	4	6	2	0.5	Moderate
D	1.5	3	4.5	6	21	7	Very High
G	1.5	3	4.5	6	10	6	High
L	3	0.5	3.5	6.5	2	0.5	High
H	2	3	5	7	7	3	Moderate
I	2	3	5	7	21	10	Very High
E	1.5	4	5.5	7	3	1.5	Moderate
C	1.5	5	6.5	8	2	1	Moderate
B	3	4	7	10	3 days	1 days	Moderate

are 1 less than in the **"Earliest Failure (at TA)"** column. That's because the turnaround is still one year away. At the turnaround, it's one year less.

Note that there are only 10 equipment, as shown in Table 17.5. Per the discussions in the earlier chapter, equipment A and F are less than five years old, so they will not make it to the turnaround on average. Therefore, we removed them from consideration since they will definitely be restored.

The middle columns show the earliest failure for 3, 2, and 1 standard deviations, respectively. This comprises 49.85% of the failures less than the mean value – or for practical purposes, all of them.

This table also includes three columns on the right side of the graphic. The first column details the production loss if the equipment should fail, labeled as **"FOFI Time,"** which is shorthand for "Feed Out – Feed In." The second column, **Repair Time**, is the amount of time that it would take to repair the equipment if it should fail during the production run. The last column, **Repair Costs**, is a qualitative measure of the cost of repairs if it should fail outside the turnaround.

17.2.3 Crude B Turnaround Data

Crude B is a lower sulfur crude, so if there is any effect at all, it will generally tend to lengthen failure times. As shown in Table 17.6, some equipment mean times have been affected, while others are not at all. It was assumed standard deviations remained the same.

If we now show the same information for Crude B, as illustrated in Table 17.7, as we have already shown for Crude A, we will see some noticeable differences.

Most importantly, we see that the values are often longer than compared with crude A. As can be expected, these differences may change what equipment gets restored in which turnaround.

17.2.4 Solution Approach

Problem solutions in this chapter were modeled in the same manner as already done for the turnaround-only problem earlier in the book. That is, the Solver was used as the main optimization

Table 17.6 Crude A/B Failure Information Comparison

Equipment	Mean Time Failure (Crude A)	Mean Time Failure (Crude B)
B	17	17
C	14	18
D	14	20
E	10	11
G	12	12
H	13	16
I	15	17
J	16	18
K	12	15
L	18	20

Table 17.7 Crude B Rearranged Information

Equipment	Earliest Failure 3 Std Dev	Earliest Failure 2 Std Dev	Earliest Failure (at TA)	FOFI Time (Days)	TA Repair Time (Days)	TA Repair Costs
G	3	4.5	6	10	6	High
J	4	6	8	2	0.5	Moderate
E	5	6.5	8	3	1.5	Moderate
L	2.5	5.5	8.5	2	0.5	High
K	5.5	7	8.5	1.5	0.5	High
I	5	7	9	21	10	Very High
B	4	7	10	3	1	Moderate
H	6	8	10	7	3	Moderate
C	9	10.5	12	2	1	Moderate
D	9	10.5	12	21	7	Very High

algorithm while including aspects of Monte Carlo analysis through the use of regressed equations. Monte Carlo was not used directly.

The solution approach required modeling equipment on/off switches, and subsequently turnaround costs, as well as estimating production run 1 and 2 availability.

It's important to note that all the equipment in the example is in series from a reliability perspective. In such an arrangement, the availability of a "system" of this equipment is simply the multiplication of the individual availabilities. Thus, if all 12 equipment in the example had 99.75% individually, then the overall "system" availability would be 97.5%. So, even though most of the equipment had exceedingly high availability individually, when multiplied together, the system availability total dropped significantly. Keep that in mind as we work through the examples since it will play a significant role in our thought process.

17.2.4.1 Discrete vs. Continuous Optimization

The decision to repair a piece of equipment is obviously a discrete decision. You either decide to repair the equipment, or you don't.

This is a classic Integer optimization problem, with the optimizer adjusting the combinations of Repair/Not Repair for each of the 10 pieces of equipment in each of the 2 turnarounds. That's 20 different decisions. Unfortunately, that could result in many millions of different combinations, which is why integer optimization is so difficult. There are literally many different possible combinations to test.

Normally, these types of problems are handled in different steps. One approach is to reduce the problem size by eliminating as many of the combinations as possible. Fortunately, there are quite a few of the decisions that don't change across a wide spectrum, so many can be eliminated before the combination testing begins.

This problem really does require a discrete optimizer to solve properly. However, because it is such a difficult mathematical problem, we will take some shortcuts in our solution approach. We will start by treating the problem as a continuous problem rather than purely discrete. That way, we can continue to use Excel's Solver since it works much better with smooth, continuous functions.

Thus, we modeled equipment on/off switches as a linear value from 0 to 1. However, it's clearly not a linear function. To make the function more closely approximate a discrete choice, we used the optimizer-chosen linear value and modified it within a weighted exponential function, which was referred to as "adjusted age." The exponential function was so steep that it did a reasonably good job of approximating a discrete choice. Adjusted age was then used to calculate turnaround costs.

17.2.4.2 Modeling Production Run 1 and 2 Availabilities

The solution approach required us to estimate both production run 1 and 2 availabilities for all equipment individually. We accomplished that through a modified Monte Carlo analysis. We didn't incorporate Monte Carlo directly in the optimization setup; but rather, those concepts were included based on regressions built using Excel's Data Tables functionality. We were forced to use that approach since the Monte Carlo analyses results are noisy, and the Solver requires repeatable derivatives, which isn't possible when using Monte Carlo directly.

There are significant differences and difficulties in calculating production run 1 and 2 availabilities. Production run 1 availability is solely a function of the decision to repair that equipment, or not, during the turnaround proceeding production run 1.

For example, equipment D's age before production 1 run begins is 6.5 years old. It has a mean failure time of 14 years with a standard deviation of 1.5 years. If we don't repair it, then the equipment will be 11.5 years old at the end of the production run. It would have a decent chance of failing sometime during that particular production run. However, if we did repair it in the proceeding turnaround, its age at the end would only be five years old, and it's not likely to fail.

Compared with production run 1 availability, estimating production run 2 availability is considerably more complex since it involves two different repair decisions for each piece of equipment. That is, the decision to repair the equipment before the first production run and the same decision before the second production run. There are three distinct possibilities.

The first is relatively straightforward and is exactly similar to that described for turnaround 1. If we chose to repair that equipment during turnaround 2, then determining its availability would be similar to what was described for turnaround 1. That's because the equipment will just continue to age until turnaround 2 and then be repaired.

However, there are two other possibilities related to turnaround 1 equipment repair decisions.

The first is the case when we repair during turnaround 1, but not turnaround 2. That's OK, but not as good as when we repair all the equipment before turnaround 2. The other case is when we don't repair the equipment in either turnaround 1 or 2, which would have the poorest availability of all the cases.

What we learn is that estimating production run 2 availabilities is a function of two different variable choices and encompasses three completely different cases as just described. Modeling those accurately requires a lot of data since the regressions use "adjusted age." In addition, it became obvious that we needed to include many more terms to adequately describe the production run 2 availability dataset compared with the simple correlation for production run 1.

While we had mentioned earlier that the independent variables were a linear value from 0 to 1, we should note that we ended up modifying that to be from 0.01 to 1. That's because the log terms in the regressed equations for production run 2 could not directly accept a value of 0 for numerical purposes. However, that tradeoff did not materially affect solution quality.

17.2.5 Trading Off Increased Production vs. Turnaround/Crude Purchase Costs

At a high level, this optimization problem features a tradeoff between two important and opposite aspects of the objective function – increased production against increased turnaround costs and/or crude costs.

We can achieve higher production by choosing higher quantities of crude B since its yield structure is more favorable. However, we incur higher crude purchase prices to achieve this.

Another way to increase production, once a crude slate has been determined, is to increase turnaround costs. What makes this problem so interesting is that the choice of turnaround costs may be much different between crude A/B.

To ensure we don't consider solutions with poor production capability, we will impose a 95% availability constraint limit for both production runs. In fact, that's probably too liberal, but we will use it for illustration purposes. However, since we are using a 95% limit, that implies that we only have a 5% range maximum (100–95%) where we could possibly increase production. That's a limited area to work within. Plus, there's a point of diminishing returns such that once we get above a certain availability, for example 99%, it may take a considerable amount of turnaround spending to increase equipment availability much, if at all, beyond this figure.

In cases where there are higher production margins, it's logical to assume that we may want to increase turnaround spending, but that runs up against the point of diminishing availability returns. On the other hand, if turnaround costs are higher, we may back off slightly on production.

So, given that introduction, let's proceed to solve this particular optimization problem.

17.2.6 Independent Variables

The independent variables for this problem are:

- Equipment A–L → Should they be restored in the upcoming turnaround?
- Equipment A–L → Should they be restored in the following turnaround?
- Percentage of crude A being run

While we mention equipment A–L as our choices, we have already eliminated two pieces of equipment from our list – equipment A and F, so there will realistically be only 10 pieces of equipment participating in the optimization.

Once the percentage of crude A has been determined, we know the percentage of crude B since they must sum to 100%. Therefore, crude A and B are not independent of one another.

17.2.7 Dependent Variables

Profit will be the revenue of the next two run cycles added together, minus turnaround and incremental crude purchase costs.

17.2.8 Constraints

As noted above, an important constraint is minimum process unit availabilities for both production runs averaged over their cycle. In addition, the problem will include a maximum turnaround budget constraint.

17.2.9 Objective Function

The goal of this optimization is to determine the maximum profit, subject to meeting all constraints.

17.2.10 Results

Many optimization runs were made to examine different aspects of the problem.

Crude A was the base crude with an assumed crude purchase price of $60/bbl for all runs, while Crude B's price was allowed to vary throughout the runs, as noted below. Product prices were chosen to be such that the unit margin was roughly $100,000 per day. That might be a bit low, but it is consistent with the nominal values used in an earlier chapter. However, we should note there are differences between the two problem definitions since margin was used directly in the turnaround-only problem, whereas in this problem definition, individual product revenues were calculated using the best estimate of product prices. Nonetheless, the overall margin was roughly $100,000.

While turnaround costs for each equipment scale in exactly the same manner as in the turnaround-only problem, they are much higher in this example, a factor of 8 to be specific, since the product revenue increases in moving to run a higher quality crude dwarfed the original turnaround costs. So, we increased those costs to make the problem results more meaningful.

In general, a two-step process was used.

First, we ran the Solver model explicitly and examined the results. In many cases, the optimizer pegged the results as either a decisive "Repair" or "Do not Repair." In some cases, the answer was quite close to one of the two decisions, so it was fairly obvious what to do. However, in other cases, it wasn't quite as straightforward. In those instances, we manually changed values and observed the effect on the objective function. There was never perfect alignment between the two approaches, but it seemed to work decently well when used in the sequence described.

17.2.10.1 Crude Selection

Many different runs were made to determine the optimal crude slate and how that changed as a function of price. It turns out that it wasn't a smooth transition from one crude selection to another. Rather, there was a sharp breakpoint at approximately a $12 differential, or when crude B price increased beyond $72/bbl (assuming a crude A price of $60). Profitability is being driven primarily by the yield increase from crude B relative to crude A – not crude price differential – except when the crude purchase price differential becomes too great.

What is really interesting is that turnaround decisions can be dramatically different, depending on which crude is being run. That makes sense because the yield structure is so different between the two crudes.

What will be noticed in all the cases is the optimizer only chooses one restoration for most equipment for the two turaround cycles – with few exceptions. That is, if the optimizer decides to restore that equipment during turnaround 1, then it will not restore for the second turnaround. The reverse is also true.

In addition, the optimizer tends to push equipment restoration to the second turnaround relative to the first. This is especially the case for crude B since, in general, all equipment has longer failure times since crude B isn't as hard on the equipment. What's also ironic is that while total profit can change dramatically due to different crude costs, turnaround spending between the two turnarounds is rather consistent.

As we were running the cases, it was observed that whether the equipment was restored in turnaround 1 or 2 was borderline for many of the equipment. In fact, the difference was often so minor that it could easily be argued one way or another.

All cases shown below will assume a crude A purchase price of \$60. Headings will refer to crude B purchase price.

17.2.10.1.1 Crude B Price = \$60

In this case, crude A and B prices are exactly equal. Therefore, there is no penalty for using the higher value crude B, so the decision was easy. It only uses crude B. An overview of the results for this run is shown in Table 17.8.

Equipment restoration decisions are shown in Table 17.9.

17.2.10.1.2 Crude B Price = \$65.0

Here, crude B's price increases by \$5. Production run availabilities remain the same, and the optimizer still only chooses to use crude B. However, profitability declines since crude B's price is more expensive, as shown in Table 17.10.

All equipment restoration decisions remain the same as shown in Table 17.11.

17.2.10.1.3 Crude B Price = \$67.5

Crude B's price is now \$67.5. The trend is the same as in the previous case, as shown in Table 17.12. Availability remains the same, and the optimizer still only uses crude B, while profitability continues to decline since crude B becomes even more expensive.

Table 17.8 \$60 Crude Turnaround – Results

Availability Production Run 1	99.99%
Availability Production Run 2	99.94%
Turnaround 1 Costs	\$22,270,887
Turnaround 2 Costs	\$55,031,276
Profit	\$2,182,659,834

Table 17.9 $60 Crude Turnaround – Equipment Decisions

	TA 1	TA 2
B	No	Repair
C	No	Repair
D	No	Repair
E	No	Repair
G	Repair	No
H	No	Repair
I	No	Repair
J	Repair	No
K	No	Repair
L	Repair	No

Table 17.10 $65 Crude Turnaround – Results

Availability Production Run 1	99.99%
Availability Production Run 2	99.94%
Turnaround 1 Costs	$22,270,887
Turnaround 2 Costs	$55,031,276
Profit	$1,269,012,527

Table 17.11 $65 Crude Turnaround – Equipment Decisions

	TA 1	TA 2
B	No	Repair
C	No	Repair
D	No	Repair
E	No	Repair
G	Repair	No
H	No	Repair
I	No	Repair
J	Repair	No
K	No	Repair
L	Repair	No

Table 17.12 $67.50 Crude Turnaround – Results

Availability Production Run 1	99.98%
Availability Production Run 2	99.94%
Turnaround 1 Costs	$21,441,510
Turnaround 2 Costs	$55,860,653
Profit	$809,729,168

Compared to the previous case, equipment restoration decisions flip for both equipment B and J, as shown in Table 17.13.

17.2.10.1.4 Crude B Price = $70

Crude B price increases further to $70. Once again, as shown in Table 17.14, availability remains the same. The optimizer is still using only crude B, but its price is getting even more expensive, so profitability continues to decline.

This is one of the exceptions where the optimizer decides to restore equipment H in both turnarounds. All other decisions remain the same as in the previous case, as shown in Table 17.15.

17.2.10.1.5 Crude B Price = $72.5

As Table 17.16 shows, we've finally reached the price where the optimizer flips and only runs crude A. Apparently, the crude price differential became too high. Interestingly enough, production run availabilities are considerably lower.

Since the crude slate flipped, what you'll notice is that equipment restoration decisions have changed significantly – compared to the previous case. Results are shown in Table 17.17.

Since the optimizer is now choosing only crude A, further increases in crude B price won't affect the optimizer's decision.

Table 17.13 $67.50 Crude Turnaround – Equipment Decisions

	TA 1	TA 2
B	Repair	No
C	No	Repair
D	No	Repair
E	No	Repair
G	Repair	No
H	No	Repair
I	No	Repair
J	No	Repair
K	No	Repair
L	Repair	No

Table 17.14 $70 Crude Turnaround – Results

Availability Production Run 1	99.98%
Availability Production Run 2	99.91%
Turnaround 1 Costs	$21,479,209
Turnaround 2 Costs	$55,860,653
Profit	$340,327,911

Table 17.15 $70 Crude Turnaround – Equipment Decisions

	TA 1	TA 2
B	No	No
C	No	Repair
D	No	Repair
E	No	Repair
G	Repair	No
H	Repair	Repair
I	No	Repair
J	No	Repair
K	No	Repair
L	Repair	No

Table 17.16 $72.50 Crude Turnaround Results

Availability Production Run 1	96.77%
Availability Production Run 2	98.93%
Turnaround 1 Costs	$17,181,529
Turnaround 2 Costs	$60,120,635
Profit	−$85,523,190

Table 17.17 $72.50 Crude Turnaround Equipment Decisions

	TA 1	TA 2
B	No	Repair
C	No	Repair
D	No	Repair
E	No	Repair
G	No	Repair
H	No	Repair
I	No	Repair
J	Repair	No
K	No	Repair
L	Repair	No

17.2.10.2 Constraining Turnaround Costs

We also conducted a couple of runs to examine how the optimizer altered equipment restoration decisions as turnaround budget constraints changed for a given crude selection slate.

Since crude wasn't changing, results are similar to those presented in the earlier chapter. In essence, it's the same problem.

For this problem, we started with the base case of crude B valued at $70/bbl crude purchase price. In that case, the optimizer always chose the higher-quality crude – crude B.

Table 17.18 shows the results for that case, where there are no active constraints.

The following are the decisions for each equipment for both of the turnarounds, as shown in Table 17.19.

So, let's start constraining turnaround costs and see how the results change. If we limit turnaround 2 costs to under $50,000,000, the results are as shown in Table 17.20.

In this case, as shown in Table 17.21, the equipment decisions for E flipped from being restored in turnaround 2 to turnaround 1. The optimizer also removed equipment H from being repaired in both turnarounds and decided to restore it only during turnaround 1. On the other hand, there was a small incremental value in adding B to turnaround 1. All other decisions remained the same.

Table 17.18 No Turnaround Cost Constraints – Results

Availability Production Run 1	99.98%
Availability Production Run 2	99.91%
Turnaround 1 Costs	$21,479,209
Turnaround 2 Costs	$55,860,653
Profit	$340,327,911

Table 17.19 No Turnaround Cost Constraints – Equipment Decisions

	TA 1	TA 2
B	No	No
C	No	Repair
D	No	Repair
E	No	Repair
G	Repair	No
H	Repair	Repair
I	No	Repair
J	No	Repair
K	No	Repair
L	Repair	No

Table 17.20 $50 MM Turnaround 2 Cost Constraint – Results

Availability Production Run 1	99.98%
Availability Production Run 2	99.81%
Turnaround 1 Costs	$27,322,547
Turnaround 2 Costs	$49,979,617
Profit	$302,010,271

Table 17.21 $50 MM Turnaround 2 Cost Constraint – Equipment Decisions

	TA 1	TA 2
B	Repair	No
C	No	Repair
D	No	Repair
E	Repair	No
G	Repair	No
H	Repair	No
I	No	Repair
J	No	Repair
K	No	Repair
L	Repair	No

Table 17.22 $30 MM Turnaround 2 Cost Constraint – Results

Availability Production Run 1	99.99%
Availability Production Run 2	99.66%
Turnaround 1 Costs	$48,056,970
Turnaround 2 Costs	$29,245,193
Profit	$245,768,966

Now, let's further constrain turnaround 2 spending to $30,000,000 with results shown in Table 17.22.

In this case, as shown in Table 17.23, equipment C and D flipped from turnaround 2 to turnaround 1. It wasn't possible to further reduce spending since I, J, and K were required to be restored to prevent unacceptable reductions in production run 2 availability.

At this point, we can't further reduce turnaround 2 spending without violating the production run 2 minimum availability constraint.

17.2.10.3 Turnaround Cost Sensitivity

One item of interest is to determine how sensitive the problem results were to changes in turnaround costs. So, we made some runs increasing turnaround costs by 50% for crude B runs of $65, $67.5, $70, $72.5, and $75.

What we observed was that there was no effect on the $65, $72.5, and $75 runs.

However, for the $67.5 case, the optimizer chose to restore equipment J in turnaround 1 instead of turnaround 2. That choice resulted in an approximate increase of $5,000,000 in the objective function.

For the $70 case, the results were more significant. There, the optimizer was able to increase the objective function by $12,000,000 by choosing to restore equipment H in turnaround 2, rather than turnaround 1.

Table 17.23 $30 MM Turnaround 2 Cost Constraint – Equipment Decisions

	TA 1	TA 2
B	Repair	No
C	Repair	No
D	Repair	No
E	Repair	No
G	Repair	No
H	Repair	No
I	No	Repair
J	No	Repair
K	No	Repair
L	Repair	No

For the lower turnaround cost cases, there was only one change, and it was very minor. That is, for the $75 case, the optimizer flipped its decision for equipment C to restore in turnaround 2, rather than 1. However, the change in objective function was only $1,000,000, which is close to flipping a coin on that decision.

17.3 Conclusion

Up until this chapter, this book has considered a series of optimization types and presented them as individual and distinct optimization problems. However, that was done only to facilitate introducing the topics. In reality, these different optimization types can be combined in whatever fashion that makes it possible for the user to answer business problems of importance.

Out of many possibilities, two combined optimization problems were presented in this chapter:

- Optimal improvement Project (Optimal Process Unit Availability) and crude A/B mix
- Turnaround equipment restoration selection and crude A/B mix

What's interesting about these choices is that one can return to previous chapters to compare and contrast solutions to the combined optimization problems with the single optimizations to see how combining variables may have changed the final results.

It's important to note that there's nothing that forces the user to just these choices. One should always choose the proper combination of independent and dependent variables, as well as constraints that solve the business plan at hand. There's never a need to artificially constrain the problem. This chapter opened our minds as to what some of those possibilities might be.

18

Mapping Models to Optimization Problems

At this point, the reader has been introduced to extended discussions of a series of major business/technical problems and seen examples of how optimization can be used to generate solutions for those problems. The next logical step is to map each of those optimization problems to the required model(s) necessary to solve those problems. Then, in the next chapter on creating a Master Plan, we will use the mapping concepts developed here to create a unique master plan for each enterprise.

There have been seven important optimization problems presented so far:

1) Turnaround Timing/Duration and Equipment Selection
2) Selection of Optimal Improvement Projects (and ultimately optimal process unit availabilities)
3) Maintenance Plan Options Optimization
4) Spares Optimization
5) Storage Optimization
6) Task Optimization
7) Asset Design Optimization

This breakdown is somewhat artificial, since there can be significant overlap between the problems highlighted above. For example, Storage Optimization can easily be formulated as a subset of the Optimal Improvement Project determination, if desired. However, for now, we'll leave each of these as their own individual optimization problem and not worry about those nuances.

In addition, we didn't include Process Conditions as a separate problem, even though we dedicated a complete chapter to that topic. In general, while accounting for how process conditions affect asset decisions can be its own problem, it is often a complexity within one of the problems already on the list. In other words, it's a detail required to aid in solving one of the other problems. Yes, there will be times when it is its own problem, but let's not worry about it for now.

18.1 Mapping Between Optimization Problem and Model(s) Required

Table 18.1 shows what types of models are required for each of the different optimization problems. They are sorted into four logical groupings.

The **Optimal Improvement Project** and **Storage Optimization** are both primarily plant-level optimization problems. The only issue is where their process unit-level information originates from. Early in your program, it would likely be based on high-level assumptions. As the program matures, it would more likely come from submodels that have already been built.

Plant Optimization in the Process Industries: Incorporating Equipment/Assets in the Decision-Making Process, First Edition. Marty Moran.

Table 18.1 Optimization Problem – Model Mapping

	Plant Model	Process Unit Model	Multiple Process Units	Multiple Failures Modes
Improvement Project Optimization	×	Maybe		
Storage Optimization	×	Maybe		
Turnaround Timing/Duration Optimization		×	×	×
Maintenance Plan Options		×	×	×
Spares Optimization		×	×	×
Task Optimization		×	×	×
Asset Design Optimization	×	×		×

Maintenance Plan Options, Spares, and Task Optimization are primarily done at the Process Unit level or sometimes across multiple process units. Often, these models will be more detailed, requiring some equipment to be abstracted with multiple failure modes instead of just assigning failure and repair information at the equipment level.

That leaves **Turnaround Timing/Duration** and **Asset Design Optimization** as their own separate categories.

Turnaround Optimization is potentially the most flexible optimization problem. For most applications, it would be detailed to the process unit level or across multiple process units. In addition, it occasionally requires multiple failure modes.

The majority of **Asset Design Optimization** will generally be done at the Process Unit Level. It sometimes requires multiple failure modes. However, we will often need to iterate between running the plant-level model and the new process unit model required when a new unit is designed.

Now, let's discuss each one of the individual optimization types, row by row.

18.2 Selection of Optimal Improvement Projects

There's a direct relationship between the optimal improvement projects in a plant and optimal process unit availability targets. By definition, the determination of the process unit availability target needs to be made in concert with all other process units in the plant at the same time – at least, those that make the major products. That's because there are usually intermediate units that influence units surrounding it. In other words, no unit is an island, by itself.

Thus, this optimization problem will, by definition, require a plant model. In general, there are two different ways to populate the basic data for each process unit used in the plant model.

The first approach is to abstract each process unit as a single block and use past availability data to estimate the unit's "reliability." Repair time would be the historical estimate of how long it takes to bring the unit back up after it has come down. The benefit of this approach is it can be used to quickly generate results. However, it creates a disconnect with the process unit models that we ultimately want to create.

The second approach is to use the submodel feature inherent in most Monte Carlo simulation packages that allow us to use the results directly from individual process unit models. For our plant model, instead of a single block with no underlying details, each process unit would be a submodel, whereby one could bring up a detailed model after clicking on that process unit.

Obviously, that would require more detailed process unit models to have already been developed. At the start of your program, that won't be the case, so it's likely that you will initially proceed with the first approach. However, over the longer term, you'll want to transition to the second approach since it would connect the detailed model with the higher-level plant model.

18.3 Storage Optimization

The comments made already for the Optimal Improvement Project will also apply to Storage Optimization, since these problems are highly intertwined with one another. Both are a plant-level type of analysis.

Thus, we would need to decide whether to use approach 1 or 2 for this analysis. As already discussed, that would likely be determined based on the maturity of our program. Early on, we will normally use approach 1, whereas as the program develops, we could transition to using approach 2.

18.4 Turnaround Timing/Duration and Equipment Restoration Selection

Determining when a piece of equipment is on the critical path will require a process unit model that is detailed to at least the equipment level. Since turnarounds often span multiple process units, this will require detailed models for each of the process units participating in the turnaround or units possibly being considered.

18.5 Maintenance Plan Options Optimization

Maintenance Plan Options Optimization can be performed on a single piece of equipment within one unit. On the other hand, it could consider several maintenance plans within one unit or across many different process units. Thus, it would require at least one process unit model. However, in other cases, many different process unit models would be involved.

The majority of the time, we would only need to detail the equipment level. However, it's possible to abstract to the failure mode level. However, that's not ordinarily the case.

18.6 Spares Optimization

The scope of spares optimization could potentially be enormous. Under normal circumstances, an optimization would not be needed to determine the optimal holdings for a single equipment in a single unit, unless constraints are a significant part of the problem definition. On the other hand, optimization is quite common for multiple spares within either one process unit or across several units.

In addition, spares optimization could be used for a common part allocated across many different process units. For example, this might be the case for instrumentation, where the same instrument is used in many different process units.

Spares optimization will require at least one process unit model, though there could be several. Also, it would be quite common to require models detailed at the failure mode level.

18.7 Task Optimization

Task optimization is normally performed across one or more process units. Sometimes, we may only require models abstracted to the equipment level. However, models built to the failure mode level would be more common.

As we consider how granular the model needs to be, keep in mind we will often want to sum the labor and material costs. Therefore, we would need to detail the model so that it is possible to capture this information.

We won't need to include more mundane tasks or those with smaller costs. Also, we don't need to include fixed activities that won't change, such as is the case with many safety-related activities.

In addition, as noted in a previous chapter, this optimization could be combined with spare parts optimization to more accurately mimic reality since many tasks require consuming a certain quantity of spare parts as a part of their activity.

18.8 Asset Design Optimization

Optimizing equipment within an upcoming Debottlenecking, Improvement, or a Grassroots project for a new process unit(s), would definitely require the associated process unit model. There may be times when that model would need to be defined at the failure mode level.

Depending on the size of the project, we may also need to run the plant model. In fact, we might find that there is a need for an iterative process alternating between running the high-level plant model and the individual process unit models.

18.9 How to Kickstart Your Program

Now that we understand the relationship between the different optimization types and the models required to execute them, let's brainstorm on how one can kickstart your program. We'll explore options to piggyback on other activities that might already be occurring in your organization.

18.9.1 Upcoming Turnaround

If an organization is planning an upcoming turnaround, whereby they are trying to make decisions on either timing, duration, or what equipment to service during the turnaround, then this may be the best place to start your program.

Performing this optimization would require process unit models for every unit participating in the turnaround. At a minimum, those units would need to be detailed to the equipment level.

While a considerable amount of work initially, starting your asset optimization program on a turnaround problem could be advantageous in that it would demonstrate the value of the program. In addition, it would help generate the data for the process units that would be part of the program on an ongoing basis.

18.9.2 Asset Design Optimization

If there is a Debottlenecking, Improvement, or Grassroots project pending, then Asset Design Optimization might also be a great place to start.

18.9.3 Optimal Process Unit Availability/Storage Optimization

If none of these are near-term possibilities, then Process Unit Availability and Storage Optimization might be attractive. These analyses require a minimal amount of information – just some high-level guesstimates outside of Improvement project estimates.

The Improvement project estimates should be relatively easy to generate since they only require information the organization should already be tracking, irrespective of an asset optimization program. So, this would generally be an easy way to start your program.

18.10 Standard Models or Not?

Now that we have discussed the mapping between optimization problems and the models required, we should address what models need to be maintained on an ongoing basis. Based on the discussion above, the answers might be obvious, but let's articulate a general strategy.

Since Monte Carlo simulation technology is inherently general, the first question one might ask is whether the models we will build should be "standard" models or whether they should be built on "as needed" or "ad hoc" basis.

While all programs will need some "ad hoc" models, the workhorse models will be process unit models. Once models have been created at the process unit level, they can easily be combined together to form a plant model through the submodel feature inherent in most modern Monte Carlo simulation software packages. The great news is that submodels provide an easy mechanism to granularize the problem to any level that makes business and technical sense.

For example, it is easy to start with a plant model and "drill down" into the process unit model. However, this could be continued by then drilling down into equipment details within the process unit model.

The asset registry hierarchy in a typical plant is broken down by functional locations, which are then segmented further by plant or site, process unit, and ultimately equipment. The submodel feature allows an organization to create a model structure that adheres quite closely to the plant's asset hierarchy structure.

18.11 Process Unit Models

Every day, all around the world, process engineers use simulation models to model the energy and material balance of process units. For that reason, it makes sense to mimic and mirror our asset management models with the models that engineers currently use and maintain.

While we can, and will, create models at other levels of granularity, this will be our "bread and butter" model. That's because process units have well-defined management and organizational structures to support their operation. There are operations engineers, process engineers, reliability and maintenance personnel that already exist for a process unit.

So, who owns the asset management model at the process unit level? To a large degree, this depends on organization specifics and will likely vary between companies.

Personally, I tend to think that reliability engineers are the obvious choice. I would assign them the responsibility of maintaining all model data inputs for the unit with information updated on a periodic basis. Yearly should be the minimum, but, if possible, more often is even better.

The Reliability Engineer (RE) would be tasked to organize meeting(s) to update all data assumptions. Since this is an incremental update, the RE should review all pertinent asset management data from the past year and then convene a meeting with experts from all applicable areas to validate and update input data.

Now, the harder question is, who is responsible for the overall model?

We advocate for the process engineer here for several reasons. First, and most important, process plants already have had existing organizational structures in place for years. We will want to build upon that structure, and then augment it. Process engineers already maintain the process simulation model for that unit, so adding another, similar model seems like a logical choice.

Plus, it will help break down existing barriers that have existed in plants all around the world. That is, they have historically been split into chemical engineers, who manage the operational aspects of the unit – what feeds to process, expected yields, how to set process unit variables required to achieve those product slates, and mechanical engineers, who have largely been tasked with managing individual equipment. While theoretically efficient, this task split results in compartmentalizing knowledge, whereby one group is not always cognizant of what the other group does.

I can personally relate to this split. I'm a chemical engineer and feel more comfortable discussing process issues, whereas my equipment knowledge is more limited.

Of course, I know many competent senior engineers who come from the opposite viewpoint. While very knowledgeable from a mechanical standpoint, they often lack a thorough understanding of process unit operation, outside of some high-level specifics.

Often, this isn't much of a problem since a thorough understanding of the process is usually not necessary to keep equipment running at peak efficiency. However, while organizationally expedient, we lose an important opportunity to integrate opposing worldviews together.

What asset management technologies have now allowed us to do is to marry these two approaches together. In other words, it's now possible to combine two totally opposite paradigms together into one overall comprehensive worldview. It's not just a "process" or "equipment" view, but rather a "plant" view.

That's important because the equipment doesn't exist in isolation as some engineers might suppose, but rather is part of a comprehensive whole that exists to make product. A combined viewpoint allows engineers, who are not as equipment-centric, like me, to better understand the tradeoffs that equipment engineers struggle with. In the past, this would never have shown up on my radar screen.

From a skill standpoint, requiring process engineers to take on added modeling responsibility shouldn't be too much to ask. That's because learning the mechanics of asset modeling is much easier than the analog for process simulation modeling, where you really have to be a chemical engineer to be able to use the technology successfully. For that reason, while the reliability engineer will own the data, and can be used to run the model for a majority of use cases, the process engineer will ultimately be the model owner.

The process engineer will use a two-pronged approach. First, they will run the first principles thermodynamic process model to calculate unit material and energy balances, along with the required process variables to meet production targets. Immediately afterward, they would run the asset model to see how much product can actually be produced.

One main difference between the two models is that the thermodynamic model assumes equipment is always available to produce products. Of course, that's unrealistic. On the other hand, the Monte Carlo model corrects that theoretical assumption with equipment reality, using the estimated failure and maintenance rates that will help calculate actual production values.

18.12 Site or Plant Models

While we will likely spend a considerable amount of time building and running process unit models, the plant model can also be helpful to answer certain important business questions. This is especially the case for larger, more complex plants such as a refinery, or any plant, where there are significant dependencies between units.

As illustrated earlier, the plant model is a prerequisite for determining the optimal availability for each individual process unit, as well as for storage optimization. In some instances, it will also be required for turnaround planning.

Initially, when a company first embarks on an optimization program, the plant model will be based on uptime estimates, as noted earlier in the chapter, and the time to bring the unit back online if the unit should shut down. However, as we build out the Process Unit Models, each block in the Site Model would eventually be replaced with a submodel that represents the overall availability of that process unit through detailed availability calculations at the process unit level.

18.13 Equipment Models

Within the process unit model, there may be times when we want to further detail a piece of equipment, instead of just assigning one failure/repair time for the entire piece of equipment. This will especially be the case for equipment participating in Task and Spares optimization.

These more detailed equipment models would be built as submodels within the process unit model framework and would not ordinarily be expected to be used outside this model. This allows a user the ability to click on the equipment submodel and see all the failure modes associated with this equipment.

However, there may be other cases where we decide to build individual equipment models, where they are intended to live on their own. Certainly, they could also function within the process models as the other individual equipment models currently do, but it's just that they could also be used on a standalone basis.

For example, this might be the case when there is large rotating equipment within the unit where it is not practical to spare that equipment. Typical examples might be the wet gas compressor in a refinery Fluid Catalytic Cracker (FCC), the recycle gas compressor in a refinery catalytic reformer, or the cracked gas compressor in an ethylene unit. The reason we might want to build these models is that engineers would like to run more detailed technical analyses on this equipment, since they are so critical to operation.

It's important to note that equipment models should only be built when it makes business sense to do so. For a significant portion of equipment in a plant, they are not necessary. One must always keep in mind there is overhead required to design, program, and maintain any model. Thus, models should only be created when the business gains from producing detail in a model exceed the effort required. That's another reason why individual equipment models that exist on their own should be incorporated into the process unit models. It avoids duplication and extra effort.

18.14 Responsibility for Equipment Models

Earlier in this section, we proposed that process engineers maintain the process unit asset models. So, should they also maintain detailed equipment models?

In this case, it would be better if the reliability engineer were the official owner of the detailed equipment model, since that person would likely need to obtain specific equipment information from other internal experts. For example, if there is a Rotating Equipment group that is responsible for, or has significant knowledge of this equipment, they would likely play a significant role in defining failure and repair information for the important subsystems inherent in this model. An example might be lubrication experts, who provide advice on the frequency of oil changes. To make the data collection process easier and more efficient, the reliability engineer would be the central focus for assembling the proper information for equipment models.

On a periodic basis, the reliability engineer would update any model changes into the same library where all other asset models reside. That way, when the process engineer runs the process unit models (or plant models, for that matter), they execute the underlying equipment models using the latest information.

18.15 Conclusion

In previous chapters, we introduced different business/technical problems that could be solved through asset optimization. In this chapter, we mapped those optimization solutions against the type of model that would be necessary to solve the problem – process unit, multi-process unit, or plant models.

There were four groupings that made the most sense. **Optimal Improvement Project (Optimal Process Unit Availability)** and **Storage Tankage Optimization** were grouped together since they were both inter-unit optimization problems. On the other hand, **Maintenance Plan Options, Task**, and **Spares Optimization** were grouped together since they are intra-unit problems. **Turnaround Duration/Equipment** selection was its own unique category, as well as **Asset Design Optimization**.

Then, there was a discussion of how to kickstart an asset optimization program by latching onto important activities that were already occurring within the organization.

Last, for several important reasons, it was determined that process unit models are the program's backbone. First, they are necessary since they are required for both intra-unit problems, as well as plant-wide challenges. Second, maintaining models at this level makes sense because process engineers already maintain other process simulation models at this level.

We also determined that it was necessary to maintain a plant model. These could start as simple high-level estimates but would be replaced with more detailed process unit models as the program matures, using the submodel feature inherent in modern Monte Carlo simulation packages.

It's possible to create and maintain equipment models, but those should only be done when it makes economic sense to do so, since all models require time to build, as well as to maintain. Many of these equipment models would be used to support Task and Spare optimization. However, some equipment models could be built by other specialties within the plant for their own use, such as the Critical Rotating Equipment group maintaining models for large compressors. Once built, these models should be incorporated into their associated process unit models.

Reliability Engineers would be responsible for collating the data for both process unit and equipment models.

19

Creating a Program Master Plan

So, if you have gotten to this point in the book, and you are either curious, or convinced, that a program like this is desirable, the natural question is, "How do I begin?"

This chapter will answer that question by helping create a master plan for your organization. It's important to recognize that each plant's priorities will be different. For that reason, we will want to draft a program tailored for each plant suited for its own individual needs.

What's really important is that the program starts off on the right foot. Therefore, we need a solid win at the start to ensure that the program endures.

As part of this discussion, we will touch on the following areas:

- Opportunity assessment
- Project selection
- Project phases
- Types and quantities of resources required?
- How to execute the consultation process
- Data and its implications
- Technologies to use/acquire?
- Work processes
- Should I develop standard models?
- Training

One important aspect of the master plan will be to lay the foundation so other parts of the program can be pieced in, like a puzzle that eventually needs to be completed, but at a pace specific to your organization. In other words, the common features of an optimization program will be the same, but the order in which they are assembled can vary from company to company.

Our starting point is to determine the business problems we want to focus on to create value. This is especially important for the first project. For that reason, we are going to introduce an Opportunity Assessment process, which would be of significant value to an organization, even if the company never embarks on an optimization program. After the opportunity assessment discussion is complete, we will delve into other important aspects of the program – all of which are necessary if the program is to endure.

Plant Optimization in the Process Industries: Incorporating Equipment/Assets in the Decision-Making Process, First Edition. Marty Moran.

19.1 Opportunity Assessment

The most defensible reason for wanting to start an asset optimization program is that it will improve business outcomes in a relatively short time frame. In order to achieve that goal, let us introduce an assessment methodology to help unearth your best opportunities. It's similar in nature to a sales process we often used when I worked at vendor companies.

In those roles, a significant portion of your work is continually finding new opportunities. That requires an effective process to be able to achieve that goal. However, that's not all. Once you have a candidate problem, you need to demonstrate that solving the problem will create value, usually financial, as well as convincing the client that your company has a solution that will solve their problem. It is, by no means, a trivial process.

What I've noticed working in the process industries that a client often learns to live with a problem they have – even if it's painful. It's familiar. It's their normal, so they often do not question if there is a better way. Even if they do, they often do not know how to proceed.

What most clients do not realize is that vendors often have solutions to those nagging problems. Our job, when working as a vendor, was to use the sales process to find those problems and propose solutions.

One particularly effective technique for finding potential opportunities, especially when working with senior management, is to structure a series of questions that asks them to rank their organization on various aspects of their business performance on a scale of 1–10. For example, you might ask, "On a scale of 1–10 (1 = very poor; 10 = excellent)", how well does your organization select the proper turnaround interval, and what equipment to include in the turnaround that properly balances all the competing objectives?

The great thing about this technique is that it's rare that an executive would rank themselves a 10. There are many reasons for their modesty. The majority of the time, it's the honest truth. Consider the previous question. Managing turnarounds is an exceedingly difficult area, in general, and I am not aware that anybody would claim that they do it perfectly.

After asking the initial question, the follow-up question is *always*, "What issues would you need to address to move from your current value of X to 10?"

This follow-up question allows the executive to start discussing issues that are unique to their organization. These are details that few outside the organization would know about before engaging in this discussion. The questioner's job is to let the executive talk and keep asking follow-up questions to elicit more details. After the conversation has run its course, both parties can jointly decide if a problem has been voiced that the vendor might be able to solve, as well as its priority to the organization.

The turnaround question was just an example to illustrate the concept of finding opportunities for optimization. However, this technique is an excellent way for organizations to assess the strength of their programs across the board, as well as where improvement opportunities might lie. This assessment process is not relegated to just asset optimization.

As one asks these questions surveying many different facets of the organization, the differing answers give both parties the ability to discern what the most important areas are to explore further. Once a range of questions has been asked, the answers become relative and remove the "absolute number bias" which is possible if only one question is asked. In other words, after a series of questions, one can compare and contrast different responses, and the explanation for their responses from senior management, so one knows where to focus one's efforts.

The following is a complete listing of questions that every organization should ask themselves as they search for opportunities.

"On a scale of 1–10 with 1 being poor and 10 being perfect, ..."

- How well are you achieving your profitability goals?
- For each process unit in the plant, how well do you achieve the unit's goals?
  - (This question is repeated for each major process unit in the plant and should be answered from an asset viewpoint for this exercise.)
- How well does your maintenance department achieve its goals while always considering and respecting the plant's production goals at the same time?
- How well does your reliability department achieve its goals while always considering and respecting the plant's production goals at the same time?
- How well does the process engineering department do in incorporating asset performance as part of their work?
- How well, when executing a debottlenecking, improvement, or grassroots project, are you able to adequately balance the initial capital costs against the longer-term operational and maintenance costs? Or are project metrics heavily tilted toward meeting the capital budget?
- Do you have confidence that you are executing enough PM tasks at exactly the right frequency while not overspending on labor costs?
- To what degree do you believe that you have spare parts optimized such that you are not carrying too much inventory, but always have enough so that when a piece of equipment has an issue, you have parts so that you do not suffer a significant loss of production?
- Given that you have a limited capital budget, to what degree do you believe you are spending those dollars on the right projects to maintain or increase production, if that is your business goal?
- How much confidence do you have that you are setting each individual process unit's availability to an optimal target value for the business? In other words, approaching it from a scientific basis, and not just plucking numbers out of thin air.
- What level of confidence do you have that equipment is replaced at the appropriate time knowing that sometimes it makes more sense from a financial perspective to replace equipment than to keep repairing it?

A couple of questions that I always ask when I am in the company of senior management, which is not specifically 1–10 questions, but still none-the-less are important are the following:

- Where areas do you feel that you are overspending in?
- What is on your plant wish list?
- Is there any area where you'd like us to concentrate on?

All of these questions should be formatted into a word or excel table that has six columns. An example is shown in Table 19.1.

Table 19.1 Question Summary Format

Question	Rating	Optimization Candidate	How Improve	Value, If Improved	Implement Ease	Implement Cost
Question 1	8	Yes		High	Difficult	High
Question 2	6	No		Moderate	Moderate	Very High
Question 3	9	No		Low	Difficult	Low
Question 4	6	Yes		Very High	Easy	Moderate
Question 5	7	Yes		Moderate	Very Easy	Moderate Low

Some notes on the table:

- The first column is the question itself.
- The second is the quantitative answer.
- Column 3 is simply a checkbox that indicates if this is an area that could benefit from asset optimization. In some cases, an opportunity may have been unearthed, but it is outside asset optimization. Nonetheless, it should be noted as such.
- Column 4 represents a written summary of how the plant could improve the area to merit a 10.
- Column 5 is a qualitative estimate of the value to the organization if the problem could be solved.
- Column 6 is a value that indicates how easy it would be to implement a solution for the problem.
- Column 7 is a high-level estimate of the cost to implement the project.

Capturing this information does not have to be in a fancy format or even incredibly detailed. What we are looking for is an easy way of consolidating the most important issues that an organization deals with, so that we are able to rank them in terms of our ability to help solve that problem, along with an estimate of the value to the organization if the problem can be solved.

19.2 Project Selection

Once the opportunity assessment process has been completed, the results need to be evaluated.

With the information collated, the table should be ordered into two main buckets. One is questions that could be improved with optimization, with the other being those that cannot. While this book is focused on optimization-related items, we do not want to lose the other ideas from the opportunity assessment process. This contains valuable information, even if these cannot be solved via optimization.

So, the first cut would be two different tables, as shown below with the original column 3 now eliminated (it is no longer relevant since the tables have been separated). The first table represents projects that optimization cannot solve (Table 19.2), while the second table shows those that are candidates for optimization, as shown in Table 19.3.

Table 19.2 Projects Not Suitable for Optimization

Question	Rating	How Improve	Value, If Improved	Implement Ease	Implement Cost
Question 2	6		Moderate	Moderate	Very High
Question 3	9		Low	Difficult	Low

Table 19.3 Good Project Candidates for Optimization

Question	Rating	How Improve	Value, If Improved	Implement Ease	Implement Cost
Question 1	8		High	Difficult	High
Question 4	6		Very High	Easy	Moderate
Question 5	7		Moderate	Very Easy	Moderate Low

Now, we can take each of the tables and prioritize them based on those that have the highest value to the organization, but balanced against the ease of implementation and the associated cost. There are many different ways that an organization could take the data in the tables to decide on the best project.

In essence, there are four different columns that need to be considered – **rating; value, if improved; implementation ease; implementation cost**. When confronted with similar problems in the past, I would normally convert the qualitative ranks into quantitative rankings using an interim table. For example, Table 19.4 shows the conversion table for "Value, If Improved."

Table 19.5 might be the conversion table for "Implementation Ease."

Table 19.6 shows the conversion table for "Implementation Cost."

Let us revisit the table for those questions that we highlighted where we could use optimization to solve. With quantitative values now substituted for qualitative values, it becomes as shown in Table 19.7.

There's still one detail that needs to be taken into consideration. That is, the original rating does not necessarily imply that we can get all the way to 10 from our current ranking. For example, for Question 1, if we implement the solution for it, there's no guarantee we would necessarily be ranked now at 10. Therefore, we would need to make an additional estimate of how much we could improve based on implementing the solution. For that reason, let us add an additional column for this new ranking shown in Table 19.8.

We can now combine the ratings column into one overall column, which is a delta ratings improvement. For example, for Question 1, there would be a 1.5 Rating improvement by moving from 8 to 9.5. Let us make that change, as well as delete the "How Improve" column since it's a text field. Table 19.9 shows those changes:

Now, let us multiply all columns together for each question, adding a column at the end to accommodate the final value. Table 19.10 shows the results after computing the final value.

Table 19.4 "Value, If Improved" Quantitative Ratings

Qualitative	Quantitative
Very High	8
High	7
Moderate – High	6
Moderate	5
Low – Moderate	4
Low	3
Very Low	2

Table 19.5 "Implementation Ease" Quantitative Ratings

Qualitative	Quantitative
Very Easy	8
Easy	7
Moderate – Easy	6
Moderate	5
Difficult – Moderate	4
Difficult	3
Very Difficult	2

Table 19.6 "Implementation Cost" Quantitative Ratings

Qualitative	Quantitative
Very High	2
High	3
Moderate – High	4
Moderate	5
Low – Moderate	6
Low	7
Very Low	8

This process has demonstrated one way to rank projects. However, it is hardly the only way. Each organization could change this approach by incorporating weighting changes as we have suggested, or by using an entirely different methodology to rank each of the attributes more suited to its own needs. The important thing is we create a defensible mechanism for the rating process.

Table 19.7 Substituting Quantitative Values into Main Table

Question	Rating	How Improve	Value, If Improved	Implement Ease	Implement Cost
Question 1	8		7	3	3
Question 4	6		8	7	5
Question 5	7		5	8	6

Table 19.8 Add Improved Rating

Question	Rating	Improved Rating	How Improve	Value, If Improved	Implement Ease	Implement Cost
Question 1	8	9.5		7	3	3
Question 4	6	8		8	7	5
Question 5	7	8.5		5	8	6

Table 19.9 Delta Improvement

Question	Delta Improvement	Value, If Improved	Implement Ease	Implement Cost
Question 1	1.5	7	3	3
Question 4	2	8	7	5
Question 5	1.5	5	8	6

Table 19.10 Final Ranking

Question	Delta Improvement	Value, If Improved	Implement Ease	Implement Cost	Final Value
Question 1	1.5	7	3	3	94.5
Question 4	2	8	7	5	560
Question 5	1.5	5	8	6	360

19.3 Project Phases

Once the master plan has been developed, and the priorities understood, the first project can be executed. Typically, projects will require the following phases:

- Kickoff meeting
- Initial RBD development
- Main consultation meeting
- Model development
- Model review
- Model documentation
- Implementation

Each of these project steps is discussed in further detail below.

19.3.1 Kickoff Meeting

The kickoff meeting should involve the major project participants, as well as important company stakeholders. Project participants should include the internal technical lead, facilitator (often this person is also the technical lead), Monte Carlo modelers, equipment experts, and the important people from Operations, Process Engineering, Maintenance, and Reliability. In addition, senior managers who are sponsoring the program and understand the financial incentives of the plant and organization should also attend. If an external vendor is participating in the project, they obviously should be there as well.

This meeting is meant to introduce the project, what the phases will be, and to inform the team how they will personally be involved. Timelines will be discussed, as well as who is responsible for each deliverable.

During the meeting, operations will also lead a high-level discussion of the process, important equipment in the process flow, and the unique challenges they have faced within the unit. Maintenance and Reliability will provide high-level input concerning equipment details. This portion of the meeting is not meant to discuss significant details, but rather just an introduction to the unit so that the next phase – Initial RBD Development – can proceed smoothly.

Any necessary data requests are also discussed at the meeting.

19.3.2 Initial RBD Development

Soon after the kickoff meeting, the technical lead will work with various internal departments – Operations, Process, Maintenance, and Reliability – to gather the appropriate documents with the goal of understanding the process flow and major equipment within the unit. They will use this information to prepare material required for the main consultation meeting, which will include a preliminary reliability block diagram (RBD), as well as the spreadsheet for capturing failure/repair data.

19.3.3 Main Consultation Meeting

The facilitator will drive the consultation meeting. Participants will include equipment experts, as well as Operations, Maintenance, Process, and Reliability personnel. The goal of this meeting is to obtain a consensus on the level of detail required for each major piece of equipment, initial failure and repair estimates, any condition monitoring tasks for the equipment, as well as estimates of what the production impact might be if the equipment fails. Information from this meeting is documented in the spreadsheet initially prepared during the previous step.

19.3.4 Model Development

Once the Consultation meeting has been completed, Monte Carlo modelers will start creating the model based on information documented in the Consultation meeting spreadsheet.

The first step in this process is to finalize the reliability block diagram initially prepared in a previous project step. Once completed, the next task is to transform the high-level failure/repair from the consultation meeting into appropriate mathematical distributions that can be used directly in the models. Any condition monitoring tasks are added, as well as how production is affected, if the block should fail.

At this point, the modeler runs the models to generate the system availability probability distributions, along with Pareto charts, tank fill/empty events, in addition to any other miscellaneous information deemed important for the project.

19.3.5 Model Review

A meeting is then held to discuss the initial model results with all prior participants. This includes a summary of system availabilities using the probability distribution, along with a Pareto chart that documents the largest contributors to unavailability. The results are discussed with the team, and agreement is reached on what changes might be needed to further evaluate the system, including provisions for any further model runs.

If additional runs are necessary, the modeler performs those runs and generates the appropriate results.

19.3.6 Model Documentation

A final documentation report is prepared and distributed throughout the organization. The spreadsheet is updated, as well as any other database that houses that information.

19.3.7 Implementation

The last step may well be the most important. Once a conclusion has been reached as to what should be improved, a plan should be put into place to achieve that goal. How this happens can vary dramatically depending on what is being implemented.

19.4 Resources

Executing an optimization program requires access to certain important skill sets. Those who will be most needed are a facilitator, Monte Carlo simulation modelers, and equipment experts. Of course, one will need occasional access to Operations, Maintenance, Reliability, Process, and management personal during different phases of the program.

Of these positions, most roles already exist. However, there will be new positions required for the facilitator, Monte Carlo modelers, and potentially a spares optimization engineer.

19.4.1 Asset Optimization Group

The Asset Optimization group would be a new entity in the organization.

19.4.1.1 Roles

Initially, it would only consist of individuals identified to be the leaders for the facilitator and Monte Carlo modeling roles. Of course, it would expand over time – especially to accommodate new Monte Carlo modelers. How many are required will depend on how mature the program has become.

The facilitator would be the logical choice for the group manager, since this individual must be the most seasoned and cognizant of how the technology improves business outcomes.

There's also a role for a spare parts optimization engineer at some point in the future once that aspect of the program has gained enough momentum.

19.4.1.2 Reporting Structure

In general, I think that it makes more sense to organize this group in the Process Engineering department. There are several reasons why.

First, this would allow all modeling activities at a unit level to be consolidated into the same group. Second, it forces the Process Engineering department to begin thinking more about the performance of individual assets than they currently do. Moreover, the entire design process, from feed input, process flow, and now individual equipment operating characteristics would reside in the same group.

Plus, if these roles ended up being in the Reliability department, I'd be worried this would perpetuate the myth that better asset decisions "isn't my job." The thing that we want is for the group to get pigeonholed in some corner of the organization and forgotten about. That would defeat all the program hopes to achieve.

19.4.1.3 Facilitators

Facilitators will play multiple roles during the project and can be either internal or external resources – at least, initially. I'd imagine that many companies will use external facilitators at the beginning of the program and identify an individual to mirror them throughout a project(s) with the goal they can successfully assume that role later.

The facilitator will help drive the Opportunity Assessment (OA) process highlighted above. Not only will that person play the central role in the actual OA facilitation, but they will also need to have enough process and equipment skills to be able to interpret the results as they are monitoring the discussion.

They will also analyze the OA results and prioritize opportunities. Normally, the facilitator is the technical lead for the project – at least on the business/process/asset side. In some instances, companies may anoint a different individual for the technical lead and facilitator roles. The facilitator will also lead the Consultation process, which was described in detail in a previous chapter. Given all these roles, the facilitation role will require senior-level expertise.

I would also envision that a facilitator-type person would be the manager of the Asset Optimization group.

19.4.1.4 Monte Carlo Simulation Modelers

Another important skill set is Monte Carlo simulation modelers. Obviously, one will need to have experience in how to use the particular software package chosen by your organization. The good news is that those skills are easily teachable. In fact, one can train an engineering student how to use the mechanical aspect of Monte Carlo simulation modeling.

However, when evaluating Monte Carlo modeling expertise, the organization needs to consider individuals, or groups of individuals, that go beyond the pure "mechanical" skills in using the software package but also possess knowledge of how to leverage the Monte Carlo simulation modeling technology to help articulate solutions to the business problems facing your organization.

In other words, it's not just about the mechanical aspect of running a model, but rather how you can efficiently solve the business and technical problems your organization faces. The Monte Carlo simulation modeling expertise within your organization should eventually be able to "guide" your company down the right path. It will almost certainly take some time to develop that set of skills within your company. Of course, there will be significant "on the job" training that will occur as one participates in one or more projects. That's where the right vendors can be a great help.

Many operating companies will choose to use a vendor for initial projects. In those instances, your company will not require immediate Monte Carlo skills. However, over time, you will want to establish full-time positions once the program begins to gain traction. What most companies will probably want to do is identify individuals to be groomed for that role at the onset of the program.

Then, have them mirror those on the vendor side responsible for Monte Carlo simulation modeling during a project(s) until you feel that they are ready to assume responsibility.

As the group grows, it will require individuals of different skill levels. The first person in the group should be groomed for a long-term role and will need to be more experienced – especially with regard to how to use this technology to solve important business problems.

As the group grows, it could be staffed with younger technical resources, such as new engineers, since learning the mechanical aspects of the technology is quite easy for engineering types. The organization could use these junior roles as a mechanism of gaining exposure for them in this area and then promote them throughout the organization. However, the group must always be led by someone more experienced.

Monte Carlo simulation modelers would play a new role in the organization. A leader would be identified for the first project. Others would be added on an ongoing basis, as the need develops.

19.4.1.5 Spare Parts Optimization Engineer

Once the organization gains some experience executing spare parts optimization problems, they'll want to consider if it makes sense to identify a full-time role for just this aspect of the asset optimization program. While this person would be formally part of the asset optimization group, they would have ties into the stores group.

Spare Parts Optimization Engineer would be a new role added at a future date, if required.

19.4.2 Supporting Roles

The following section describes other roles that will be required for the program but are almost certainly part of the existing organization.

19.4.2.1 Equipment Experts

Equipment experts are an important source of knowledge as to how the equipment behaves in different operational scenarios. This group will include true equipment experts, such as a company rotating equipment expert who specializes in that particular type of equipment, or fixed equipment expertise, such as a fired equipment expert. Of course, these skill sets will be augmented by Operations and Maintenance personal that have detailed knowledge of how the equipment behaves within that particular unit. One of the more important tasks for equipment experts will be to supply estimates of the failure and repair of the major equipment in the unit.

Equipment experts already exist in most organizations. Thus, no new roles are required to support the asset optimization program.

19.4.2.2 Operating/Maintenance/Reliability Personal

Unit Operations, Maintenance, and Reliability personal will need to be involved at different points during the project phase. One of their more important duties will be to participate during the consultation process, along with equipment experts, to help generate failure/repair data.

Operations, Maintenance, and Reliability personal already exist in the organization. There are no new roles required to support the asset optimization program.

19.4.2.3 Process Engineering

Process engineering will have a limited role in the asset optimization program, at least directly. They will primarily provide information on material balance changes (especially production quantities) for incremental improvements in different process units. This would most likely be used as intermediate calculations or part of the dependent variable calculations used in evaluating the optimization function.

Process Engineering already exists in the organization. There are no new roles required to support the asset optimization program.

19.4.2.4 Senior Management

Senior management's role during the process is to provide guidance on the economic incentives driving the plant. In addition, they will sponsor the program and provide important moral support as the program proceeds.

Obviously, senior management already exists in the organization. There are no new requirements to support the asset optimization program.

19.5 Consultation Process

The Consultation Process may well be the most important part of the project. During this exercise, a facilitator will lead consulting sessions with those knowledgeable of the equipment to "extract equipment knowledge" from the experts. In many ways, this is similar to the role that an RCM/FMEA or HAZOP facilitator plays during those programs. The facilitator can be hired either internally or externally for the project.

If chosen internally, a good choice might be those that have successfully led root cause investigations in the past, or similar exercises. Not only should this individual have significant technical skills, but they also need to have excellent interpersonal communication skills to be able to keep the meeting moving. That's not a trivial skill set.

If hired externally, that may only be necessary for the first project, until the organization understands the role and can staff it properly. On the positive side, external resources often possess more advanced interpersonal skills since they perform this service all the time. So, using external resources for this role might be a good idea if the organization does not naturally possess those skill sets. However, it does not matter if the individual is an internal or external resource; it's the expertise the team must have.

Once all the resources have been assembled, the Consultation process proceeds as described in a previous chapter.

19.6 Data – and Its Implications

An asset optimization program will generate, and require access to, various sets of data.

On the input side, if it exists in the organization, "nice to have" data is failure/repair data. Normally, this data would reside in the Computer Maintenance Management System (CMMS) or in an adjacent Asset Performance Management (APM) system.

The asset optimization program will also generate data from the Consulting exercises as described in the chapter on that topic. That data will need to have a home and be managed properly.

19.6.1 Failure/Repair Data

If you believe the Monte Carlo modeling purists, they'll tell you that your program requires a set of failure and repair data. That's true, you will need data to perform the optimizations outlined in this book. However, it does not have to be in the form of data in a database honed over many years. That's great if you are one of a couple of organizations on the planet that has data like this, but it's certainly not the norm.

If you do have this data, it likely resides in your Computer Maintenance Management System (CMMS) or in an Asset Performance Management (APM) system. Depending on the organization, that data will be owned by either the Maintenance or Reliability department. Most likely, it will be maintained by the IT department.

However, I'd be weary of relying too heavily on data from these sources. It's important to remember that operating conditions play a significant role in the process industries in determining how well equipment operates. Thus, if you do not have access to what the operating variables were in the time period before the equipment failed, it's easy to misinterpret failure data. Locating the operating parameters would be difficult, at best.

That's why I'd be suspicious about data from these sources – especially for use within an asset optimization type program. I would consider having access to failure and repair data to be a "bonus." It's not the backbone of any asset optimization program.

It's not that this group cannot also use data of this type. They can. However, the equipment experts would use any available data in the same way that a doctor uses medical tests. It just confirms or refines the "gut feeling" of the equipment experts. It's not a crutch. In fact, I'd be weary of anybody that blindly just uses data. If so, they likely do not understand the equipment very well.

19.6.2 Data from Consulting Exercises

The consulting process will generate a significant quantity of data that will need to be managed. For the first project, it probably makes sense to simply store project data within Excel. However, as time moves on, and additional projects are executed, the company will need to decide on the best place to store the data, so that pertinent personnel have access to it.

Since each company uses different IT systems, there is no right answer for each company. The best choices are probably the CMMS or higher-level asset management system database. If those are not available, then it might make sense to create a SQL or Access database. However, standalone programs are often hard to manage.

One also needs to understand the required data structure, as described in the consulting approach chapter.

The consulting exercise data owner would be the asset optimization group, while the IT group would maintain that data.

19.7 Technologies

This book has shown that there are two technologies necessary to enable plant optimization based on better asset decisions – true optimization and Monte Carlo simulation technology, often called reliability, availability, and maintainability (RAM) modelling within our industry.

This book will remain agnostic on which vendor(s) a company should engage for these technologies. However, it's easy to locate the major vendors for these products by just doing a simple internet search. But there are certain comments to keep in mind as one considers different choices.

19.7.1 Monte Carlo Considerations

With regard to Monte Carlo modeling, there are a handful of major vendors – all with certain plus or minuses. Virtually all of their packages can be made to work, and the author has worked with several of them over the course of his career. However, it may take more time on a particular problem when working with one package versus another. Nonetheless, the problems can usually still be solved.

One area where there are differences is storage tankage. Not all packages allow for the "accumulation" of material between equipment or processing units. That's a problem for the plant-level problems where there are storage tanks, such as Improvement Project or Storage Tankage Determination/Optimization.

Another difference is whether your chosen package allows for "custom coding" when an event occurs. For example, is it possible to execute certain logic through the use of custom code when a failure or repair occurs (though it could also be another event)?

That's a nice feature to have since we often want to take action when certain events occur. For example, if a certain piece of equipment fails, from an operational perspective, we might reduce rates. Without the ability to perform "custom coding" when an event occurs, we would not be able to mimic what operations actually do in a given situation. The author has worked with packages that did not have this feature, but it's clearly nice to have.

19.7.2 Monte Carlo/Optimizer Integration

Since the Monte Carlo simulation packages and optimizer are not likely to come from the same company, another important consideration is how easy it is to integrate the Monte Carlo package with the optimizer technology. During a particular optimization run, the optimizer will frequently choose new sets of variables, and those need variables to be seamlessly passed to the Monte Carlo program.

We were able to make it work when we put together the 2004 NPRA example, but it did require some effort on our part. It may well be much easier today.

However, it's important to ask these questions of any potential RAM package or optimizer. More than likely, the optimizer will already have some "hooks" available to perform this duty since, by definition, they are manipulating variables elsewhere on a regular basis. It's the Monte Carlo packages that one needs to investigate further, since they have historically operated solo.

19.7.3 Optimizer Considerations

For most clients, the optimizer is easier to choose. When selecting an optimization technology make sure it is a constrained, nonlinear optimization technology. Both words are important.

The optimizer must allow dependent constraints to be calculated, since we will frequently be concerned about limits. For example, we might be interested in ensuring that we do not violate either the minimum or maximum amount of an intermediate or finished product component. Those values would ordinarily be calculated by some intermediate calculations. However, there are situations when even values directly calculated by the Monte Carlo program might need to be constrained. Dependent variable constraints are a necessity.

The optimization must also be nonlinear, since the underlying system is clearly not simple, and certainly not linear. Therefore, simple linear optimization (LP) programs are not acceptable.

Another optimizer consideration is whether the optimization can handle mixed integer problems. These are generally much more complicated programs. For example, in our earlier discussion of optimizing process unit availabilities, we showed that the optimizer would be choosing from a set of discrete, individual projects. If we used these projects directly without trying to smooth them out, then, for sure, we'd have to use an optimization algorithm that could handle mixed integer problems. However, if we converted those to a smooth function, then we could get away with not using a mixed integer algorithm. The downside of the smooth approach is that it is not how our world actually works, so by necessity, that would be an approximation. However, if we can just use smoothing, it certainly widens our choice of optimization algorithms.

19.7.4 Organizational Choices

Of course, what choices we ultimately make for the Monte Carlo package or optimizer must align with our organization's ability to manage and maintain those software packages. In other words, the IT department will need to be involved as we consider what package to purchase. However, we will not delve into any further IT requirements for the packages since individual IT organizational limitations tend to be quite company-specific and beyond the scope of this book.

19.7.5 Technology Owner

The Asset Optimization group will be the owner of both major pieces of technology. Regarding the optimizer, if this is the same technology used by the Advanced Control and Optimization (ACO) group, then the asset optimization group would need to share ownership responsibility with ACO. Of course, IT will maintain the software implementation.

19.8 Work Processes

An asset optimization program touches your organization in many different ways – some of which might not be fully appreciated. For example, it's likely that many different internal work processes in your organization will be affected. Since each company has its own internal processes, this book will not touch on the specifics of any given work process. Rather, our discussion will be confined to a higher level.

Table 19.11 shows the optimization problems defined in this book and what departments are primarily and secondarily affected.

Certain work processes, such as **Turnaround Timing Duration And Equipment Selection**, and **Optimal Improvement Projects** are so broad-based that they touch virtually every department in the organization.

On the other hand, a few optimization problems only affect selected departments in the organization. For example, **Maintenance Plan Options** and **Task Optimization** are more local and principally affect only the Reliability and Maintenance departments. Another example is **Storage** and **Asset Design Optimization**, which concerns primarily the Process Engineering department, though Operations is sure to weigh in on any storage tankage modifications.

Spares Optimization is unique in that it affects the store department, as well as Reliability.

Table 19.11 Who's Affected by Project Type

Issue	Primarily Affected	Secondary Affected
Turnaround Timing/Duration and Equipment Selection	Everybody	
Optimal Improvement Project	Everybody	
Maintenance Plan Options Optimization	Reliability	Maintenance
Spares Optimization	Stores	Reliability
Storage Optimization	Process Engineering	Operations
Task Optimization	Reliability	Maintenance
Asset Design Optimization	Process Engineering	

19.9 Training

Each organization will want to pursue a vendor-supplied training class for each major technology supplied. For example, all the important players in your organization should attend Monte Carlo and optimizer training classes.

However, in addition to vendor training, you'll want to take advantage of any system modeling training classes provided by universities in your area, since it would be nice to have a better theoretical background for this topic, if possible. As this is an emerging area, classes that were not available in the past might now be.

19.10 Conclusion

The thrust of this chapter was how to create a master plan for an optimization program based on better asset decisions. The plan would include a list of suggested projects, project methodology, resources required for both a given project, as well as for future program development and sustainment.

An Opportunity Assessment approach was presented that detailed how to use a series of probing questions, usually tailored toward a senior management audience, which ranked an organization on a scale of 1–10 (1 being poor; 10 excellent) in a series of categories. The results from these questions were then analyzed in a structured process to determine the best project priority order.

After success on the first project, the organization would need to put into place an infrastructure to sustain the program. These would include decisions on the following:

- Which individual Monte Carlo simulation and optimization technologies should be purchased?
- Where would program data from the consulting process reside?
- Which work processes, and to what extent, should internal work processes be modified to account for the discussion in this chapter?
- Addition of new organization roles. At a minimum, this would include a facilitator, as well as Monte Carlo modelers. As the program matures, it would also be advisable to choose one person responsible for spares optimization.
- What training should be provided for the internal staff? For the most part, these would be primarily vendor-supplied, but system modeling classes from a local university would also be desirable.

20

Conclusion

Optimization of a process plant involves many different facets.

First, we must recognize that refining is a business. So is petrochemical manufacturing. In fact, the same is true for all other process industry businesses. Thus, our optimization objective will always be to maximize profit, balanced across all the relevant time frames – short, medium, or long term.

Second, generating profit for a process plant means we must convert raw materials into saleable products. This requires process technology consisting of a process flow, equipment tailored for that specific process, correct operating conditions, as well as the proper catalyst if a chemical reaction is part of the process flow. For larger companies, they can occasionally develop this technology internally. However, most companies must license it through a company that specializes in this particular type of process technology.

One aspect of maximizing profit is to generate the optimal product slate and overall production amounts for the plant. This involves a thorough understanding of the process and how to adjust operating variables to improve production quantities and product split. The good news is our industry has put into place technology, methodologies, and tools during the last generation to achieve those goals.

Lastly, generating profit requires a significant capital investment for the equipment necessary to make the product.

20.1 The Need for a Complex Asset Base

Speaking of equipment, it's important to remember that process equipment exists for one and only one purpose – to help make product. All equipment within a process unit is part of a collection of, and interconnection with, other equipment. Even though it co-exists with a significant amount of other equipment, each individual piece of equipment has a specific and well-defined role within a process unit. But that does not trump its purpose to help make products. Equipment does not exist on its own.

Unfortunately, in our business, plants end up being complex because that's what is required to achieve our business goals. While we may not have wanted a complex manufacturing environment, it's necessary to meet our business objectives. In fact, if we could get away with building a process unit with fewer equipment, we would. However, that's not our reality.

Plant Optimization in the Process Industries: Incorporating Equipment/Assets in the Decision-Making Process, First Edition. Marty Moran.

Therefore, since we do need to spend vast sums of money on pumps, compressors, distillation towers, furnaces, and reactors in order to help us achieve our goals, the real challenge is how we manage this vast asset base to help generate the optimum profit for our facility. That's been the subject of this book.

20.2 High-Level Business Goals

In an earlier chapter, we stated that the high-level business goals of most process plants are as follows:

- Maximize production when we are able to do so, using the existing asset base.
- Keep the plant running as long as possible between turnarounds. However, when choosing turnaround timing, duration, and what equipment to be restored, balance the increased production of longer run rates against exposing the plant to unacceptable production or safety risks.
- Carefully manage capital dollars to earn the best return on those investments.
- When new capital projects are proposed consider designs with higher reliability if they can be cost-justified, subject to budget constraints.
- Perform day-to-day maintenance that balances the short- and long-term production needs in the most cost-efficient manner.
- Maintain enough spare parts to prevent any production disruptions due to a lack of spare parts, but do not overstock.

Let's see how well we were able to bolster those goals by examining the proposed solutions in our Asset Optimization Program.

20.3 Asset Decisions that Can Drive Optimal Profit

In previous chapters, we detailed optimization solutions that can improve plant profitability from an asset viewpoint by solving various operational, maintenance, and design problems. These included solutions for those generic problems that virtually every plant deals with, as well as how to approach unique challenges.

In general, we would like to minimize the cost of our assets, both their initial purchase price, as well as their ongoing costs.

The same could be said for maintenance costs. In fact, if we could get away with no maintenance, and still attain our business goals, then that's what we would do. That's why, every task performed in a process plant environment must be cost-justified.

However, it's not just about the assets. We need to make asset decisions without negatively affecting the optimal product slate and overall production decisions the organization has already made. For that reason, optimization will drive us to spend the **proper amount** for the original equipment, as well as spending the **right amount** for their care over the long haul, in a way that makes the most business sense.

Balancing process and asset objectives at the same time is not trivial, but it is our ultimate business objective.

The main problems we outlined solutions for in this book were the following:

- Turnaround timing/duration, as well as what equipment should be serviced in a given turnaround
- Determination of Optimal Improvement Projects (optimal process unit availability)

- Maintenance Plan Options Optimization
- Spares Optimization
- Task Optimization
- Storage Tankage Determination/Optimization
- Asset Design Optimization

20.3.1 Optimal Improvement Project Selection (Optimal Process Unit Availability Targets)

For most process units, it's possible to assemble a group of unit experts – process engineers, operations, maintenance, and reliability personnel to determine a list of recommended projects that will increase process unit availability. However, how much each project costs and how it will ultimately impact overall plant production can vary significantly from one process unit to another. In addition, the ease of implementation, and when the project could be executed could also vary. For example, some projects might be relatively easy to implement, and done immediately, while others may require a shutdown.

In an integrated facility with several interconnected process units, it's not always trivial to figure out which improvement project provides the most incremental value for a limited amount of improvement dollars. Prioritizing these projects is an optimization problem.

There's no business reason why each process unit should have the same availability target. Since costs increase as availability improves, it makes more sense to consider lower targets consistent while still meeting business and technical goals. That's especially true since costs increase exponentially as one approaches 100% availability.

When I go on vacation, I only spend the amount of money that ensures I have an enjoyable time. However, I don't try to maximize my spending (at least, I hope not). So, why do we have to shoot for 100% availability for our process units when we know that it will cost us a significant amount of money and not potentially increase business outcomes?

What availability target for each unit is a classic optimization problem. In fact, that choice is a direct result of what improvement projects we decide to implement.

20.3.2 Turnaround Timing/Duration and Equipment Selection

Turnarounds are expensive and disruptive. Not only is production impacted, but hundreds, if not thousands of people, must be employed by the company to execute the turnaround. In addition, we must spend significant amounts of money on equipment and repairs. It is not trivial to determine when to shut down for a turnaround and what equipment should be restored during the turnaround.

The reasons for a turnaround can vary. Sometimes, the supply chain has been disrupted, so this unit or other related units must shut down. The catalyst might need to be replaced, the company might want to make critical repairs, a critical inspection needs to take place, or an improvement project needs to be implemented. Sometimes, equipment limits production, and turnarounds are scheduled to make repairs to improve availability.

Detailed turnaround decisions have historically been a combination of art and science. One place where science can be substituted for art is what equipment needs to be restored during this turnaround and what can be left until a future turnaround. Of course, if we decide to postpone work on a given piece of equipment, there's always the risk that the equipment will fail before we reach the next turnaround. Therefore, we need to understand the probability of that happening, so risks can be adequately managed.

What equipment to work on and the decision of when to shut down units for a turnaround are classic optimization problems.

20.3.3 Maintenance Plan Option Optimization

There's a mistaken belief that the only way to increase a process unit availability is to improve equipment reliability. While that is certainly one approach, another is to improve the "maintainability" of equipment. That refers to returning failed equipment back to service sooner after a failure has occurred.

Anything that can decrease the time from when the failure occurred to the instance when the equipment is back online producing on-spec products will improve process unit availability. There are certain steps after a failure occurs – such as cooling the plant down, isolating the equipment, returning it back to service, and time to heat up to process temperature – where we have little ability to make changes.

However, we may have options to reduce the time required for the repair step itself. For example, there may be options for repairing the equipment "in situ" versus pulling the equipment and repairing it in the shop. Plus, we might be able to preposition equipment or have plans in place ready to execute if we know that there is some likelihood that the equipment may soon fail.

It's sometimes possible to distill those choices down to a set of different maintenance plan options that have different costs, schedules, and production consequences. This is especially true for the more significant or complicated repair jobs. When we can do that, the best choice is sometimes obvious.

However, there may be other times when it isn't. In those cases, optimization can evaluate a series of options to help identify the best economic choice. The quickest return to service would be the most expensive option, cascading down to other options that are cheaper, but take longer to complete.

The choice of the best maintenance plan from a set of options is a classic optimization problem.

20.3.4 Spares Optimization

How many spares are stocked is often determined initially based on a vendor's recommendation and rarely altered thereafter. This is suboptimal and leads to excessive spare parts spending, or the real possibility that a part may be unavailable when it is really necessary. In the worst case, this could lead to a loss of production. Spare optimization is an attractive option to address either one of these two extremes.

A plant's decision on what spare parts participate in an optimization program could also be driven by vendor issues. For example, if the lead time for a critical spare part suddenly increases, the plant may need to evaluate if they should now hold a spare part when they didn't in the past, or if they need to increase their holdings. In addition, if a preferred vendor can no longer supply a critical part, the plant may need to hold additional spares to offset the possibility they will not be able to obtain the spare part from an alternate vendor on the same timetable, or at the same price point.

20.3.5 Task Optimization

All plants execute many different maintenance tasks on a daily, weekly, or monthly basis. Usually, task frequencies were determined years ago, and rarely questioned after the initial decision. This is suboptimal.

There are many reasons why tasks should be reevaluated. One is that technology could have improved allowing the interval to be extended – without any impact on production or equipment. Or labor costs to execute these tasks may have changed.

However, reevaluating task frequencies rarely happens. Not only does this potentially waste labor resources, but it can also consume more spare parts since certain equipment, such as rings, must be replaced every time that equipment is repaired. Thus, how we manage tasks affects the quantity of spares we must hold.

Task optimization is an area where organizations could more effectively balance production, labor, and spare costs.

20.3.6 Storage Tankage Determination/Optimization

Increasing storage tankage provides a buffer for downstream units when an upstream process unit is unavailable. While Operations will generally always argue for more since it provides flexibility, tankage costs capital to build and maintain, plus it can have tax implications.

What often isn't thought about is that there are times when it is more cost-effective to increase overall plant production through increased storage capacity. That's because the cost of an incremental availability improvement for a process unit improvement project is sometimes more expensive relative to the same storage capacity investment. In order to determine the best expenditure of funds, one must look at possible improvement projects and storage capacity at the same time in order to make the best economic decision.

The question then becomes, "What's the best balance between all these differing objectives?"

Storage tankage determination/optimization can determine the proper amount of tankage or how to allocate that storage capacity throughout a plant.

20.3.7 Asset(s) Design Optimization

There are many design decisions required for each asset, or collection of assets, in a plant. These decisions potentially affect the performance of the equipment, the cost required for a given level of reliability, as well as whether there needs to be redundancy for that equipment or a collection of assets, such as in a train, and the level of spare parts required to support the asset(s). When undertaking an improvement, debottlenecking, or grassroots design project, one should consider if any of the design selections are candidates for optimization. The optimizer can then evaluate a series of possible options to determine the best combination that maximizes profitability.

20.4 A Side Benefit → Combining the Process and Equipment Views of the Business

The optimization modeling solutions just described focus on assets, or collection of assets, and permit our industry to achieve something that has not been possible in the past. That is, we can combine the process and asset sides of our business into one comprehensive whole.

This is a positive development and will eventually allow us to improve how we run and manage our businesses. It's important to understand the history behind why this is now possible, and why we should take advantage of it.

20.4.1 Two Different Views of the Business

The process and asset sides of the business have historically been the domain of different engineering types. Typically, chemical engineers have been the process side masters, and less interested in equipment details. On the other hand, mechanical engineers have managed the equipment, yet were not usually cognizant of process details.

While there have always been knowledge overlaps between the two groups, there have also been significant gaps in the understanding of each other's domain. After working on both sides of the business, I noticed instances where these groups don't work well together, or where there could be improvements. We can now significantly bridge this gap by combining the two different simulation technologies for each domain.

20.4.2 Process Simulation Has Been Very Successful

Chemical engineers have been using industry-standard process simulation products for well over 30 years. These models utilize first-principle models grounded in thermodynamics, with the strength of these packages being the accuracy of their physical property algorithms.

Using those tools, engineers specify basic unit operations, necessary product specifications, and feed streams that will transform raw materials into final products. Reactor yield data is often supplied via internal or process licensor data and manually entered into the simulation. The package then calculates missing stream compositions and flow rates based on thermodynamics and conservation of mass principles.

In addition to determining the material, energy balance, operating conditions, and necessary equipment, the simulations allow engineers to size the equipment properly. Unfortunately, the equipment in process simulation models is generic. The downside of this is that operating characteristics and the real-time limitations of actual plant equipment are never considered.

In addition, these models always assume equipment is available. That is, pumps always work, furnaces never trip, and that process units never shut down. Of course, this is unrealistic. We all know that a real-life plant never achieves 100% equipment availability.

20.4.3 Asset Simulation Is Now Available

Whereas process simulation has been successfully used for many years, our industry has lacked the natural process simulation complement that should proceed immediately after process simulation – asset modeling. This technology would predict individual equipment availability. This would not be done on "generic" equipment, but rather on specific equipment with known failure characteristics.

Asset modeling is a system-wide analysis, often across the entire plant, normally focused on a particular process unit that estimates the availability and/or asset utilization over an extended period of time. It incorporates failure and repair characteristics of the specified individual equipment, as well as including tankage capacity and the layout and interconnection of the process equipment.

Asset modeling is different from process simulation models because models are created based on actual operating data, or assumptions of how real-life equipment will perform in a given environment, rather than on first-principles models. This includes the physical locale, as well as how personnel with differing experience levels affect operation. It includes how often maintenance is done, when, and how skilled those personnel must be to accomplish those duties. Calculations are not done at a moment in time, as they are in process simulation, but rather over an extended period of time.

These packages utilize the knowledge of different individuals who have participated in the equipment selection, operation, and maintenance over the years. Fundamentally, those are the "smarts" inherent in asset modeling.

This includes the EPC vendors who made the initial equipment selection years ago when the unit was originally being designed and built, equipment experts in the organization who may have made improvements to that original design, or have experience operating and maintaining the equipment, the support of different vendors with particular expertise in one equipment area, as well as internal operations and maintenance personnel. One important job in asset management is to access all this knowledge and institutionalize it.

20.4.4 Combining the Simulation Technologies

The Monte Carlo simulation tools now available in the marketplace can marry the chemical engineering of the process simulation world with the mechanical engineering of the equipment world to produce a much more accurate estimate of actual plant production. Fundamentally, it's availability modeling, if throughput is not considered, or asset utilization modeling, if throughput can vary.

So, while the asset modeling approach is different, it tightly complements the process modeling effort. In fact, the two modeling approaches form a comprehensive whole that more accurately depicts plant operation than either by itself. Asset modeling takes the theoretical numbers from the process modeling and makes it realistic in the sense that now we are dealing with real live equipment operated and maintained by fallible human beings in a dynamic organization.

Asset modeling can also help us bridge the process-equipment gap by incorporating process conditions, as shown in an earlier chapter. This allows us to include process considerations in our asset decisions since feed rate, process conditions, and potentially even contaminants, can significantly influence equipment performance.

That's why asset modeling should become as important to an operating company as process simulation is today. In essence, it should be an equal partner with process simulation technology.

In fact, the ultimate nirvana would be to create an analogous evergreen availability/asset utilization model for each process unit, which would be maintained by the process engineering department, and synchronized with the simulation model. These models would then become available for use within plant optimization problems, as described in this book.

20.5 How to Move Forward with Your Program

Once an organization has decided that it wants to move forward with an Asset Optimization Program, it's necessary to decide how to proceed. Choosing the right project is an important first step.

To achieve that goal, I suggest using the Opportunity Assessment process described in a previous chapter to help unearth good opportunities, if one does not initially have a project in mind. This generalized approach will shine a light on important issues that the organization has become accustomed to, or never even thought about. Using the prioritized list, the organization could pick one top issue to focus on and pilot a program around that problem. Even if the organization decides not to pursue an optimization based on better asset decisions, it would still be worthwhile to undertake the Opportunity Assessment, since it still highlights issues within the organization that may be worth solving.

A key decision will be whether to pursue this effort using all internal resources or source some of those to a vendor. Many companies will find that using a vendor might make the most sense for the first project. However, there's been enough information presented in this book so a company could proceed without a vendor if they feel that they have the necessary internal expertise.

That's not to say that process companies don't have the required *asset* expertise. They clearly have that in spades. In fact, the knowledge that Operations, Maintenance, Reliability, Process Engineering, and equipment experts possess is the backbone of your program.

It's the facilitator and Monte Carlo modelers that might be challenging. Both roles are important, but of the two, the facilitator is key to project success.

On the other hand, the mechanics of Monte Carlo simulation modeling are relatively easy to teach.

20.6 Limitations of Asset Modeling

As with all technologies or methodologies, asset modeling combined with optimization is not a panacea, but rather another tool in your toolbox to improve the organization's profitability. When used properly, it can help achieve that goal. However, it can be misused if you don't have a good idea of what problems to focus on.

As a doctor once told me, if your doctor has ordered a test and they don't have a really good idea of what the issue is before the test is conducted, you should be worried. That's because medical tests should be used more for confirmation of an issue (or lack thereof), or a refinement of the diagnosis. The same is true of modeling.

A company needs to stay focused on three important aspects.

First, the Optimization problems described throughout this book are the major issues that virtually all process plants have to deal with. Solving these problems is probably the best way companies could derive value from an asset optimization program.

However, the Opportunity Assessment and Consultation processes described in earlier chapters are also important. Opportunity Assessment helps guide what projects to focus on in a prioritized order for a particular organization, while the Consultation process provides the best equipment data estimates for the project.

If an organization stays focused on these three major items, the odds of success are considerably higher.

20.7 Comparing Process and Asset Optimization

Ironically, while optimization may have first appeared on the process side of the business, it may ultimately have more applicability on the asset side.

That's because a considerable portion of true optimization on the process side occurs during the LP planning process. At least, that's true in the refining industry, though it may not be quite as important in other process industry segments. If we think about LP Planning, it is fundamentally based on a linear programming optimization algorithm. It's just that it's a linear program – not nonlinear. Nonetheless, it's still optimization.

Other optimizations may occur in the Advanced Process Control world, where we often layer Optimization on top of Multivariable Control. However, our industry has learned over the past 30 years that we only need optimization in certain circumstances. That's because if the economic optimum occurs at either a predefined maximum or minimum, for example, a certain well-known process constraint, then we could achieve that goal directly with a Multivariable Controller without the need for optimization.

However, in performing the research for this book, the author has realized that optimization is probably more natural on the asset side rather than the process side. That's because it isn't always obvious what the best improvement project to execute or what equipment to select during a turnaround should be. Spares optimization isn't trivial either, or determining the optimal storage tankage allocation.

That's even more the case when we add constraints to the problem. Without constraints, the answer is sometimes obvious. However, layer in constraints, and suddenly, it's not so easy to make that determination. But, as always, it will depend on a company's particular unique circumstances.

20.8 The Future of Optimization

Of course, real nirvana will be achieved when process and asset optimizations are truly combined together. Today, there is one package for process simulation and a totally separate one for assets.

Today, process simulation technology is extensively used. Asset modeling is not.

However, wouldn't it be great if we could break down the barriers that separate these two worlds and do a better job of combining these aspects of the business together?

It won't happen overnight and would likely need software integration between the packages to make it realistic. However, back in the late 1970s, there was no Advanced Process Control. Today, it is everywhere.

So, the question we should be asking ourselves is, "Why haven't we combined these technologies together?" instead of "Why should we?"

I hope that I live long enough to see the day when there is a seamless integration between the two technologies.

Appendix A

Nuts and Bolts of Monte Carlo Simulation

This appendix will provide a more detailed examination of Monte Carlo simulation modeling. While this discussion will improve a user's understanding, it is not meant to replace vendor or university training, which is by its nature is more comprehensive and complete.

Fundamentally, Monte Carlo simulation technology, which is often referred to as reliability, availability, and maintainability modeling (RAM) within the industry, gives users the ability to evaluate the availability and cost performance of a collection of assets over a period of time, assuming a given process/equipment configuration and individual equipment failure/repair characteristics.

Since Monte Carlo modeling is a probabilistic mathematical technique, model outputs are also probabilistic. In other words, outputs are probability distributions, rather than specific discrete values.

A.1 Example Projects

In order to ease into this discussion, it's best to start with some example projects that Monte Carlo modeling can solve that I have participated in. (No actual client information is used for obvious reasons.)

1) A large refiner was trying to determine if a second, largely redundant electrical line was required for their plant. Their hunch was that it made sense, but they wanted technical and financial justification to show to their management team to prove their point.
   a) When initially assigned to the project, the ultimate project objective was not clear. To seed the discussion, they sent some detailed one-line drawings, which were briefly reviewed before attending the kickoff meeting.
   b) Unfortunately, electrical systems can be quite difficult to model within Monte Carlo tools, because, unlike process flow, which typically moves in only one direction, electricity always flows from high voltage to lower voltage. For that reason, it can seemingly flow backward on the reliability block diagram (RBD), making modeling difficult within a typical RBD. Because it is such a specialized problem, only certain vendors have the capability to model such a difficult problem. Unfortunately, we weren't one of them.
   c) Fortunately, after we talked through the problem at the kickoff meeting, we all agreed that the problem really didn't require that level of detail. In fact, we could answer their question with a vastly simplified model.

Plant Optimization in the Process Industries: Incorporating Equipment/Assets in the Decision-Making Process, First Edition. Marty Moran.

   d) We then sat together, agreed on some important assumptions about the failure characteristics of certain high-level blocks, and quickly iterated to an answer.
   e) The exercise validated their hunch. That is, we were able to justify the cost of a second redundant electrical line with the model providing the business rationale to proceed with the project.
   f) This project was a great example of why a more detailed model is not always better. In fact, it's usually better to create the simplest model that answers the business/technical question.
2) The second project was actually a series of similar projects, where we were trying to determine if availability would be adversely affected if the plant increased production. In other words, could the plant, as currently designed, accommodate the increased production without a pronounced impact on availability, or would the plant need to be modified?
   a) These were natural resource plants located in the southern hemisphere that were desperately trying to increase production due to a phenomenal increase in Chinese demand.
   b) To start, we imported some rough failure data into a spreadsheet and used that information to estimate failure and repair characteristics of major plant equipment. The data wasn't perfect, but it was good enough for our analysis. We then linearly extrapolated that data to determine what would happen to failure rates if production increased, then reran the models and examined the results.
   c) We determined that, while availability might decrease a small amount, the plant could safely meet the increased production goals without major modifications.
3) The third project was a large coal gasification project in the Southern US.
   a) The client was trying to determine if the plant, as built, would be able to meet the original project design availability targets.
   b) A previous model had been built, but they had doubts as to how accurate it was. So, to avoid biases, the project team decided it would be better if a brand-new model was built by people not previously involved with the project.
   c) This was a large plant on the order of a medium-sized refinery.
   d) We didn't have any failure or repair data, so we used the consultation methodology as described earlier in this book.
   e) In essence, we built RBD components for all the major equipment within each process unit.
   f) We used submodels extensively using three major levels of detail to portray the plant accurately.
      i) The top level was the entire plant.
      ii) The second level was usually individual process units, though there were a few instances where, we grouped a collection of units together – especially when there were parallel trains.
         1) In those cases, level 3 was the actual individual process unit.
   g) The most important part of the project was the consultation process. In this step, the consultants, working under my guidance, held a series of meetings with the company's operations and maintenance personnel. During those meetings, they talked through each unit's major equipment and assigned estimates of equipment failure rates, along with their best guess of repair time, as well as production impacts.
   h) The agreement reached by the group for the failure and repair estimates were so well done that it validated the consultation approach. In fact, one could peruse the tabs in the spreadsheet for each process unit and fairly easily determine the availability for that unit by examining the team's original estimates.

i) Those results confirmed my gut feeling about the Monte Carlo simulation. That is, it's just a large mathematical engine. What's much more important was the failure and repair data fed into the model.

i) Estimates from the consulting meeting were then transformed into failure and repair distributions, and the models were formally run.

j) The completed model showed that the plant would not meet the project's goals and highlighted many issues with the design. It was able to shine a light where improvement projects could address major, underlying issues.

k) This project is a great example of how an organization could proceed with Monte Carlo technology.

These three projects are a sampling but should give the user a flavor of how Monte Carlo technology can be used to solve difficult, real-world problems successfully. In other words, how could this technology be used to answer important business questions that were not part of the more general business problems delineated in Chapter 3.

A.2 Monte Carlo Modeling Steps

The project sampling above highlights the process of creating an accurate RAM model:

- Accurately describe the business/technical problem one is trying to solve
- Determine how detailed the model needs to be to adequately solve the problem
- Create reliability block diagram(s)
- Assign throughput information
- Assign failure/repair information for each piece of equipment or failure mode
- Run the models and review their results

Let's discuss each of these steps in more detail.

A.2.1 Defining the Problem to be Solved

One must explicitly understand the business/technical objective of the problem at hand and why you are even considering using Monte Carlo simulation modeling. You should be able to state the objective in a simple sentence, such as:

- Determine the economic optimum of whether there should be a common fractionation section or individual trains.
- Determine the economic optimum of whether we should store special equipment onsite or purchase them when required at market prices and availability.
- Determine the economic optimum quantity of spare parts for a given application.

The objective can either be a technical or business objective. In this book, it has normally been a business objective with the goal of determining the optimal profit.

What's especially important is understanding how detailed the model should be to answer the business/technical problem adequately. That's where your organizational process and equipment knowledge become paramount in the form of equipment experts from operations, maintenance, process engineering, etc.

In fact, that knowledge is much, more important than the specific skills required to use the RAM modeling software tools. In reality, any competent engineer, or for that matter any college-level engineering student, can be trained to use the Monte Carlo software tools. What isn't easily taught is the experience needed for the front end of the analysis process. That comes from knowledge of the process and equipment, usually gained over many years in the industry.

The bottom line is that the real "art" in effectively using Monte Carlo modeling is not the use of the tool at all, but rather the front-end problem analysis. The problem must be explicitly and accurately defined.

A.2.2 Model Complexity Possibilities

Commercially available Monte Carlo RAM tools software tools are quite flexible and can be structured to build models of any degree of complexity. For example, one could create a much more generalized model of an entire plant with each process unit modeled as a simple block with no underlying detail.

Or one can proceed to build a model, similar to the last example highlighted above, where we abstract the major equipment in a unit and only assign generalized failure and repair distribution to each major piece of equipment. These process unit models can then be aggregated into one larger, plant model. Using this structure, one can run the models at either the process unit level only or aggregate them together to run the entire plant model.

On the other hand, one could create an equipment compressor model that, in addition to the normal mechanical components inherent in any compressor, includes detailed submodels that adequately describe the oil lubrication system, cooling system, and supporting instrumentation.

Unfortunately, the flexibility inherent in the software tools has been both a blessing and a curse.

It's a blessing when technology can be used to solve a difficult problem. However, the ability to create overly complicated models is a sin that many a RAM modeler has succumbed to.

Even though the tools allow one to create models to any degree possible, the general guidance is that one should build a model with only enough complexity to answer the business or technical questions that one is attempting to ascertain.

As the next section implies, simpler is almost always better.

A.2.2.1 Simpler Is Better

When setting up Monte Carlo problems, the level of detail required to solve the problem should naturally flow from the problem definition. However, engineers, by their nature, are technical people, often very technical, and are usually biased toward adding more detail rather than less. In the Monte Carlo world that has often led to models that are *far* too technical.

In fact, one of the reasons that Monte Carlo modeling has failed to gain a stronghold in the process industries is that there is a false belief that one has to build a very detailed model of the equipment in a processing unit in order to adequately analyze it. Not true.

What we learned from the three examples described earlier in this chapter is we can construct relatively simple models to answer the problem at hand. The important lesson is that the model does NOT have to be complicated to add value to the problem that one is trying to solve.

However, that doesn't mean that all models will be simple. There may be times when one has to build more detailed models. That's OK. But one should take that approach only when detail is absolutely necessary to answer the problem at hand.

One reason for simplicity is that each failure point requires the definition of how often it fails, along with how quickly it can be repaired. This requires either hard data or an estimate.

Second, the amount of time to create the model initially, as well as the long-term maintenance to support it, increases significantly as the model becomes larger. Save yourself the hassle, not to mention work, as well as numerous headaches, and only abstract a problem to the level of detail necessary to obtain an answer for the problem at hand.

A.2.3 Create Reliability Block Diagrams

Once the exact business problem has been decided upon and the level of detail required for that analysis, the next step is to create a reliability block diagram (RBD).

RBDs are graphical representations of a system that depict equipment, or groups of equipment, that could potentially fail such that the system, or subsystem, becomes unavailable. It highlights the interconnection among the equipment, or groups of equipment, but is different from a typical process flow path found on a process flow diagram (PFD). Each block on an RBD contains failure and repair distribution data, along with production and maintenance interval information.

The system is assumed to be "available" if enough blocks are "running" such that one can trace a complete path through the system from start to finish. If the system is available, the results show whether it is in full or partial production. An example of a RBD for a catalytic reforming unit is shown in Figure A.1.

Individual blocks, usually equipment, can be connected in any combination of series and parallel paths. While the blocks often follow a process flow for a particular plant, that can be misleading.

For example, most plants have backup pumps for liquid service, which are shown as parallel paths on an RBD. In general, whenever there are parallel paths in an RBD, only one of the paths needs to be active in order for the system to be considered "up and running."

All modern Monte Carlo simulation packages have special provisions for redundancy so that if one component fails in a parallel arrangement, the software will naturally transition to its backup, as would occur in actual production. This is the normal modeling practice for backup pump systems.

In general, blocks should only be as detailed to answer the business/technical question being sought. Often, the majority of blocks on a RBD are individual pieces of equipment level. However, there is nothing that requires this to be so. A block can be anything that could cause the unit to shut down.

A common rule of thumb is to only add blocks to a RBD for equipment that realistically could fail. For example, instrumentation rarely needs to be added for most practical problems, since it doesn't often lead to a unit shutdown. Therefore, there's no need to include it in a model.

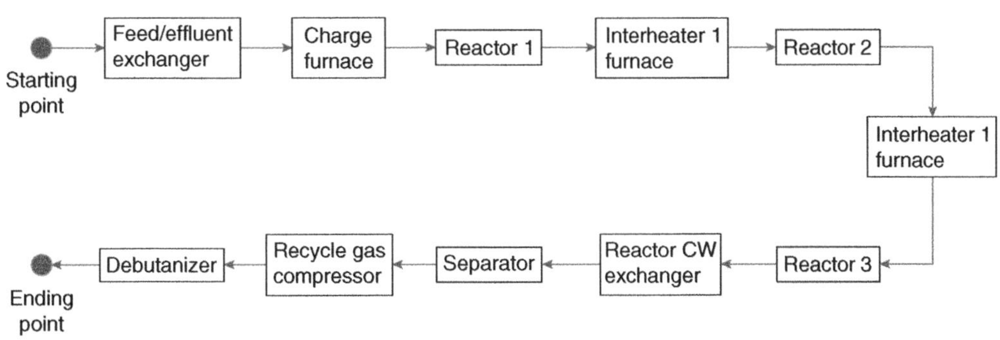

Figure A.1 Catalytic Reformer Reliability Block Diagram

A.2.3.1 Submodels

Submodels, or subdiagrams as they are sometimes referred to as, are an important part of modern Monte Carlo simulation packages. You will use them extensively. These allow a user to build complete RBDs using different levels of RBDs, hence supporting detail, yet access them seamlessly within one overall, higher-level RBD.

Using this feature, one can build relatively simple models at the highest level and let the user "drill down" into the details only when one needs access to those details. For example, it's common practice to construct a process unit model that shows the interconnection of the major equipment, as shown for the catalytic reformer unit.

Some of the process unit blocks could be constructed as a submodel, instead of a single block. This allows the user to "drill down" into the process unit to view a detailed RBD model of individual equipment for that unit, if one desired. The submodel for the debutanizer after the user clicks on it is shown in Figure A.2.

Within the "submodel," the user would likely abstract the majority of equipment as single blocks with no supporting detail. However, there may be equipment even at that level that the user wants to detail further.

There isn't a compressor on the debutanizer RBD, but if there had been one, then the user could build a detailed compressor model that would allow a user the ability to drill into compressor details, such as the lube oil system. The basic process of "drilling down" into sublayers can go on indefinitely.

When Monte Carlo projects with submodels are executed, the lowest level is calculated first, and the results are passed up to the next higher level. This process continues until the highest level is run.

The great thing about submodels is that the lower-level RBDs can always be executed separately. For example, one could run the debutanizer, independent of the catalytic reformer, if it's desired to understand the results at only that level. Submodels give users the best of both worlds.

It's important to note a potential downside with submodels. That is, as more blocks are added to the overall RBD (whether it be at the highest level or one of the lower RBDs), the model will require more processing time to execute. In fact, it's possible the user will reach a point of diminishing returns regarding model size when running complex models due to unrealistically long processing times.

In those instances, the user is forced to break the models apart. At that point, they would likely create and manage entirely separate lower-level models. They would run those individual models separately and then manually key in model results from the lower-level models into the

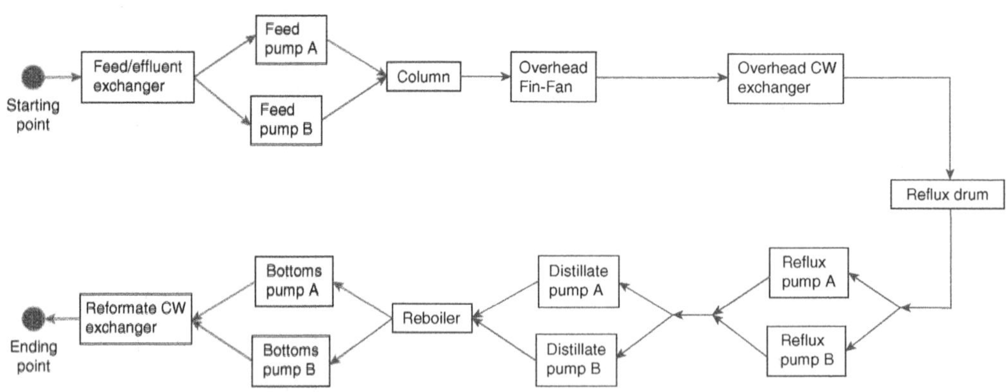

Figure A.2 Debutanizer Reliability Block Diagram

higher-level model. As can be imagined, this is a laborious process. For that reason, I personally start with submodels in the appropriate spots and only use totally separate models when calculation time exceeds some critical threshold.

The great thing about submodels is that they also make the management of Monte Carlo projects considerably easier than creating a series of models as separate Monte Carlo projects. That's because submodels and higher-level models are all part of one Monte Carlo project.

A.2.4 Assign Throughput Information

Once RBDs are completed, "Throughput" values should be assigned for each block. These depict the quantity of total feed rate that the block can process as a percentage of total feed. This becómes important when one is trying to evaluate how throughput changes as various equipment (blocks on the RBD) fail.

Sometimes, when a block fails, that failure will result in a total loss of production from that system. However, there may be other times when a block fails that another part of the system can still process some of the total feed, resulting in only a partial loss of production, rather than a total loss.

Whether equipment failures result in a total or partial loss of production becomes important when evaluating Monte Carlo results. One important metric is system availability, which is simply the amount of time that the entire system is mechanically available. However, another important measure, and particularly important in our optimization calculations, is system throughput over a period of time.

If all equipment failures result in a total loss of production, then total system throughput is simply the availability multiplied by the nominal feed rate. However, if it is possible to have partial production when certain equipment fails, then total system throughput will be larger by some amount over the case when all equipment failures result in a total loss of production.

A.2.5 Assign Failure/Repair Information

The "heart and soul" of any RAM modeling project is the assignment of the failure and repair data for the individual pieces of equipment or failure modes. In fact, the success or failure of a RAM modeling project is highly tied to the accuracy of the failure and repair information.

In some industries, such as electric power, automobiles, and aircraft, credible industry-wide sources of failure data exist. An excellent example would be the failure rate of transformers in the electrical power distribution industry. However, the process industries are unique in that there are a wide variety of operating conditions, plant configurations, and feedstock differences that have made compiling accurate industry-wide data difficult.

For that reason, the best source of failure and repair data in our industries are the people knowledgeable about specific equipment and processes in YOUR plant. That's why we suggest using the consulting process described earlier in this book. The work product from the workshops is an initial estimate of the failure frequencies and repair time for each piece of equipment throughout the complex. While these are estimates from the workshops, they can be updated over time as real data becomes available.

A special thought about repair time. It should be noted that when most people are thinking about "repair time," they are thinking about wrench time to actually repair the equipment. However, from a modeling standpoint, what is really required is the total amount of time the unit is down for a particular incident. So, we are interested in the total time to cool the equipment down, isolate the equipment, repair the equipment, return it to service, and bring it up to operating conditions. You will often hear this referred to as FOFI – "Feed Out Feed In." Since it encompasses many different steps, it is usually a much larger number than simple wrench time. That reflects the fact that a unit needs to be cooled down before one can even consider gaining access to the equipment. Plus, we are not producing on-spec products until they have had time to be brought back up to temperature and lined out.

A.2.6 Run Models/Review Results

The last step in the process is to run the simulations and review the results. The two key analysis results are the probability distribution and the Pareto chart. Using the probability distribution, we can tell if the system can achieve the desired availability target.

On the other hand, the Pareto chart details what equipment has the greatest contribution to "unavailability." Thus, if one needs to increase availability, these are possible equipment that one should concentrate on. As shown in the chapter on increasing availability, it's one way to proceed (but as the chapter highlighted, it's not the only path forward).

An example of a probability distribution is shown in Figure A.3.

Examining the graph can tell you the maximum and minimum availability values. In other words, availability MUST be within that range and can't possibly be greater or less than those values. In this particular example, the minimum availability is 94%, while the system will never achieve availabilities greater than 98.5%. The mean value is 96.4%.

The shape of the curve is also important. Often, they are not bell-shaped. In fact, most curves are relatively similar to Figure A.3 since that shape indicates it's harder to achieve higher availabilities because one starts approaching a point of diminishing returns. On the other hand, the tail on the left-hand side of the curve is often more pronounced since it's much easier to lose availability than gain it.

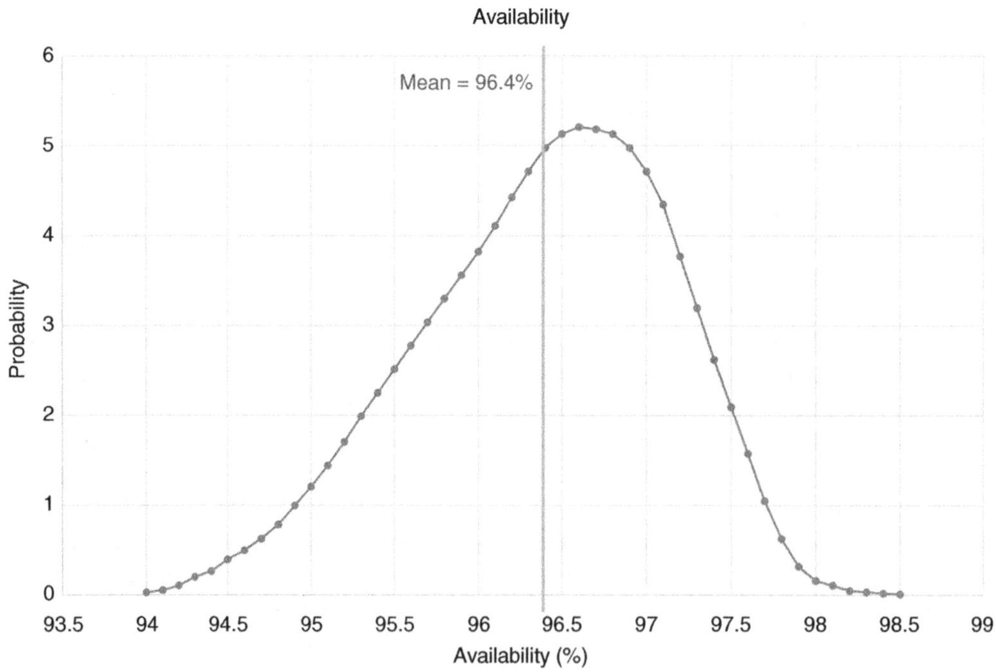

Figure A.3 Availability Probability Distribution

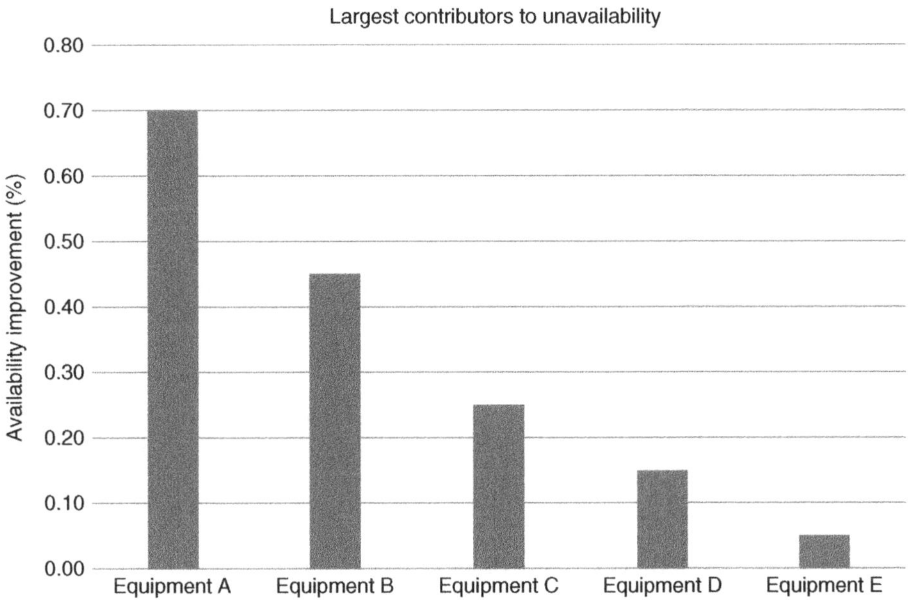

Figure A.4 Largest Contributors to Unavailability

The Pareto chart shown in Figure A.4 highlights equipment that contributes to unavailability. The largest contributors are always shown on the left-hand side of the chart and decrease in importance as one moves toward the right.

This chart shows that equipment A contributes 0.7% to unavailability. If it is practical to restore that equipment, then total system availability could be increased by that amount. In a similar fashion, if equipment B's issues could be addressed, then system availability could increase by 0.45%.

It's important to note that there is no requirement that one start with equipment A before proceeding to equipment B. In fact, sometimes it's impractical to correct issues with a piece of equipment farther to the left in the graphic, yet it is possible to implement equipment changes farther to the right. For example, equipment A might require a turnaround to implement, whereas equipment B does not.

A.3 Using Monte Carlo Modeling in Absolute vs. "Delta" Sense

Monte Carlo simulation modeling can be used for many different possible problems, but in its broadest sense, Monte Carlo technology is a decision-making tool that allows one to evaluate the pros/cons of different alternatives. When used in that manner, results are reported as the difference between two different case runs. We refer to that as a "delta" mode.

However, it is also possible to run Monte Carlo models in "absolute" mode.

Let's discuss the pros/cons of these two different approaches, starting with the absolute case.

A.3.1 Using Monte Carlo Modeling in Absolute Mode

Monte Carlo modeling can be used in an absolute sense, though one has to be careful when interpreting those results.

That is, one can start by creating an RBD to mimic the equipment in a process unit or plant. The components within the RBD are then assigned failure and repair distributions representing their real-world performance.

However, the accuracy of the results is directly related to the confidence level of the input reliability and maintenance distributions. So, one needs to be quite careful when interpreting absolute availability results, since they are entirely dependent on input data. In fact, a good rule of thumb is "the confidence in the absolute availability result is in direct proportion to the quantity and quality of ALL distributions in the model."

Thus, as the sheer quantity of equipment in the model goes up, hence the total number of distributions, the availability confidence level goes down just due to the total quantity of distributions needed to depict the model. That's because all distributions have some uncertainty associated with them.

So, the user should be weary when interpreting absolute Monte Carlo modeling results. Directionally, the models will provide the correct results, but caution should be used in assessing the exact absolute value.

A.3.2 Using Monte Carlo Modeling in "Delta" Mode

Most of the issues with absolute Monte Carlo results disappear when used in "delta" mode. That's because the quality of failure and repair distributions becomes much less important since the majority of distributions are used in both the "base" case and the "delta" case. That allows one to focus solely on how the delta input changes affect the delta final availability determinations. In fact, when used in "delta" mode, models don't necessarily need to be overly complicated, and sometimes not super accurate, in order to generate useful results.

Let's now take a look at how Monte Carlo can be used in "delta" mode for some typical scenarios. We'll examine a design, maintenance, and operational case separately.

A.3.2.1 Using Monte Carlo Modeling in "Delta" Mode for a Design Decision

First, let's evaluate a design scenario.

One problem where Monte Carlo has been used successfully is to evaluate whether parallel trains with common components should be used or whether it makes more sense to build two individual trains. Fundamentally, one is trying to evaluate if one spends additional money to eliminate the common components, how much will the availability improve? Is it cost-effective to make this change? One might evaluate a number of different scenarios before arriving at the most economical choice.

For example, one could ask, "Should we double the entire plant or just a portion of it?" One scenario might be comparing two reactor sections with a common fractionation section or two separate plants with each having a reactor and its own fractionation section.

The fundamental tradeoff between these two scenarios is that there will be fractionation capital savings from a common section. However, there will be times when the fractionation section shuts down taking both reactor trains with it.

In the two-plant scenario that wouldn't have happened. Fifty percent of production would have continued unabated.

To model these scenarios properly, the user would construct an entire plant model – one reactor section and one fractionation section. Then, run the simulation, noting the availability results. This would allow them to calculate the net margin across the unit, since they would know both feed costs and product volumes. They also would have determined the capital costs to construct the unit. They would use both sets of information, doubling them, to calculate the final rate of return on that investment.

To properly evaluate the second case, we need to modify our model. The reactor section wouldn't change. But there would now be two of them. On the other hand, the fractionation section would double in size and consolidate into one in our model. We would then make any failure/repair distribution changes to the fractionation section since there might be differences due to size changes. Then, we would rerun the model to evaluate the availability results for the entire system.

Once again, we know feed costs and total product volumes. We also know the total capital costs. The reactor section design costs don't change, but there is some net change in the fractionation section. We roll all that information together to arrive at a final rate of return.

The two cases now allow us to determine which results in the best economic alternative.

A.3.2.2 Using Monte Carlo Modeling in "Delta" Mode for a Maintenance Decision

That's the design scenario for a common plant section. Let's say that we have decided from a design standpoint that a common fractionation section was the right answer. Could we now use that model in a "delta" sense to make important maintenance decisions? That question becomes important because we are now more vulnerable to fractionation downtime than previously.

For example, could we evaluate whether it makes sense to store special equipment onsite or make contractual arrangements to expedite its delivery, as opposed to trying to obtain it through the normal market? We could evaluate all three of those scenarios in "delta" cases.

We could also extend that thought to spare parts holdings. Do we need to alter our spare part holdings since we are more vulnerable now?

We won't talk through the exact modeling aspects of either of those scenarios. However, the process is the same. We start with a base case, run the alternatives, and examine the financial difference between the cases.

A.3.2.3 Using Monte Carlo Modeling in "Delta" Mode for an Operational Decision

We can also use the model to evaluate different operational scenarios.

For example, it's well known that how equipment is operated can have a dramatic effect on availability. If one runs the equipment harder or at higher temperatures, then the failure rate can be affected. So, if the reactor is operated at 30 °F higher than the design, how will this affect plant operation and economics?

We would first need to evaluate how the reliability characteristics of the equipment change as temperature is increased. Plus, we would need to understand how the reaction yield changes to fully evaluate product distributions. This example is similar to the one we covered in the section on Process Conditions in an earlier chapter.

Another scenario is where the model could be used to evaluate how multiple startup/shutdown cycles affect plant economics.

Given that most failures occur during startup, there can be additional availability penalties because one is starting and stopping so frequently. For example, if one unit shuts down twice a year for a mean time of two days down each time, we might think that this can be linearly extrapolated.

For example, if the plant were now to shut down 6 times in a year, we might assume that the plant would now be down for $(6 \times 2) = 12$ times. However, when we actually run the model, we learn that if a plant has this many shutdowns, the total number of lost days might increase to 15. That's because every time that you start up, there's the possibility of equipment failing. Once again, we can easily model this in a "delta" sense.

Appendix B

Refinery Example Process Description

This section will provide a process description for the refinery example that has been used throughout this book. The majority of the discussion in this appendix section will be to describe each of the 11 process units in our example refinery, highlighting important features of the unit, along with the production changes of key streams as they are processed in each downstream unit.

Like all businesses, oil refining exists to satisfy certain market demands. As is obvious in any American city, there is an exceedingly high demand for transportation fuels – primarily gasoline for cars, kerosene for jet travel, and diesel for trucks and machinery. While a refinery can produce any number of smaller volume products, such as waxes and lubricating oils, the bulk of the products are transportation fuels. Thus, this section will only report the major products – and not the many other miscellaneous products.

B.1 Crude Oil

The feedstock to any oil refinery is crude oil. What isn't widely understood by those not part of the industry is that not all crude oil is created equal.

The two main parameters used to characterize crude oil are crude density, expressed as API gravity, and the percentage of sulfur in crude. While these are the two main indications of crude oil quality, and hence the price they command in the marketplace, there are many other variables required for a complete characterization.

One major difference between crude oils is the proportion of gasoline, kerosene, diesel, and heavier material that a particular crude oil can produce. API number is a general indication of distilled products. As the API number increases, the quantity of the more valuable lighter boiling materials increases. Thus, higher API crudes are more valuable.

A crude with a sulfur content of less than 1% is considered to be a "sweet" crude. On the other hand, crudes with sulfur values greater than 1% are termed "sour." The lower the sulfur value, the more valuable the crude since higher sulfur levels require more processing to remove sulfur compounds.

To make it easier for buyers to evaluate crudes on an "apple to apple" basis, the industry has standardized the practice of crude "assays," which is an independent evaluation of crude oil by a third-party entity. The assay catalogs the percentage of crude that falls into ten different boiling point ranges distilled at atmospheric pressure (normally called "cuts"), and another five cuts distilled under vacuum. In addition, sulfur levels and other miscellaneous parameters are reported as part of the assay.

Plant Optimization in the Process Industries: Incorporating Equipment/Assets in the Decision-Making Process, First Edition. Marty Moran.

B.2 Crudes Used in Problem Examples

There were two different crudes used as examples in this book: crude A and crude B. Most of the examples used crude A, though some later examples utilized both crudes. It should be noted that the material balance for crude A was largely developed (with some minor modifications) based on the material balance for the process units from the book, *Petroleum Refining, Technology and Economics, 2nd Edition, Revised & Expanded* by James H. Gary and Glenn E. Handwerk, and are used with permission. All crude B examples are fictious.

The best way to characterize the crudes without getting into the details of crude assays is to show their product distribution after distillation in both the crude and vacuum units. That basic material balance, assuming a crude run rate of 100,000 barrels per day (BPD) for both crudes is shown in Table B.1.

There are a couple of main differences between these two crude oils. The first is that crude B produces considerably more naphtha material and less crude resid than crude A. That implies it is a lighter crude and has a higher API number. The other difference is the amount of vacuum residue produced.

Another important characterization is sulfur content. In our example, we will assume that Crude A is a "sour" crude and has a sulfur content of 2%, while Crude B is a "sweet crude" and has a sulfur content of 0.25%.

The material balance and sulfur content have important considerations for processing beyond the crude unit. For a refinery to be economically viable, a considerable amount of processing beyond the crude and vacuum units needs to exist. While both crudes will be processed in exactly the same units, there will be differences in feed rates between various units. Those differences are shown in Table B.2 and will affect each unit's material balances, as will be demonstrated throughout the rest of this chapter.

Table B.1 Crude A/B Material Balances

	Gas	Naphtha	Kerosine	Diesel	AGO	Crude Resid	LVGO HVGO	Vacuum Resid
Crude A	1300	18,000	12,900	11,900	13,400	42,300	18,697	23,519
Crude B	1600	25,000	11,000	11,933	12,400	38,000	19,000	19,000

Table B.2 Process Unit Feed rates

	Crude A	Crude B
Crude	100,000	100,000
Sat Gas	1,300	1,600
Vacuum	42,300	38,000
Naph Splitter	18,000	25,000
Isom	5,562	8,750
Reformer	17,941	21,570
FCC	32,097	31,400
Alkylation	7,858	7,557
Coker	23,519	19,000
Hydrocracker	21,879	19,066

One will notice that crude B has a higher Reformer feed rate due to the higher amount of gasoline produced in crude B. On the other hand, there are higher crude A feed rates for both the Coker and Hydrocracker since more material will need to be "converted" in those units into more valuable products. Details behind how those units operate follows below.

B.3 Crude Unit

Crude oil is not valuable in its native state and cannot be used without further processing. Thus, the first unit in the processing train within a refinery is always a crude unit. This is a large, multiproduct distillation tower that separates streams by boiling point with lighter cuts gravitating toward the top of the tower.

After removing water, salts, and other impurities in the desalter, crude oil is heated to nearly 700 °F in the crude furnace before entering lower in the crude tower. Major product streams from the crude unit are fuel gas, naphtha, kerosene, diesel, atmospheric gas oil (AGO), as well as crude resid.

Product distribution data shows we can't produce enough gasoline, kerosene, and diesel for the refinery to be economical based solely on running these crudes in the crude unit. In fact, only 18–25% of these two crudes are naphtha after processing in the crude unit. (Naphtha is gasoline-grade material that has not been reformed.)

Unfortunately, the largest product from the crude tower for both crudes is resid, a low-value, heavier material that boils in a range much higher than the desired products. Thus, we will need to convert some of that heavier material into lighter products for our oil refinery to be profitable. That's the crux of a modern oil refinery and the main purpose for the conversion units described later in this section.

The product distribution after processing in the crude unit for both crudes is shown in Table B.3.[12]

Where the Crude Unit fits logically into the flow schematic is shown in Figure B.1.

B.4 Saturated Gas Plant

The Sat Gas Plant exists to separate out valuable propane and butane from the overhead gas stream. That's done because those products can command a higher sales price than the remainder of this

Table B.3 Crude Unit Products

	Crude A	Crude B
Fuel Gas	1,300	1,600
Naphtha	18,000	25,000
Kerosene	12,900	11,000
Diesel	11,900	11,933
AGO	13,400	12,400
Crude Resid	42,300	38,000

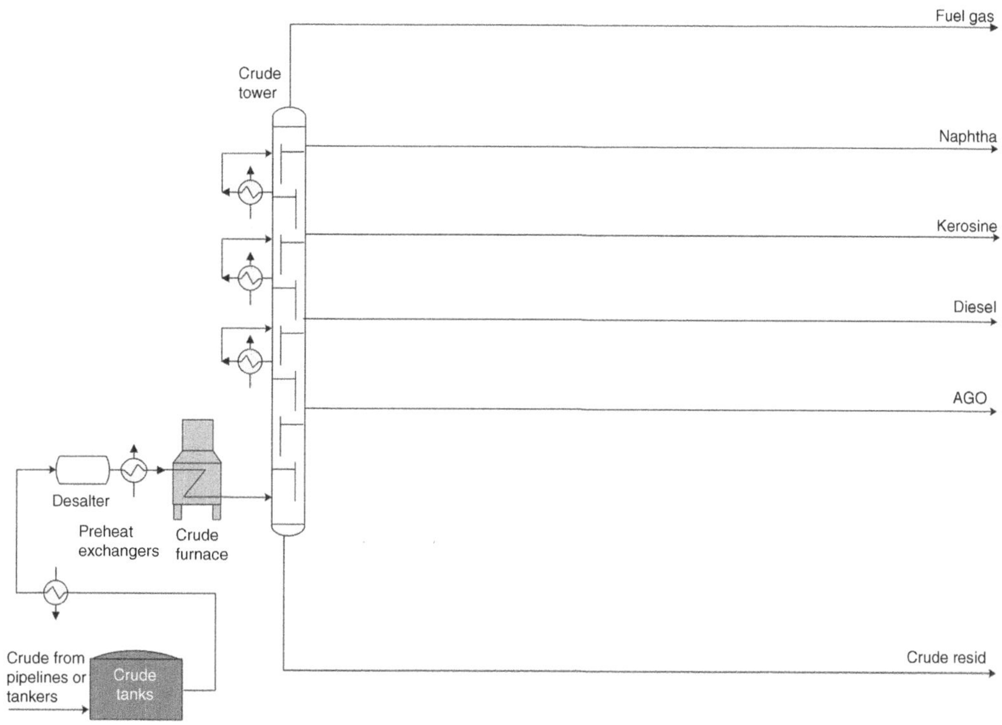

Figure B.1 Crude Unit

Table B.4 Material Balance after Processing in Sat Gas Plant

	Crude A	Crude B
Propane	400	560
Butane	900	1,040
Naphtha	18,000	25,000
Kerosene	12,900	11,000
Diesel	11,900	11,933
AGO	13,400	12,400
Crude Resid	42,300	38,000

stream, principally methane, which is valued as natural gas. The product material balance after adding processing in the Saturated Gas plant is shown in Table B.4.[13]

Where the Sat Gas Plant fits logically into the flow schematic is shown in Figure B.2.

B.5 Catalytic Reformer

Octane number, which we are all familiar with from filling up our car, is a relative comparison of the "knocking" tendencies of a fuel relative to burning isooctane in a combustion engine. To prevent knocking, a modern engine requires fuels with at least a 90-octane number. Unfortunately,

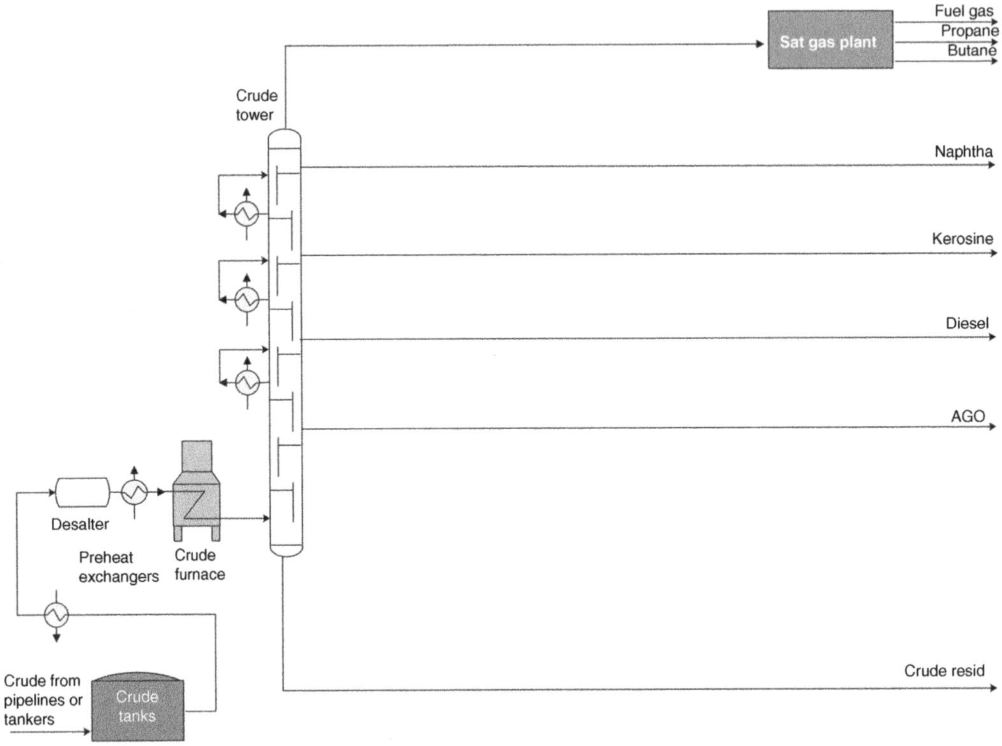

Figure B.2 Add Sat Gas Plant

the octane number of the naphtha stream from raw crude is normally in the 70s. Thus, the naphtha stream must be upgraded so the final blended gasoline has an octane number high enough to be saleable.

That's the purpose of the catalytic reforming process, which increases the octane number of naphtha feed by converting cyclic hydrocarbons to aromatics in a series of low-pressure reactors over a platinum catalyst. The gasoline stream from the reformer is "Reformate."

A huge side benefit of the dehydrogenation reactions is that they create huge quantities of hydrogen, which are used in either other refinery hydrotreating processes, or in the Hydrocracker.

The product material balance after processing in the Reformer is shown in Table B.5.[14]

Table B.5 Cumulative Material Balance after Processing in Reformer

	Crude A	Crude B
Propane	938	1,207
Butane	2,156	2,550
Gasoline	16,147	19,413
Kerosene	12,900	11,000
Diesel	11,900	11,933
AGO	13,400	12,400
Crude Resid	42,300	38,000

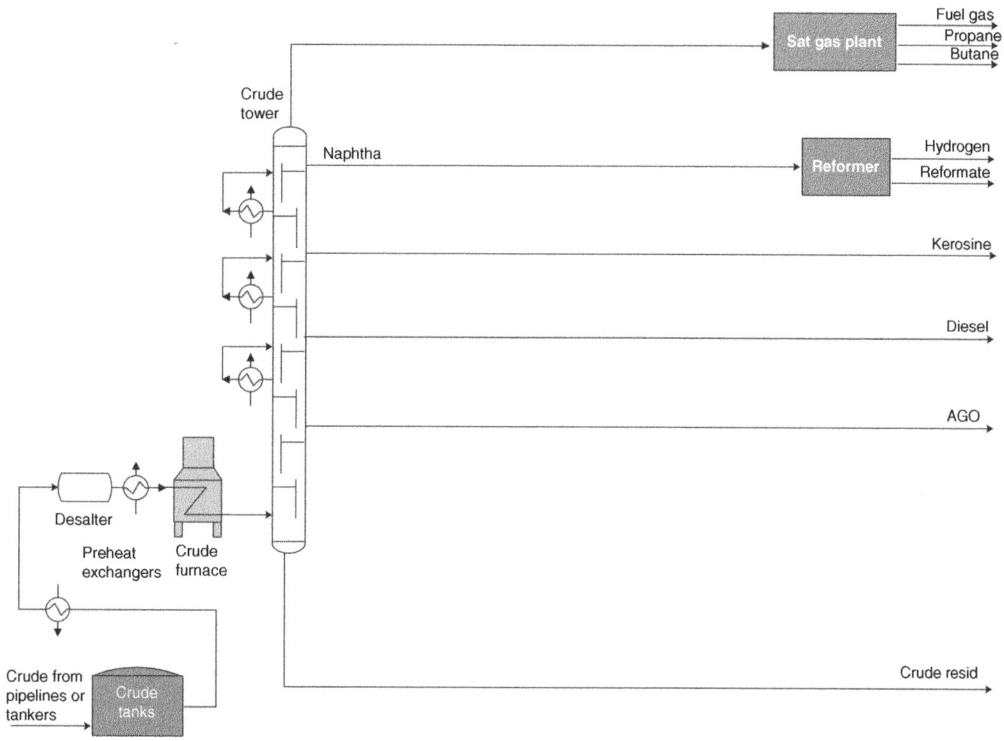

Figure B.3 Add Reformer

Note that naphtha is now termed "Gasoline" since it can sold as final product.

Where the Reformer fits logically into the flow schematic is shown in Figure B.3.

B.6 Naphtha Splitter/Isomerization Unit

If a refinery has a C5/C6 Isomerization unit, which ours does, then there is an optimal split of the naphtha from the crude unit into a C5/C6 paraffin stream and a bottoms C6 aromatics and heavier cut. The Naphtha Splitter is the distillation column that performs that fractionation. The top distillate stream is the feedstock to the Isomerization unit, while the bottom is Reformer feed.

The octane number of straight-chain C5 and C6 molecules is significantly less than branched isomers of the same number of carbon atoms. Thus, the C5/C6 Isomerization unit attempts to convert n-C5 into iso-C5, as well as to convert nC6 into one of four C6 isomers to produce a slight boost in overall stream octane. The resultant gasoline product stream from the Isom unit is called Isomerate.

Since the Isom stream is relatively small and yields are so high, we won't update the material balance, since it won't change appreciably for our purposes.

Where the Naphtha Splitter and Isom unit fits logically into the flow schematic is shown in Figure B.4.

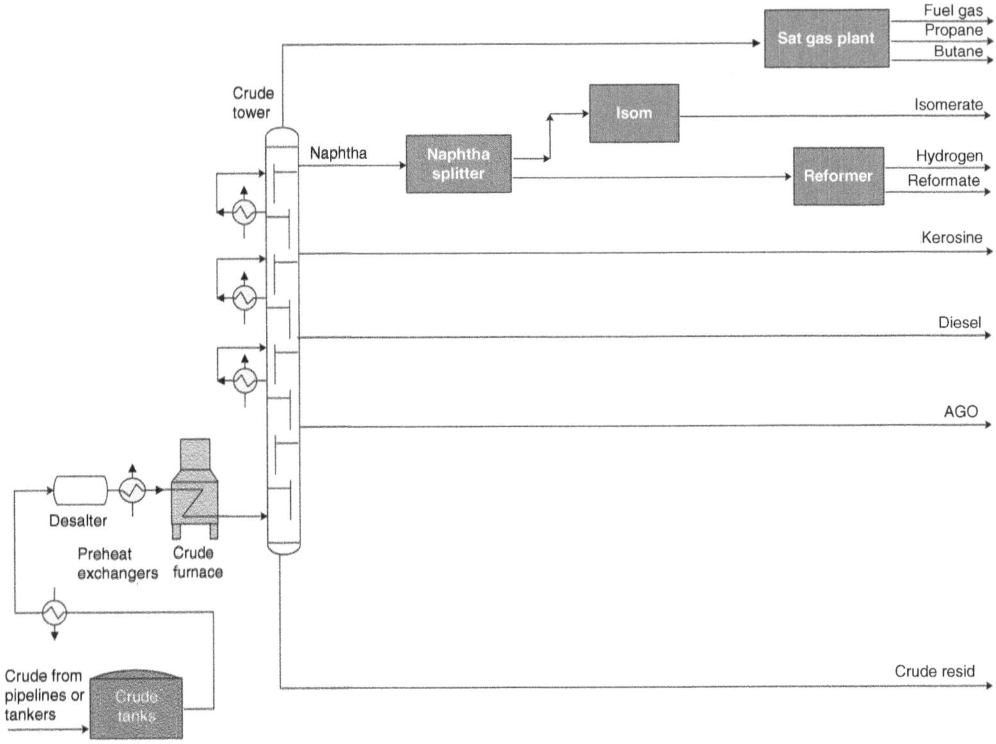

Figure B.4 Add Naphtha Splitter and Isom

B.7 Diesel Hydrotreater

To prevent the formation of sulfur dioxide when diesel fuel is combusted, government authorities all over the world have tightly regulated the quantity of sulfur allowable in diesel fuels. Since the sulfur in crude oil tends to concentrate in the heavier cuts, the diesel product from the crude unit needs to be further treated to remove sulfur compounds and other objectionable materials. This is achieved by processing diesel at moderately high pressures with hydrogen in several beds over either a Cobalt/Molybdenum or Nickel/Molybdenum catalyst.

For our purposes, there is no significant net change to the material balance after processing in the diesel hydrotreater.

Where the Diesel Hydrotreater fits logically into the flow schematic is shown in Figure B.5.

B.8 Vacuum Tower

If we examine the product distribution from our original crude distillation, over 40% of the initial crude feed ends up as crude tower bottoms, called resid. In other words, this is a heavier material that did not distill into one of the more valuable lighter fractions.

The high quantity of resid is partially due to the fact that crude tower feed material cannot be heated over 700 °F. That's because higher temperature causes crude oil to "crack" into lower-value lighter gases.

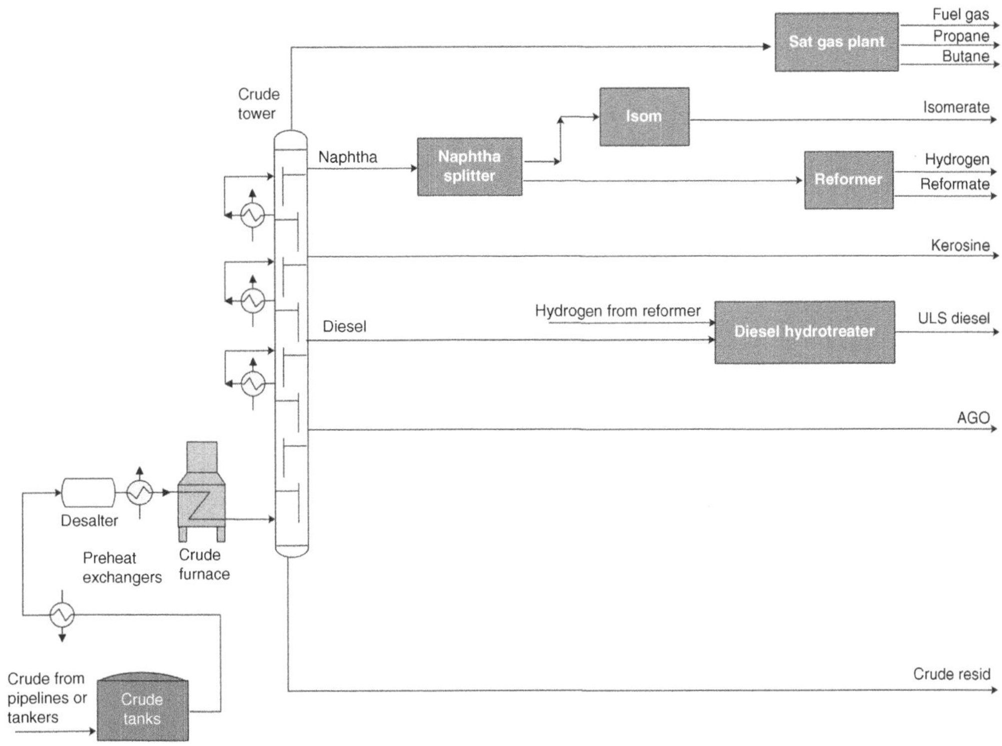

Figure B.5 Add Hydrotreater

In order to reduce the amount of crude resid, and increase the quantity of the more valuable fractions, we need to perform another distillation. However, this time the distillation will need to be done under reduced pressure, since it is well-known that boiling points decrease as the pressure is reduced. This will not only recover higher valued material but will also avoid any cracking issues.

Thus, the vacuum tower takes resid from the crude tower and fractionates it into a Light Vacuum Gas Oil (LVGO) and Heavy Vacuum Gas Oil (HVGO) stream. Any material not captured in those streams ends up as Vacuum Tower bottoms resid. The low-pressure vacuum required in the tower is produced by steam flowing at high velocities through a series of jet educators.

The net result of the vacuum tower operation is the significant reduction of crude resid into LVGO and HVGO streams, which ends up being fed to the Fluid Catalytic Cracking Unit. (They are drawn as separate streams in the Vacuum Tower solely for heat recovery purposes.)

The product material balance after adding processing in the Vacuum tower is shown in Table B.6.[15]

Where the Vacuum Unit fits logically into the flow schematic is shown in Figure B.6.

B.9 Conversion Processes

Up to this point in time, we have largely only discussed distillation processes, in addition to one unit that can normally be considered a cleanup process – the Diesel Hydrotreater, as well as the Reformer and Isom units, which largely rearrange molecules to improve octane number.

The four units discussed next are "Conversion" processes, which create higher-value lighter products through catalytic or thermal processes. In our case, the principal product is gasoline.

Table B.6 Cumulative Material Balance after Processing in Vacuum Unit

	Crude A	Crude B
Propane	938	1,207
Butane	2,156	2,550
Gasoline	16,147	19,413
Kerosene	12,900	11,000
Diesel	11,900	11,933
AGO	13,400	12,400
LVGO/HVGO	18,697	19,000
Vacuum Resid	23,519	19,000

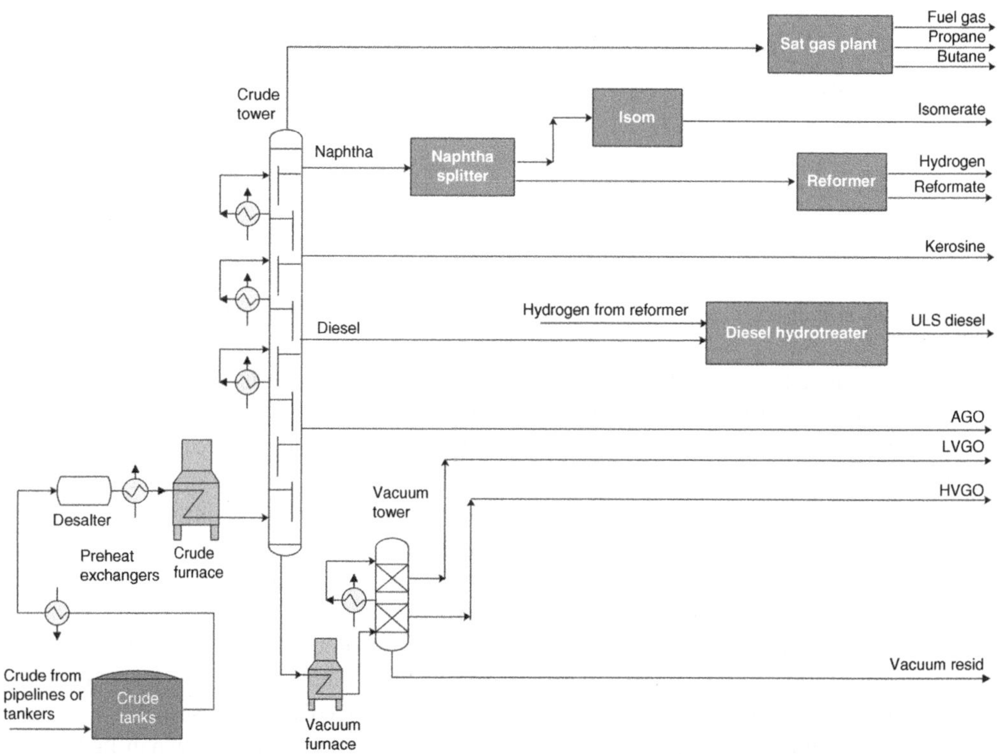

Figure B.6 Add Vacuum Tower

Let's start by discussing what most consider to be the most important unit in all of refining – the Fluid Catalytic Cracker (FCC).

B.9.1 FCC

The FCC is often considered to be the "heart" of most oil refineries. In our example, there are three feedstocks to the unit – the lowest cut in the crude tower, atmospheric gas oil (AGO), and the main vacuum tower products – LVGO and HVGO.

The principal product from the FCC is a high-octane gasoline stream. However, the FCC also produces important lighter C4 streams that will be processed further in the Alkylation unit.

Byproducts from the FCC include lower-value cycle oils. This heavy fraction is normally blended into fuel oil or alternatively, as in the book example, used as Hydrocracker feedstock.

Gasoline from the FCC is called FCC gasoline. The product material balance after processing in the FCC is shown in Table B.7.[16] Gasoline quantity increases significantly compared to processing in other units.

Where the FCC fits logically into the flow schematic is shown in Figure B.7.

Table B.7 Cumulative Material Balance after Processing in FCC

	Crude A	Crude B
C3	2,005	2,130
C4	3,436	3,552
Gasoline	33,824	40,700
Kerosene	12,900	13,000
Diesel	11,900	12,000
FCC GO	11,202	11,618
Vacuum Resid	23,519	19,000

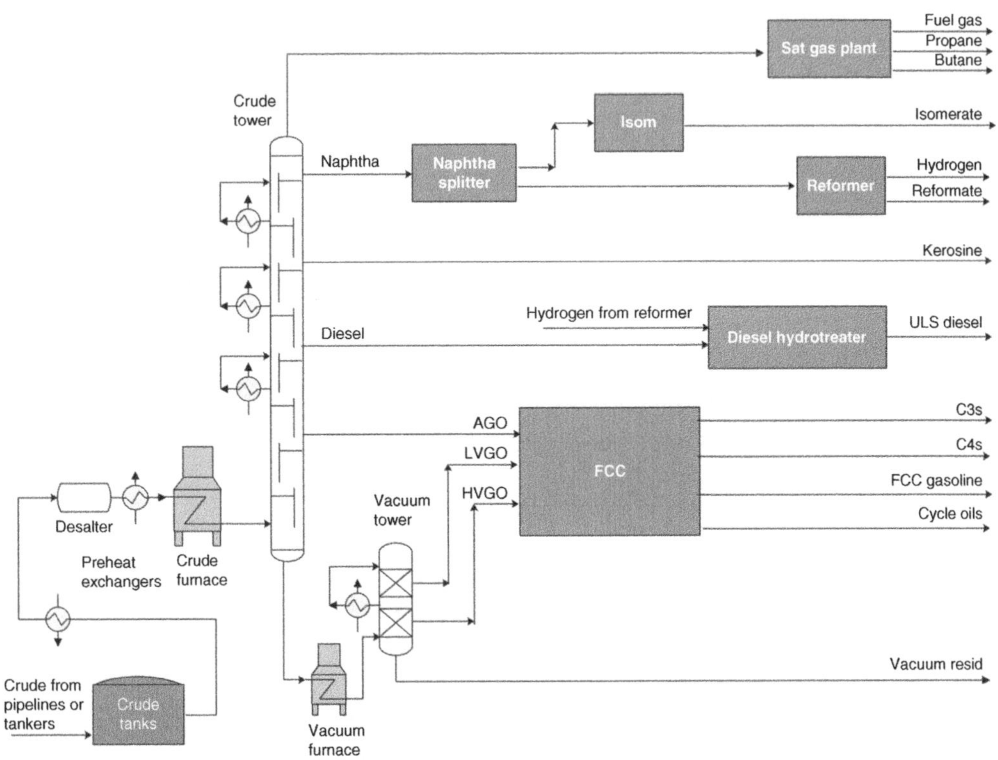

Figure B.7 Add FCC

B.9.2 Alkylation Unit

The FCC unit produces a significant quantity of several different C4 materials, principally isobutylene and isobutane. While not valuable by themselves, they can be combined catalytically in the alkylation unit to produce a gasoline-grade material – isooctane. This product is valuable not only because it boils in the typical gasoline range but because it has a very high octane number.

The one significant downside of an alkylation unit is it requires either sulfuric or hydrofluoric acid as a liquid catalyst. Unfortunately, both of those materials are highly acidic. Thus, anybody working near the unit must wear special PPE (personal protective equipment).

Gasoline from the Alkylation unit is called Alkylate. The product material balance after processing in the Alkylation Unit is shown in Table B.8.[17] Note the significant increase in gasoline quantity.

Where the Alkylation fits logically into the flow schematic is shown in Figure B.8.

B.9.3 Coker

The vacuum tower bottoms material is a highly viscous material that can plug pipes if it cools suddenly. For this reason, it must be further processed. Therefore, most refiners send this hot material to a Coker, which essentially "cooks" the feed stream to thermally crack as much material as possible into a gasoline-grade stream. Whatever material doesn't vaporize into either gasoline or a gas oil stream is used to make petroleum coke – a material similar to charcoal.

Since it is a highly olefinic stream, it must be further processed in the Reformer before it can be blended into the gasoline pool. Gasoline from the Coker is called Coker gasoline.

In addition, the Coker produces Coker gas oil, which will be a feed stream to the Hydrocracker in our example refinery.

The product material balance after adding processing in the Coker is shown in Table B.9.[18]

Where the Coker fits logically into the flow schematic is shown in Figure B.9.

B.9.4 Hydrocracker

The nine process units described previously are standard units found in most American oil refineries. However, some refineries also have a Hydrocracker, which can produce high-quality gasoline, kerosene, and diesel, depending on operating severities and catalysts. This unit operates

Table B.8 Cumulative Material Balance after Processing in Aklylation Unit

	Crude A	Crude B
C3	750	750
C4	0	0
Gasoline	42,860	49,390
Kerosene	12,900	13,000
Diesel	11,900	12,000
FCC GO	11,202	11,618
Vacuum Resid	23,519	19,000

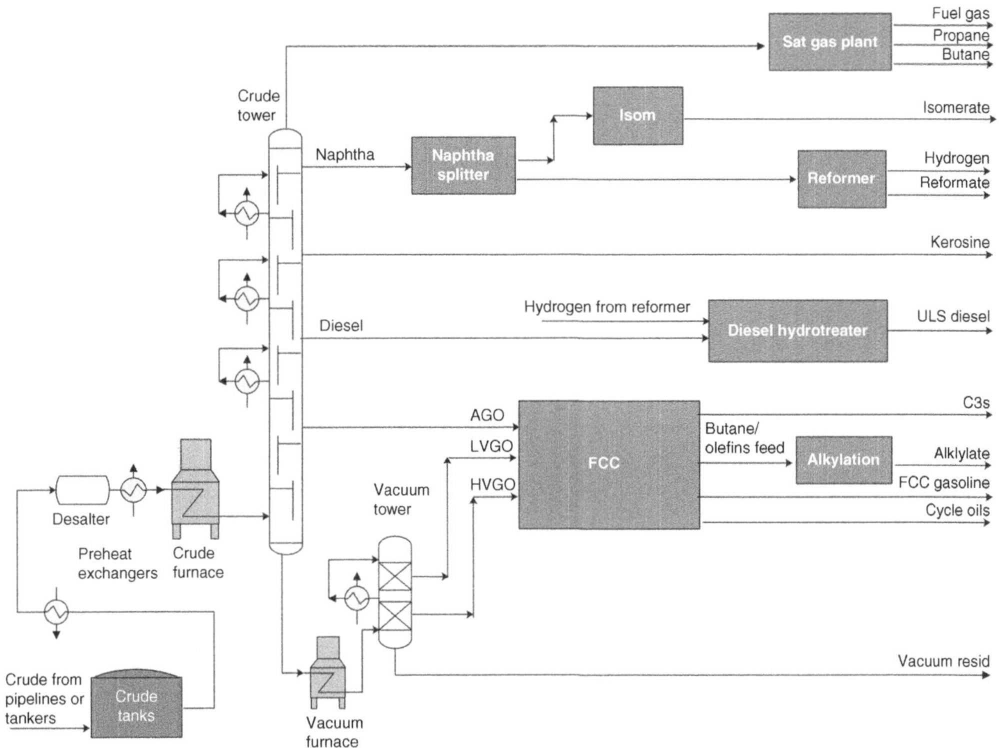

Figure B.8 Add Alkylation

Table B.9 Cumulative Material Balance after Processing in Coker

	Crude A	Crude B
C3s	1,691	1,700
C4s	1,176	760
Gasoline	48,364	54,710
Kerosene	13,000	13,000
Diesel	11,900	12,000
FCC LCO	11,202	11,618
Coker GO	10,678	7,448

at high pressure and is highly exothermic, but unfortunately, there is always the possibility of runaway reactions if the unit is not operated properly. In addition, it consumes significant quantities of hydrogen, which can often be a bottleneck since hydrogen is in high demand in a modern refinery.

Hydrocracking has significant flexibility to crack feed material to kerosene and diesel or all the way down to gasoline-grade material. The final product slate is a function of the catalyst and operating conditions.

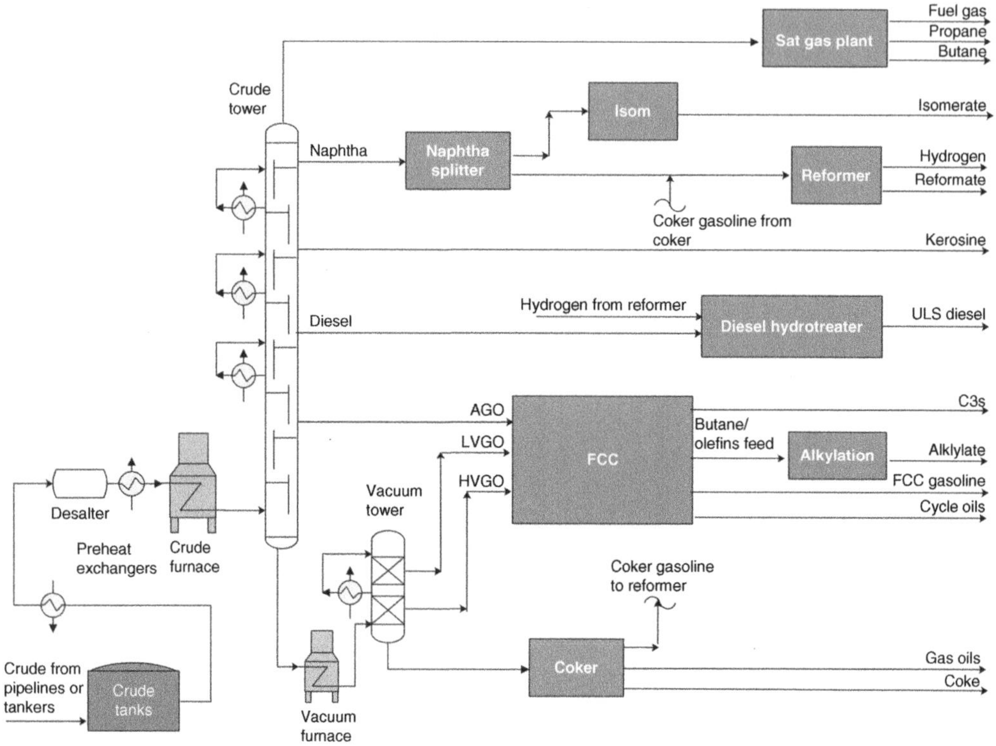

Figure B.9 Add Coker

Table B.10 Cumulative Material Balance after Processing in Hydrocracker

	Crude A	Crude B
C3s	1,691	1,700
C4s	3,167	2,911
Gasoline	68,252	72,041
Kerosene	13,000	13,000
Diesel	11,900	12,000

In our case, the hydrocracker has been designed to maximize gasoline production. The feed to the Hydrocracker is FCC light cycle oil (LCO) and Coker gas oil. Gasoline from the Hydrocracker is called Hydrocracker gasoline.

It should be noted that many refineries do not operate Hydrocrackers due to their high initial cost, the requirement for substantial quantities of additional hydrogen, which cannot be supplied by the Reformer, and the potential risks to both personal and facility inherent when operating a Hydrocracker. The product material balance after adding processing in the Hydrocracker is shown in Table B.10.[19]

Where the Hydrocracker fits logically into the flow schematic is shown in Figure B.10.

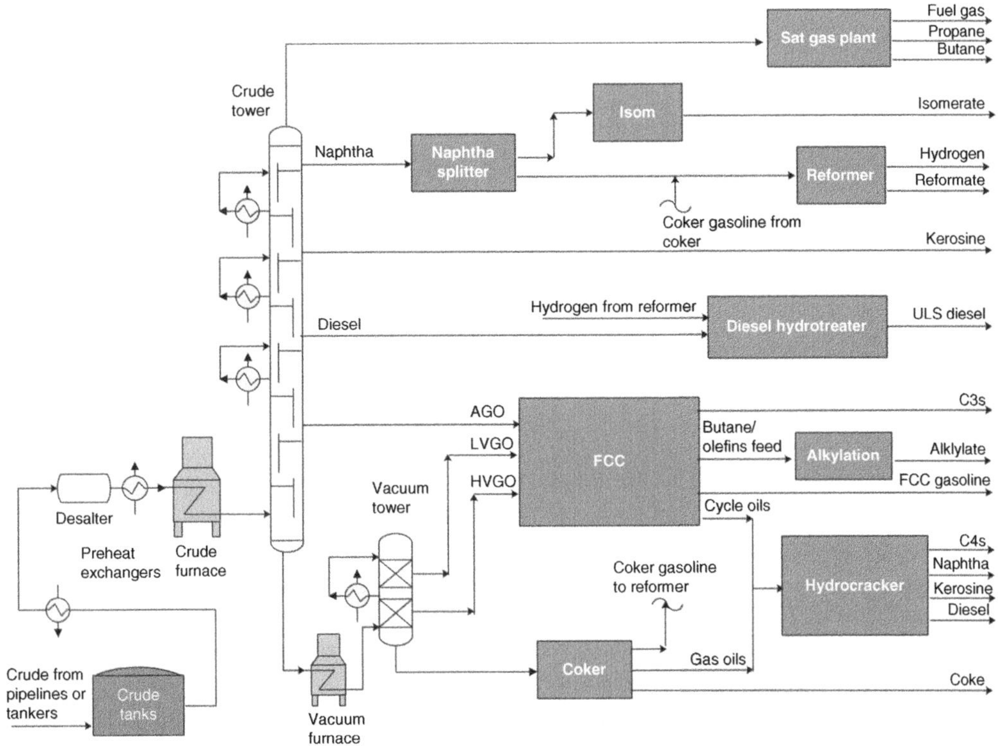

Figure B.10 Add Hydrocracker

Notes

About the Author

ix US patent number 5,488,561, Multivariable Process Control Method and Apparatus.

x Moran and Bond, "The Tradeoff Between Gasoline Pool Availability and Capital Expenditures in a Complex Modern-Day Refinery."

xi Moran and Bond, "Tradeoffs: unit availability vs capital expenditure."

xii National Science Foundation Proposal number, PD 05-5371, submitted by Onstream Optimization. This proposal was submitted for Overhaul Optimization, which is essentially the same ideas as Turnaround Optimization detailed in this book.

xiii https://www.slideshare.net/JoelBarger/meridiumpas55, "PAS 55 Compliance using Meridium APM", Barger, Joel, Feb 5, 2017. This presentation details important aspects of the Meridium Value Assurance Program (MVAP). This includes the MVAP Value Map which shows the relationship between Business Drivers, Asset Management Objectives, and APM Work Processes. In addition, it shows the complete Maturity Model, including the 10 key APM work processes – one of which was Reliability in Design. The PowerPoint presentation shows the APM work process diagram for Strategy and Risk Management. However, there were complete work process diagrams for all the other 9 work processes, such as Reliability in Design, though they are not shown in this presentation.

xiv Created complete, detailed, plant-wide RAM model under a subcontract to World Oil Services.

xv Moran, Martin, "Winning the Race among Nations for Economic Superiority", *Amazon*, Nov 2016

xvi Moran, Martin and Hayek, Fareed, "https://www.hydrocarbonprocessing.com/magazine/2019/march-2019/special-focus-petrochemical-technology/how-did-asset-utilization-modeling-save-sadara-1-b-in-capital-cost/"How did asset utilization modeling save Sadara $1 B in capital cost?," *Hydrocarbon Processing*, March 2019.

xvii Moran, Marty, "How are leading organizations implementing operational excellence?", *Hydrocarbon Processing*, August 2014

xviii Ayral, Tom and Moran, Marty, "Quantifying the ROI of an Asset Management Program", *Hydrocarbon Processing*, May 2007

xix Moran, Marty and Pompa, Valerie, "Real World Experiences Implementing Refinery Expert Systems", *Hydrocarbon Processing*, May 1999

xx Perino, Joe and Moran, Marty and Gomez J. P., "Process model-based multivariable control alternatives", *Hydrocarbon Processing*, 1994

Plant Optimization in the Process Industries: Incorporating Equipment/Assets in the Decision-Making Process, First Edition. Marty Moran.

Chapter 3

1 Moran, Marty and Bond, Hollis, "The Tradeoff Between Gasoline Pool Availability and Capital Expenditures in a Complex Modern-Day Refinery," NPRA Annual Meeting, San Antonio, Texas, March 21–23, 2004.
2 Moran, Marty and Bond, Hollis, "Tradeoffs: unit availability vs capital expenditure," *Petroleum Technology Quarterly*, April 2004.
3 National Science Foundation Proposal number, PD 05-5371, submitted by Onstream Optimization. This proposal was submitted for Overhaul Optimization, which is essentially the same idea as Turnaround Optimization detailed in this book.

Chapter 4

4 Moran and Bond, "The Tradeoff Between Gasoline Pool Availability and Capital Expenditures in a Complex Modern-Day Refinery."
5 OptQuest was the name of the commercial optimization technology used in the 2004 NPRA paper. OptTek Systems, Inc. is the name of the company that owns this technology.
6 US patent number 5,488,561, Multivariable Process Control Method and Apparatus.

Chapter 7

7 https://www.theguardian.com/environment/2018/mar/02/clean-coal-america-kemper-power-plant#:~:text=These%20construction%20snafus%20and%20planning,plant's%20%E2%80%9C clean%20coal%E2%80%9D%20design. The author built a complete plant-wide RAM model under subcontract to World Oil Services.

Chapter 13

8 Moran and Bond, "The Tradeoff Between Gasoline Pool Availability and Capital Expenditures in a Complex Modern-Day Refinery."
9 Ibid.

Chapter 14

10 Moran and Bond, "The Tradeoff Between Gasoline Pool Availability and Capital Expenditures in a Complex Modern-Day Refinery."
11 Ibid.

Appendix B

12 Gray, James H. and Handwerk, Glenn E., Petroleum Refining, Technology and Economics, 2nd Edition (Marcel Dekker, 1984), pg. 51 – Table 4.7.

13 Ibid., pg. 51 – Table 4.7.
14 Ibid., pg. 96 – Table 6.5.
15 Ibid., pg. 51 – Table 4.7.
16 Ibid., pg. 126 – Table 7.4.
17 Ibid., pg. 126 – Table 7.4.
18 Ibid., pg. 181 – Table 10.12.
19 Ibid., pg. 156 – Table 9.5.

Index

Plant Optimization in the Process Industries: Incorporating Equipment/Assets in the Decision-Making Process, First Edition. Marty Moran.

Printed and bound by CPI Group (UK) Ltd, Croydon, CR0 4YY
17/09/2026
14964324-0001